Pharmaceutical and Nutraceutical Potential of Cyanobacteria

Muhammad Aamer Mehmood • Pradeep Verma •
Maulin P. Shah • Michael J. Betenbaugh
Editors

Pharmaceutical and Nutraceutical Potential of Cyanobacteria

Springer

Editors
Muhammad Aamer Mehmood
Department of Bioinformatics and Biotechnology
Government College University Faisalabad
Faisalabad, Pakistan

Pradeep Verma
BPBEL, Department of Microbiology
Central University of Rajasthan
Ajmer, Rajasthan, India

Maulin P. Shah
Industrial Wastewater Research Lab, Division of Applied & Environmental Microbiology
Ankleshwar, Gujarat, India

Michael J. Betenbaugh
Department of Chemical and Biomolecular Engineering
Johns Hopkins University
Baltimore, MD, USA

ISBN 978-3-031-45522-3 ISBN 978-3-031-45523-0 (eBook)
https://doi.org/10.1007/978-3-031-45523-0

This Springer imprint is published by the registered company Springer Nature Switzerland AG
The registered company address is: Gewerbestrasse 11, 6330 Cham, Switzerland

Paper in this product is recyclable.

This book is dedicated to all those who are trying to make this world a better place for the living kind.

Preface

Cyanobacteria are the extraordinary microbes that are believed to have started life on Earth. It's been billions of years since they colonized this planet. Now, the question arises, what has made them so successful? The answer to this question lies in their remarkable potential to survive unfavorable environments, their substantial ability to fix the atmospheric carbon through photosynthesis, and their amazing secondary metabolites having antioxidant, osmoprotectant, and stress-tolerance abilities. Research on cyanobacteria has shown that these secondary metabolites and the substantial photosynthetic rates of cyanobacteria can be exploited for environmental, industrial, nutraceutical, and pharmaceutical applications.

This book "**Pharmaceutical and Nutraceutical Potential of Cyanobacteria**" is a collection of 14 book chapters that have covered almost all aspects related to the opportunities, challenges, and potential applications while employing cyanobacteria as feedstock for various industrial and environmental applications with a special focus on pharmaceutical and nutraceutical applications. Some sections have also covered the enhanced biosynthesis, extraction, storage, and marketing of the cyanobacterial bioactive compounds (phycobilins, carotenoids, fatty acids, amino acids) and applications of cyanobacteria as food/feed of the future. We believe that this book will provide substantial learning opportunities to the readers including graduate students, academicians, phycologists, policymakers, environmental entrepreneurs, and industrialists.

Faisalabad, Pakistan — Muhammad Aamer Mehmood
Ajmer, Rajasthan, India — Pradeep Verma
Ankleshwar, Gujarat, India — Maulin P. Shah
Baltimore, MD — Michael J. Betenbaugh

Acknowledgments

I am glad to extend my sincere thanks to Springer Nature and its representatives, especially *Susanne Dathe* and *Bibhuti Bhusan Sharma* for their patience, guidance, and kind support throughout. I appreciate the time and sincere efforts of all authors for their valuable contribution to this book, without which this work would have never been completed.

I appreciate the scientific and technical advice of the co-editors Prof. Pradeep Verma, Dr. Maulin P. Shah, and Prof. Michael J. Betenbaugh. A very special note of appreciation to my dedicated and passionate team members Dr. Ayesha Shahid and Dr. Sana Malik for providing substantial technical input during the preparation of this document.

Muhammad Aamer Mehmood
Faisalabad, Pakistan

Contents

About the Editors

Muhammad Aamer Mehmood is a Professor and Group Leader of the BioEcoTech Research Cluster, Bioenergy Research Center, at the Department of Bioinformatics and Biotechnology, Government College University Faisalabad, Pakistan. His research group focuses on process optimization for the sustainable utilization of bio-resources, wastewater recycling, resource recovery, and biotransformation of biomass bioproducts including biofertilizers, industrial enzymes, and biopolymers employing microalgae/cyanobacteria/fungi as microbial cell factories in a multiproduct biorefinery paradigm.

Pradeep Verma is a Professor and Group Leader of the Bioprocess and Bioenergy Laboratory at the Department of Microbiology, School of Life Sciences, Central University of Rajasthan, Ajmer, India. He is working extensively in the area of microwave/biological delignification, enzyme-mediated hydrolysis, and the development of consolidated/integrated biorefineries. He has contributed significantly to the area of lignocellulosic biomass-based biorefineries.

Maulin P. Shah is Chief Scientist and Head of the Industrial Wastewater Research Lab, Division of Applied and Environmental Microbiology Lab at Enviro Technology Ltd., Ankleshwar, Gujarat, India. His work focuses on the impact of industrial pollution on the microbial diversity of wastewater and genetically engineering high-impact microbes for the degradation of hazardous materials. His research interests include biological wastewater treatment, environmental microbiology, biodegradation, bioremediation and phytoremediation of environmental pollutants from industrial wastewater.

Michael J. Betenbaugh is a Professor of Chemical and Biomolecular Engineering and a lead PI of the Advanced Mammalian Biomanufacturing Innovation Center (AMBIC) at Johns Hopkins University, Baltimore, USA. He is known for integrating systems biology with cellular, metabolic, and biochemical engineering for eukaryotic biotechnology applications. AMBIC is an Industry-University Cooperative Research Center (IUCRC) funded by NSF and 20 industrial and governmental sponsors, including major biopharmaceutical manufacturers, contract manufacturers, and suppliers.

Chapter 1
Cyanobacterial Cell Factories; Insight into Their Pharmaceutical and Nutraceutical Properties

Ayesha Shahid, Iqra Kamal, Muhammad Nabeel Haider, Muhammad Imran Arshad, Sultan Habibullah Khan, Ning Wang, and Hui Zhu

Abstract Increasing demands for healthier humans and nutritional food have compelled researchers to explore nonconventional sources with better properties to meet the market requirements. Compared to synthetic molecules, biologically derived bioactive compounds are more sustainable, reliable, and safe to use. Cyanobacteria are promising pharmaceutical and nutraceutical cell factories with the tendency to produce bioactive secondary metabolites such as peptides, fatty acids, antioxidant enzymes, dietary pigments, alkaloids, polyketides, polysaccharides, and phenols. These metabolites have antitumor, antimicrobial, antifungal, anticancer, anti-inflammatory, antioxidant, antiproliferative, and immunosuppressive properties. The market for cyanobacterial therapeutics/nutraceuticals is emerging at an exponential rate with an expected share of $6.09 billion by the year 2026. The first part of this chapter provides insights into the cyanobacterial market along with an overview of the cyanobacterial portfolio for secondary metabolite biosynthesis that showed the suitability of cyanobacteria as the "organism of choice" for pharmaceuticals and nutraceuticals. To meet commercial requirements, the bioactive potential of cyanobacteria has to be explored in detail for which techniques like adaptive evolutions, genetic source/pathway identification by omics, and genetic engineering

A. Shahid · S. H. Khan
National Center for Genome Editing, Center for Advanced Studies in Agriculture and Food Security, University of Agriculture Faisalabad, Faisalabad, Pakistan

I. Kamal · M. N. Haider
Bioenergy Research Center, Department of Bioinformatics and Biotechnology, Government College University Faisalabad, Faisalabad, Pakistan

M. I. Arshad
National Center for Genome Editing, Center for Advanced Studies in Agriculture and Food Security, University of Agriculture Faisalabad, Faisalabad, Pakistan

Institute of Microbiology, University of Agriculture, Faisalabad, Pakistan

N. Wang · H. Zhu (✉)
School of Bioengineering, Sichuan University of Science and Engineering, Zigong, China

M. A. Mehmood et al. (eds.), *Pharmaceutical and Nutraceutical Potential of Cyanobacteria*, https://doi.org/10.1007/978-3-031-45523-0_1

have been discussed in the last section. The focus has been on the research gaps and possible opportunities to improve the pharmaceutical and nutraceutical potential of cyanobacteria.

Keywords Bioactive compounds · Nutritional supplements · Therapeutics · Antioxidants · Genetic engineering

1.1 Introduction

Population spurts pose threats to human health and food security due to increasing risks of communicable diseases and malnutrition. Malnutrition and food scarcity are responsible for 3.5 million deaths annually (Nethravathy et al. 2019). Alternatively, the world is relying heavily on synthetic pharmaceuticals to improve human health, turning them from a blessing to a nuisance due to their overuse. Most of the pharmaceutical agents are persistent organic compounds whose mitigation is challenging; thus, they are contaminating the ecosystem at an alarming rate and pose a great threat to mankind (Khan et al. 2022). It has compelled researchers to explore alternative pharmaceutical and nutraceutical sources with improved properties.

Cyanobacteria are unconventional and effective sources to replace synthetic drugs and combat issues of malnutrition due to their unique cellular composition. Their use as therapeutic agents, especially in drug discovery and as food sources, has been in practice for 1500 years (Agarwal et al. 2022). They are of vital and pragmatic importance due to the production of bioactive compounds in response to varying environmental conditions. Additionally, their CO_2 mitigation abilities, conversion of nutrients into valuable compounds, utilization of nonarable land for cultivation, and tolerance against harsh environmental conditions allow their exploration for the sustainable production of bioactive compounds (Shahid et al. 2020). Based on the chemical nature of the cyanobacterial secondary metabolites, they have potent pharmaceutical and nutraceutical applications. The complexity of secondary metabolites allows us to produce a multitude of products, but only a small fraction have been characterized yet. According to an estimation, 59% of cyanobacterial metabolites are cyclic, while 69% are identified as peptides with molecular weight of 0.12–2.17 kDa (Verma et al. 2022). Generally, cyanobacteria produce most of the secondary metabolites except peptides and polyketides through (i) MEP (methylerythritol 4-phosphate) pathway, (ii) shikimate pathway, (iii) mevalonate pathway, and (iv) malonate pathway. MEP pathway is generally explored for drug discovery as it is responsible for chlorophylls and hormone production (Amadu et al. 2021; Verma et al. 2022). However, secondary metabolite production is highly species specific and depends on environmental fluctuations. In response to stress conditions, cyanobacteria tend to produce antibacterial metabolites, while under nutrient stress, pharmaceutically important storage metabolites such as lipids and starch are accumulated (Amadu et al. 2021).

Owing to the pharmaceutical and nutraceutical potential of the cyanobacterial bioactive compounds, its market is gaining attention recently (Fig. 1.1) and is

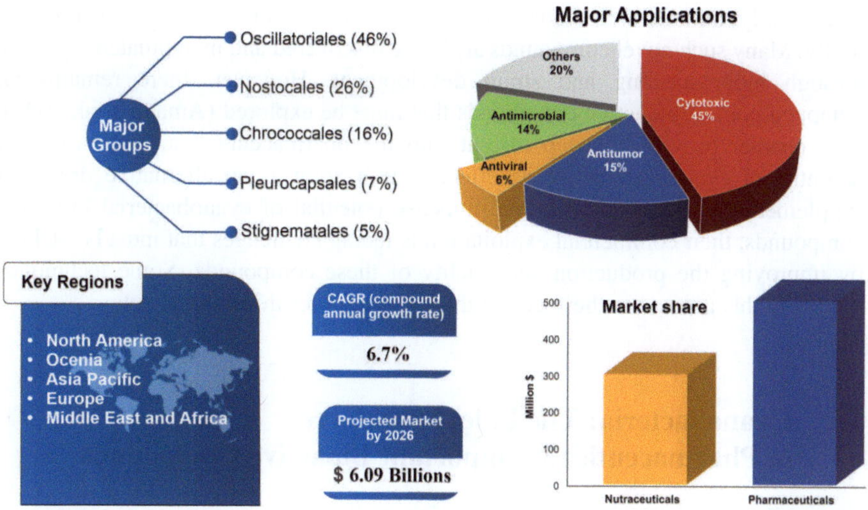

Fig. 1.1 Market analysis of cyanobacterial products in terms of pharmaceutical and nutraceutical applications

expected to reach \$6.09 billion by 2026 with a CAGR of 6.7% (Verma et al. 2022). Cyanobacterial metabolites are being explored commercially as potential sources of food, supplements, functional foods, therapeutics, and pharmaceutical cell factories for Mars habitation (Mapstone et al. 2022). Growing food demands, customization of food products, and public awareness of the health benefits of algal-based products are the driving factors of the algal product market. According to the experts of the "National Center for Biotechnology Information" by 2025, 18% of proteins will be provided by algae as their production is expected to increase during 2020–2027 by CAGR (compound annual growth rate) of 4.3% (Hachicha et al. 2022). Currently, >75% of the algal-derived products are dedicated to pharmaceuticals and nutraceuticals with market value of \$300 million and \$500 million, respectively (Levasseur et al. 2020). It is projected that the phycobilin market will increase from \$30 million to \$140 million by the year 2030, while the phycocyanin market alone will contribute to a \$409.8 million market share. The pharmaceutical-grade phycocyanin market is expected to reach from \$12.5 million in 2020 to \$35.8 million by 2030 with 10.6% of CGAR (Kamble and Deshmukh 2021). The market of omega-3 fatty acids, particularly DHA (docosahexaenoic acid), is expected to reach \$305 million from \$158 million with a CAGR of 11.6% by the year 2027 (Business Wire 2022). Considering the great industrial importance of cyanobacteria, the world is encouraged to enhance the production of these valuable organisms. Until 2011, global microalgal exploitation was 10,000 tons year^{-1} which almost doubled during 2016–2018 and is expected to reach 27,500 tons year^{-1} by 2024 (Hachicha et al. 2022). North America mainly the United States is dominating the algal product market, but Asia Pacific and Europe are expected to be the major leaders in the upcoming years (Levasseur et al. 2020; Custom Research 2022).

Presently, innumerable cyanobacterial bioactive compounds are produced industrially. Many such novel compounds are being discovered and manipulated regularly through bioprospecting and strain development. However, there remains an untapped pool of bioactive compounds that must be explored (Amadu et al. 2021). This chapter provides a deep insight into the pharmaceutical and nutraceutical potential of cyanobacteria for its effective utilization as an alternative drug and supplementary source. Despite the immense potential of cyanobacterial bioactive compounds, their commercial exploitation is facing challenges that must be tackled by improving the production and quality of these compounds. Some techniques related to this aspect are the focus of the last section of this book chapter.

1.2 Cyanobacteria: The Potential Source of Pharmaceutically Important Bioactive Compounds

During the last few years, bioactive compounds from cyanobacteria have gained considerable attention due to their remarkable pharmaceutical applications (Fig. 1.2). The secondary bioactive compounds offer protection against environmental stress conditions to cyanobacteria (Amin and Kannaujiya 2021). Cyanobacterial compounds exhibit potent antiviral, antibacterial, antiparasitic, antifungal, anticancer, antioxidant, anti-inflammatory, antiproliferative, and anticoagulant activity (Bishoyi et al. 2022). These compounds also play a crucial role in neuroprotection

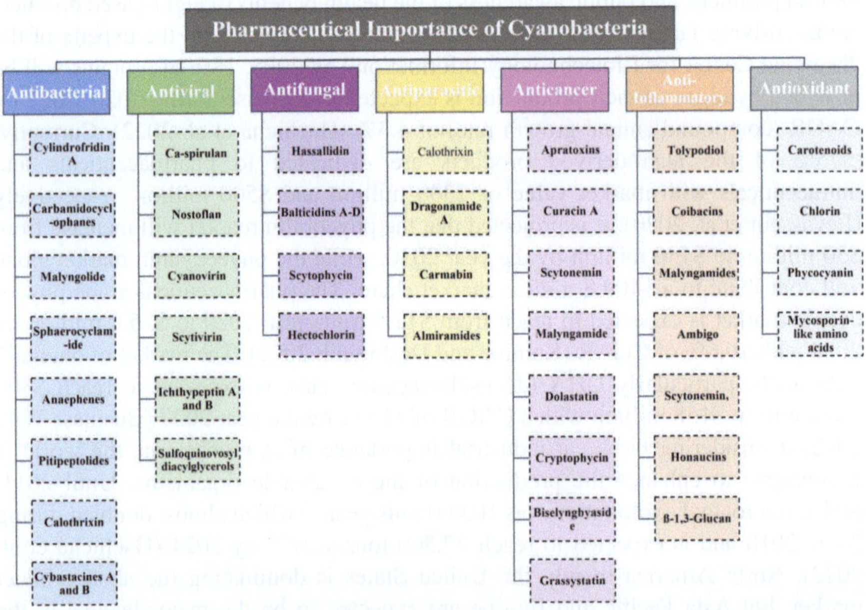

Fig. 1.2 Overview of cyanobacterial bioactive compounds with pharmaceutical potential

and immunostimulation. The biologically active compounds from cyanobacteria are alkaloids, depsipeptides, cyclic peptides, linear peptides, purines, macrolides, polysaccharide guanidines, phosphonates, polyketides, terpenoids, pigments, phenols, and lipids (Kini et al. 2020). The diversity of these bioactive compounds from cyanobacteria is remarkable, making cyanobacteria one of the exceptional pharmaceutical sources.

Cyanobacteria genera like *Nostoc, Microcystis,* and *Lyngbya* have been explored for their immense pharmaceutical potential. The important pharmaceutical compounds from these cyanobacteria strains such as curacin and crytophycin have entered the preclinical and clinical phases of drug development (Thuan et al. 2019). The analysis of a wide variety of bioactive metabolites from cyanobacteria with pharmaceutical applications against diseases is the focus of recent advances and research in therapeutics.

1.2.1 Cyanobacteria as Antidiabetic Agent

Diabetes is a progressive metabolic disorder that was the ninth leading cause of death in 2019 with approximately 1.5 million deaths (World Health Organization 2021). It demands the effective treatment of the disease. Several in vivo and ex vivo assays have been performed to investigate the antidiabetic potential of cyanobacterial metabolites. To check the hypoglycemic properties of cyanobacteria, diabetes-induced albino rats were exploited as animal models (Pandurangan and Kim 2016). Cyanobacterial bioactive compounds were administrated to rats for 60 days in a row. Cyanobacterial compounds not only regulate the level of sugar in the body, but they also elevated the level of blood insulin, C-peptide, and body mass. Here, antidiabetic effects were observed due to the increased activity of lipid peroxidase and hexokinase enzymes and the reduction in the activity of the glucose-6-phosphatase enzyme. In another study, refined pigments, and crude extracts of *Microcoleus, Lyngbya,* and *Synechocystis* sp., have been explored for antidiabetic activity. The extract from *Lyngbya* exhibits the highest antidiabetic properties. The digested cyanobacterial bioactive compounds of these strains act as strong inhibitors of α-glucopyranosidase and endoamylase glycogenase enzymes. In addition to lowering the level of blood sugar, these extracts also exhibit high antioxidant activity (Ghosh et al. 2016).

1.2.2 Cyanobacteria as Anticancer Therapeutic

According to a World Health Organization report, cancer is the second leading cause of death with approximately 18 million new cases and 10 million deaths in the year 2018. The prevalence and mortality of cancer are expected to be doubled by the year 2040 (World Health Organization 2020). Currently, all the medications for cancer treatment have some sort of side effect. Cyanobacteria have been explored as a potential anticancer source (Centella et al. 2017) as it is known to produce pharmaceutically rich anticancer metabolites (Table 1.1). The alkaloid molecules derived from cyanobacteria *Calothrix* sp., *Fischerella musicola*, *Hapalosiphon welwitschii*, *Fischerella ambigua*, *Hapalosiphon fontinalis*, *and Westiella intricate* possess cytotoxic activity against cancer cell lines (Mondal et al. 2020). The bioactive compounds like terpenes, polyketides, peptides, macrolide, nucleosides, and carbohydrates from cyanobacteria *Lyngbya* sp., *Plectonema* sp., *Okeania* sp., *Symploca* sp., *Anabaena* sp., *Tolypothrix* sp., and *Nostoc* sp. exhibit significant anticancer activity against different types of cancer cell lines (Demay et al. 2019). Apratoxin A has also shown anticancer properties as its family is involved in the arrest of G1-phase and apoptosis (Qamar et al. 2021). In silico study of bioactive compounds from *Lyngbya majuscula* exhibited their higher treatment potential against human lung cancer (Baig et al. 2022).

Out of the total bioactivities shown by cyanobacterial metabolites, cytotoxic activity against cancer cell lines is the most prevalent activity that is shown by more than 40% of the metabolite's families (Demay et al. 2019).

1.2.3 Cyanobacteria as Antimicrobial Agent

Since the beginning of the nineteenth century, viral outbreaks have drastically impacted the economics and health of people around the globe. Different cyanobacterial strains produce low molecular weight proteins called lectins. This family of bioactive metabolites consists of important compounds such as microvirin, cyanovirin-N, *Oscillatoria agardhii* agglutinin, scytovirin, and *Microcystis viridis* lectin (Pradhan et al. 2022). These lectins exhibit potent antiviral activity against simian immunodeficiency viruses, human immunodeficiency viruses, hepatitis C viruses, Ebola viruses, and influenza viruses (Wu et al. 2015). Cyanovirin-N a polypeptide isolated from *Nostoc ellipsosporum* reported antiviral potential against lentiviruses and HIV (Vijayakumar and Menakha 2015). The virucidal properties of *Phormidium* and *Lyngbya lagerheimeii* have been reported against HIV (Kini et al. 2020). The two important polysaccharides Ca-pirulan and nostoflan from cyanobacteria cause the inhibition of many viruses. Ca-spirulan isolated from *Arthrospira platensis* has broad-spectrum activity against enveloped viruses. Nostoflan is isolated from cyanobacteria *Nostoc flagelliforme*, and this compound exhibits antiviral activity against both enveloped and non-enveloped viruses (Mazur-Marzec et al. 2021; Carpine and Sieber 2021).

Extensive drug resistance and the presence of multidrug-resistant pathogens are the most prevalent issues faced by the world right now (Gómez-Espinoza et al. 2022). It is assumed that by 2050 antibiotic resistance could result in 300 million premature deaths with a 2–3.5% reduction in gross domestic product (GDP) resulting in a $100 trillion economic loss (Laundy et al. 2016). The variety of bioactive secondary metabolites from cyanobacteria has extensive antimicrobial

Table 1.1 Summary of the important cyanobacterial bioactive compounds, their metabolic activity, and mechanism

Bioactive compounds	Cyanobacterial species	Metabolic activity	Target disease/microbe	Mechanism	References
Apratoxins	*Lyngbya majuscula*	Anticancer	Multiple tumors	These marine peptides function as Sec61 inhibitors and target the HER/ErbB family proteins. They have vigorous activity toward extensive tumor cell lines in the nanomolar range	Andler and Kazmaier (2021)
Curacin A	*Lyngbya majuscula*	Anticancer	Renal, colon, and breast cancer cell lines and lung cancer cells	A ketopeptide inhibits cell proliferation by interacting with the colchicine-binding site on tubulin and hinders the process of microtubule assembly which results in disorganized cell shape. Due to its anticancer properties, curacin A has entered the clinical trials for cancer treatment	Giordano et al. (2015)
Scytonemin	*Nostoc*, *Calothrix*, *Scytonema*, *Hyella*, *Lyngbya*, *Chlorogloeopsis*, and *Rivularia*	Anticancer, antiproliferative, and anti-inflammatory	Multiple myeloma cells	Scytonemin is a serine/threonine inhibitor that inhibits the activity of the human polo-like kinase. It affects cell cycle kinesis and the development of mitotic spindles	Sugumaran et al. (2022)
Malyngamide	*Lyngbya majuscula*, *Moorea producens, and Nostoc* sp.	Anticancer	Breast cancer, colon cancer, colorectal carcinoma, lung cancer, and neuroblastoma	It suppressed tumor proliferation by inactivating the phosphorylation of Akt and FAK pathways. Three malyngamides isolated from	Lee et al. (2021, Qamar et al. (2021)

(continued)

Table 1.1 (continued)

Bioactive compounds	Cyanobacterial species	Metabolic activity	Target disease/microbe	Mechanism	References
				Moorea producens showed potent anticancer activity by activating the function of adenosine monophosphate-activated protein kinases	
Dolastatin	*Symploca* sp. VP642 and *Symploca* sp. VP452	Anticancer	Breast, colon, and lung cancer	This compound showed cytotoxic activity by disorganizing microtubule assembly and hindering the G_2/M phase of the cell cycle leading to cell death. Its anticancer activity was assessed against several murine cell lines. Several synthetic and semisynthetic analogs of dolastatins are in the clinical phase as anticancer drugs	Luesch et al. (2001)
Cryptophycin	*Nostoc* ATCC 53789 and *Nostoc* GSV 224	Anticancer	Ovarian carcinoma, breast carcinoma cells, and non-small cell lung cancer	It showed potent anticancer activity by disorganizing microtubule assembly and hindering the G_2/M phase of the cell cycle via capase-1- and caspase-3- dependent pathway ultimately leading to programmed cell death. Cryptophycin causes microtubule disorganization in multidrug-resistant cancer cells. Several synthetic and semisynthetic analogs of	Qamar et al. (2021), Shih and Teicher (2001)

				cyrptophycin have entered the preclinical and clinical phases	
Biselyngbyaside	*Lyngbya* sp.	Anticancer	HeLa S3, and HeLa, SNB 68, NCI H522, and HL-60 cells	This macrolide glycoside induces endoplasmic reticulum stress which ultimately leads to cell death. It causes apoptosis in cancer cells by nuclear condensation	Mondal et al. (2020)
Grassystatin	*Lyngbya confervoides*	Anticancer	HT29, mouse neuro-2a blastoma cell lines, and HeLa cancer cell lines	These cyclic depsipeptides obstruct the G_1 and G_2 phases of the cell cycle. Some forms of grassystatin also function as protease inhibitors and act as antiproliferative and antimetastatic agents	Al-Awadhi et al. (2017)
Bisebromoamide	*Lyngbya* sp.	Anticancer	HeLa S3, 769-P and 786-O, human renal carcinoma cell lines	It is a potent kinase inhibitor and inhibits the phosphorylation of the extracellular signal-regulated kinase (ERK) and protein kinase B (Akt) pathway. It also functions in the stabilization of the actin filament. It can function as a cytotoxic drug as it inhibits both ERK and Akt pathways	Teruya et al. (2009)
Ca-spirulan	*Spirulina* sp.	Antiviral	HIV, influenza virus, herpes simplex virus, measles, mumps, and other enveloped viruses	This compound hinders the process of virus and host cell attachment and fusion. They perform anti-HIV functions by suppressing the activity of reverse transcriptase enzyme.	Carpine and Sieber (2021), Hayashi et al. (1996)

(continued)

Table 1.1 (continued)

Bioactive compounds	Cyanobacterial species	Metabolic activity	Target disease/microbe	Mechanism	References
				These compounds also hinder the fusion between HIV-infected and non-HIV-infected white blood cells which plays an important role to lessen viral infectivity	
Nostoflan	*Nostoc flagelliforme*	Antiviral	Enveloped viruses, herpes simplex virus, influenza A virus, and human cytomegalovirus	This bioactive compound has extensive activity against enveloped viruses with carbohydrate receptors. It works by hindering the virus's attachment to the host cell	Kanekiyo et al. (2005, 2007)
Cyanovirin	*Nostoc ellipsosporum*	Antiviral	HIV and SIV, measles, herpes simplex virus, and lentivirus	It hinders virus attachment and fusion with the target cells. As an anti-HIV agent, it restricts the interaction between virus surface glycoprotein gp120 with the target cell which results in the inhibition of virus fusion	Keeffe et al. (2011)
Scytovirin	*Scytonema varium*	Antiviral	HIV virus, Ebola virus, Marburg virus, and SARS coronavirus	This peptide works by attaching itself to mannose sugar in the glycoprotein of the virus envelope and blocks the virus attachment and penetration into the host cell	McFeeters et al. (2013), Garrison et al. (2014)
Sulfoquinovosyl diacylglycerols	*Spirulina platensis*, *Lyngbya lagerheimii*, and *Phormidium tenue*	Antiviral	HIV, AIDS, and herpes simplex virus	These sulfoglycolipids inhibit the activity of DNA polymerase, P-selectin receptors, HIV	Chirasuwan et al. (2009)

				reverse transcriptase, and AIDS telomerase enzyme which in turn obstruct viral development. This bioactive compound also has activity against herpes simplex virus-1	
Cylindrofridin A	*Cylindrospermum stagnale*	Antibacterial	Gram-positive bacteria mainly methicillin-resistant *Staphylococcus aureus* and *Streptococcus pneumoniae*	–	Preisitsch et al. (2016)
Carbamidocyclophanes	*Nostoc* sp. CAVN2	Antibacterial	Methicillin-resistant *Staphylococcus aureus*, *Enterococcus faecalis*, *Streptococcus pneumoniae*, and *Mycobacterium tuberculosis*		Preisitsch et al. (2015)
Malyngolide	*Lyngbya majuscula*	Antibacterial	Gram-positive bacteria (*Bacillus subtilis*, *Staphylococcus aureus*, and *Streptococcus pyogenes*) and Gram-negative bacteria (*Pseudomonas aeruginosa*, and *Chromobacterium violaceum*)	This bioactive compound works by blocking the crucial steps of quorum sensing and blocks the communication between bacteria	Dobretsov et al. (2010)
Sphaerocyclamide	*Sphaerospermopsis* sp. LEGE 00249	Antibacterial	*Halomonas aquamarina* CECT 5000	This prenylated cyanobactin inhibits the activity of biofilm-producing marine fouling bacterium. The mechanism of action is unknown	Martins et al. (2018)
Anaephenes	*Hormoscilla* sp. *Oscillatoriales* sp.	Antibacterial	Gram-positive bacterium *Bacillus cereus* and methicillin-resistant *Staphylococcus aureus*	Mechanism unknown	Brumley et al. (2018)

(continued)

Table 1.1 (continued)

Bioactive compounds	Cyanobacterial species	Metabolic activity	Target disease/microbe	Mechanism	References
Pitipeptolides	*Lyngbya majuscula*	Antibacterial	*Mycobacterium tuberculosis*	The N-methylation of phenyl-alanine amino acid in pitipeptolides due to this cyclic depsipeptides plays a crucial role in the inhibition of bacteria	Montaser et al. (2011)
Calothrixin	*Calothrix* sp.	Antibacterial and antiparasitic	*Escherichia coli*, *Bacillus cereus*, and *Staphylococcus epidermidis*	This bioactive compound exhibits antibacterial activity by inhibiting the activity of bacterial RNA polymerase and DNA topoisomerase. As an antiparasitic compound, it inhibits the growth of malarial strains in a dose-dependent manner	Xu et al. (2016)
Cybastacines A and B	*Nostoc* sp.	Antibacterial	*Streptococcus* spp., *Staphylococcus* spp., *Mycobacterium* spp., *Enterococcus* spp., *Nocardia* spp., and other closely related species	Mechanism unknown	Cabanillas et al. (2018)
Hapalindoles	Cyanobacterium *Hapalosiphon* sp. CBT1235.	Immunostimulatory	Autoimmune diseases	This indole alkaloid inhibits the proliferation of T-cells of the immune system and its apoptosis	Chilczuk et al. (2020)
Hassallidin	*Hassallia* sp., *Nostoc* sp., *Anabaena* sp., and *Tolypothrix* sp.	Antifungal	Human pathogenic fungi (*Candida* spp., *Fusarium* spp., *Penicillin* sp., *Aspergillus fumigatus*, *Ustilago maydis*, *Cryptococcus*	This bioactive compound causes disruption and permeabilization of sterol-containing cell membranes	Neuhof et al. (2005), Vestola et al. (2014)

			neoformans, and *Acremonium strictum*		
Balticidins A-D	*Anabaena cylindrica* Bio33	Antifungal	*Candida maltose*, *Candida albicans, Candida krusei, Microsporum canis, Microsporum gypseum, Mucor* sp., *and Aspergillus fumigatus*	Mechanism of action unknown	Bui et al. (2014)
Scytophycin	*Scytonema* spp., *Nostoc* sp., *Anabaena* sp., and *Cylindrospermum*	Antifungal	*Aspergillus* spp. and *Candida* spp.	Mechanism unknown	Shishido et al. (2015)
Dragonamide A and carmabin	*Lyngbya majuscula*	Antimalarial	Chloroquine-resistant strains of *Plasmodium falciparum*	Mechanism unknown	Ye et al. (2018)
Almiramides	*Lyngbya majuscula*	Antiparasitic	Leishmaniasis	Mechanism unknown	Nguyen et al. (2021)

activity (Table 1.1) which is believed to be associated with the presence of a high percentage of unsaturated fatty acids, phenolic compounds, and carotenoids (Alsenani et al. 2020). The presence of a phenolic compound known as eucapsitrione that exhibits potent antibacterial activity against *Mycobacterium tuberculosis* has been reported in freshwater cyanobacteria *Eucapsis* (Sturdy et al. 2010). Laminarin and fucoidan polysaccharides from algae showed antibacterial activity against *S. aureus* and *E. coli* and prevent the biofilm creation of *Helicobacter pylori* (Hernández et al. 2016). The growth of *E. coli* and *Bacillus cereus* is also inhibited by the fatty acid extract of *Synechocystis* sp. (Sarada et al. 2011). *Oscillatoria redekei* is reported to prevent the growth of *S. aureus*, *Micrococcus flavus*, and *B. subtilis* due to the presence of linoleic acids and dimorphecolic fatty acids (Raveh and Carmeli 2007).

The secondary metabolites from cyanobacteria *Fischerella* sp., *Nostoc* sp., *Anabaena* sp., *Oscillatoria* sp., *Spirulina* sp., *Lyngbya* sp., *Scytonema* sp., and *Eucapsis* sp. exhibit broad-spectrum antibacterial activity against pathogenic microbes (Rojas et al. 2020). Similarly, secondary metabolite extracted from *Westeilla*, *Fischerella*, *Westiellopsis*, and *Hapalosiphon* exhibits broad-spectrum antibacterial activity against 27 pathogenic bacterial strains. In addition to antibacterial activity, these secondary metabolites also showed insecticidal, fungicidal, algicidal, and cytotoxic activity against drug-resistant cancer cell lines (Demay et al. 2019). Due to multiple modes of action and synergy among these compounds, it is difficult to develop resistance against them.

Pathogen fungal species also have a detrimental effect on human health, and this problem became worse due to the development of antifungal resistance as the number of antifungal resources is limited. The structurally diverse secondary metabolites from cyanobacteria also have potent antifungal activity (Table 1.1). Important biomolecules that have strong activity against pathogenic and opportunistic fungus include hassallidins, lyngbyabellins, microguanidines, and majusculamides. Hassallidins are cyclic glycolipopeptides synthesized by *Tolypothrix*, *Nostoc*, *Anabaena*, *Cylindrospermopsis*, *Planktothrix*, and *Aphanizomenon* (Vestola et al. 2014). They have strong inhibitory action against pathogenic fungi such as *Aspergillus*, *Fusarium*, *Saccharomyces*, *Candida*, and *Penicillium* (Humisto et al. 2019; Vestola et al. 2014). Pharmaceutically important cyanobacteria such as *Moorea* and *Lyngbya* produce cyclic depsipeptides called lyngbyabellins which are responsible for the inhibition of *Candida albicans* by disrupting the microfilament network of fungal cell (Demay et al. 2019). Glycolipopeptide hassallidin and macrolide compounds found in *Nostoc* sp. and *Anabaena* sp. proved to be effective against the pathogenic fungi (Shishido et al. 2015). The ethanolic fraction of cyanobacterial cells also exhibits antifungal properties. Metabolites from *Microcystis aeruginosa* such as hexadecenoic acid, BHT, and methyl esters showed their antifungal properties against the *Aspergillus* sp. (Mickymaray and Alturaiki 2018). Another study reported the antifungal characteristics of *Scytonema*, *Nostoc*, and *Anabaena* species due to the presence of scytophycin (Marrez and Sultan 2016). The antifungal properties of *Anabaena laxa* appeared due to the presence of laxaphicins B and C lipopeptides that are effective against *Penicillium notatum*, *C. albicans*, and *Aspergillus Oryzae* (Gautam and Mannan 2020).

Protozoal diseases such as malaria, Chagas, sleeping sickness, and leishmaniosis have caused millions of mortalities and morbidity around the world and huge economic losses as well. The development of resistant strains of these protozoal organisms has urged researchers to develop new drugs against them. Different strains of cyanobacteria produce peptides and lactams that have significant antiprotozoal activity against resistant causative agents. The cyanobacteria *Hyalidium* sp. produce companeramides that inhibit the activity of three chloroquine-resistant strains of plasmodium (Vining et al. 2015). Another important antiprotozoal bioactive compound hoshinolactam which is secreted by *Oscillatoria* sp. has inhibitory action against *Trypanosoma brucei* which is a causative agent of sleeping sickness (Ogawa et al. 2017). Dolastatins are the most important bioactive metabolites discovered from cyanobacteria *Symploca* sp. and *Lyngbya* sp. that have antiparasitic activity against resistant strains of *Plasmodium falciparum.* Despite the high cytotoxicity of dolastatins, it has not been reported as promising antiprotozoal medicine due to their unfavorable pharmacokinetic profile that targets molecular host instead of parasite (Demay et al. 2019; Saad et al. 2022). However, derivatives of dolastatin in combination with monoclonal antibodies have recently been approved by the FDA (Luan et al. 2021).

1.2.4 Cyanobacteria as an Anti-inflammatory and Antioxidant Source

Inflammatory diseases such as Parkinson's disease, atherosclerosis, endometriosis, rheumatoid arthritis, obesity, inflammatory bowel diseases, asthma, multiple sclerosis, and type 1 diabetes are the leading cause of death around the globe (Roh and Sohn 2018). Bioactive compounds from cyanobacteria exhibit potent anti-inflammatory and antioxidant activity. Important anti-inflammatory metabolites isolated from different cyanobacterial strains include tolypodiol, coibacins, malyngamides, ambigol, scytonemin, and honaucins (Rastogi et al. 2015; Demay et al. 2019).

Cyanobacterial cells have the potential to synthesize antioxidants including carotenoids, tocopherol, phycobilisomes, lutein, vitamin E, and polyphenols in response to oxidative stress (Nagarajan et al. 2012). Antioxidants are the compounds that interrupt or inhibit the formation of harmful free radicals that is the major cause of many chronic diseases (Wilson et al. 2017). Many other cyanobacterial compounds such as oleic acid, palmitoleic acid, linolenic acid, and zeaxanthin also have antioxidant features. The free radical scavenging ability of antioxidants prevents several disorders such as cancer, neurodegenerative disorders, stroke, heart diseases, renal failure, atherosclerosis, and aging in addition to its antibacterial, antifungal, and anti-inflammatory properties (Guedes et al. 2011).

Although many bioactive metabolites have been reported so far, only a few bioactive metabolites from cyanobacteria have entered clinical trials for drug development due to various limitations. Several challenges are associated with the development of drugs from active cyanobacterial metabolites. Being secondary metabolites, their development on a commercial scale is hindered by low titers and difficult processing. Associated toxicity of the intermediate or the final product and in vivo inactivation of the product is another challenge. The mode of action, targets, and inhibitory pathways of these metabolites are still unknown as the studies on their properties are mostly limited to in vivo and in vitro animal trials. Therefore, to determine the actual effectiveness of these metabolites, further human clinical trials are required.

1.3 Cyanobacteria as Nutraceutical Agents

Nutraceuticals are nutritional compounds with health benefits (Udayan et al. 2017) that provide concentrated bioactive agents from nonfood matrices. They are known to enhance health that could not be simply obtained from traditional food sources. Cyanobacterial nutraceuticals are viable sources to combat nutritional deficiencies and boost the immune response against heart-related disorders, cancer(s), autoimmune diseases, neurodegenerative disorders, etc. Cyanobacterial genera like *Spirulina*, *Nostoc*, *Microcystis*, *Anabaena*, *Lyngbya*, *Oscillatoria*, and *Tolypothrix* are more prevalent in their applications as nutraceuticals (Singh and Rajan 2022). Cyanobacterial compounds such as polyunsaturated fatty acids, a wide range of essential amino acids, glycolipids, lipoproteins, polyphenols, terpenoids, and bioactive compounds are marketed as nutraceuticals (Singh et al. 2020). This section covers the range of nutraceuticals and their impact on health in a comprehensive way.

1.3.1 Pigments

Scytonema aquatilis and other cyanobacterial species are reported as major sources of phycobilins including phycocyanin and phycoerythrin (Sivakumar et al. 2011) which can scavenge hydroxyl, alkoxyl, and peroxy radicals, thereby contributing as anti-inflammatory and neuroprotectant (Jensen et al. 2015). *Nostoc spongiforme* is a potential source of the high content of phycobilins and phenolic compounds. Other cyanobacteria-pigmented compounds like carotenoids and lutein are also associated with health benefits. Carotenoids play an important role in the reduction of reactive oxygen species. The dominant antioxidative character of carotenoids increases its value in terms of nutrition and health benefits. The effectiveness of carotenoids depends upon the number of double bonds in the molecule (Kini et al. 2020). These pigments also act as precursors of vitamin A in animals and are involved in embryonic development and its morphogenesis. Lutein has also been recommended to prevent retinal degeneration and cancer. It is reported that lutein and β-carotene are helpful in cognitive impairment to a higher extent (Landrum and Bone 2001).

There is also strong evidence that supports the anticancerous activities of cyanobacterial pigments. Astaxanthin shows antitumor features in colon and oral cancer in the post-initiation phase. The suppressive impact of astaxanthin was also reported in transplantable tumor cells with methylcholanthrene-induced fibrosarcoma (Baig et al. 2022). Investigation of the usefulness of phycobilins in the prevention of lung and colon cancer highlighted their suppressive impact against TPA-induced tumors (Manochkumar et al. 2022).

1.3.2 Polyunsaturated Fatty Acids

Polyunsaturated fatty acids (PUFA) including short-chain and long-chain ω-3 PUFA and ω-6 PUFA are known for their health benefits. Traditionally, these compounds are produced by aquatic phytoplankton and fish. Their increasing demand in the current scenario demands the identification of nontraditional PUFA-producing sources. Cyanobacteria such as *Oscillatoria*, *Chlamydomonas*, *Nostoc*, *Isochrysis*, and *Spirulina* are commercially exploited for the production of bioactive compounds mainly PUFA(s) (Hassan et al. 2022).

These compounds are reported to have antioxidant and anti-inflammatory properties. Based on these properties, they have a protective effect against diabetes, obesity, liver fibrosis, dyslipidemia, cardio diseases, neurotoxicity, tumors, chemoradiotherapy-induced bone marrow damage, and weight loss as well as tumor-associated psychological disorders (Xu et al. 2023). They are also associated with improved membrane fluidity, vision, memory, and brain development. The higher concentration of EPA and long-chain fatty acids also plays protective roles in colorectal cancer. Furthermore, the nutrient supplementation of PUFA has shown positive effects against pancreatic cancer by tumor suppression (Hassan et al. 2022). Omega-3 fatty acids help to improve the function of human arteries by lowering blood pressure as well as lowering the levels of low-density lipoproteins (LDL) in diabetic patients. The EPA deficiency in childhood causes issue related to mental health such as increased levels of depression and other heart-related problems. So, this is the best time to raise the level of EPA in the diet as it is reported that this boosts academic performance, behavior, attention, and focus while reducing aggressive behavior. The increased level of EPA in childhood also prevents conditions of dry skin, allergies, and several chronic illnesses. Omega-6 fatty acids are also important as the human body could not synthesize them. These metabolites have a significant role in the regulation of metabolism and development/growth of hairs, skin, and bones (Mathur 2019).

1.3.3 Polysaccharides and Proteins

Polysaccharides serve to prevent oxidative stress in cyanobacterial cells (Gautam and Mannan 2020). They are associated with the prevention of membrane injury and cell apoptosis, reduction in lipid metabolites, and upregulation of catalase enzyme. Cyanobacterial polysaccharides could regulate glucose and lipid levels and thus help in regulating blood pressure. Intake of these polysaccharides results in the reduced production of live-associated lipids and cholesterol esterase. In addition to these, it hampers the adsorption of pancreatic cholesterol and helps in removing excessive cholesterol from the blood. Cyanobacterial polysaccharides are reported to have prebiotic activity, but their use as prebiotics is restricted due to the presence of gelling agents and sulfur compounds which could have undesirably changed the gut microbiota, leading to bacterial outgrowth and inflammatory response (Gomes-Dias et al. 2022). Furthermore, they are known for their techno-function food properties which enables their use as food emulsifier, gelling agent, and stabilizer as well as to produce biodegradable food packaging (Muthukumar et al. 2021). The structure and properties of cyanobacterial polysaccharides have not yet been fully characterized, which limits their commercial applicability. Although they have been explored for novel products, the acquisition of polysaccharide-rich biomass and cost-effective extraction/processing pipeline are the major bottleneck in the commercial feasibility of cyanobacterial polysaccharide-related products.

The Food and Agriculture Organization (FAO) declared *Spirulina* as ideal food due to its potential health benefits as the protein extracted from it contained all the essential amino acids. *Spirulina* is cultivated extensively because of its antioxidant, antiviral, and nontoxic dietary supplement properties. It is also declared a humanitarian instrument to combat malnutrition by the WHO (Mathur 2019).

1.4 Advances in the Extraction Technologies of Nutraceuticals

Several techniques are used to extract these metabolites from the cyanobacterial biomass. However, the selection of suitable methods has a significant impact on the performance and yield of these bioactive molecules. Conditions of the extraction process such as temperature, pressure, solvent amount/quality, and pH are dependent on the nature of extracted compounds. Extraction of bioactive compounds such as fatty acids, carbohydrates, and pigments from cyanobacteria follows conventional, traditional, and modern methods. Conventional methods such as solvent extraction (liquid-to-liquid extraction) and solid-to-liquid extraction methods and Soxhlet apparatus have some downsides including large processing time, thermal denaturation of metabolites, utilization of large volumes of toxic solvents, and noneconomic approaches. In recent years, advanced sustainable methods have been devised to overcome these bottlenecks. These green technologies have higher efficiency, shorter extraction time, limited use of solvents, selective extraction that reduces the chances of contamination from nontarget metabolites, and extraction of metabolites without any effect on their activity. These modern extraction methods include pressurized liquid extraction, ultrasound-assisted extraction (UAE), microwave-assisted extraction (MAE), subcritical water extraction, and solvent-free extraction by supercritical fluid (SFE). The selection of extraction method is dependent on the size of the molecule and its chemical properties, and it is essential to find appropriate methods for optimized extraction of biomolecules.

1.4.1 Supercritical Fluid Extraction

SFE is mainly used for the extraction of high-value metabolites such as lipids, and pigments particularly chlorophylls, and carotenoids due to its low polarity by using supercritical fluids that are considered a nontoxic, cost-effective, and environment-friendly approach. This extraction methods rely on the utilization of a supercritical solvent that infuses into cells in the form of gas vesicles and dissolves different metabolites (Abrahamsson et al. 2018). SFE is generalized as green technology due to its low utilization of organic solvents and less extraction time in comparison to other extraction methods. Higher extraction of γ-linolenic acid from *Arthrospira maxima* was observed with SFE as compared to traditional organic solvent methods (Mendes et al. 2003). Another study reported enhanced supercritical CO_2 extraction in the case of *Nannochloropsis oculata* (Barba et al. 2015).

1.4.2 Wave-Base Extraction Methods

Extraction by microwave and ultrasound is classified as wave-based extraction systems. In microwave-assisted extraction, heat in the form of radiation is used that causes cell wall degradation. The only downside of this method is the thermos de-stability of pigments, but this method is suitable due to its low extraction time and temperature control under vacuum conditions that significantly increase the yields (Lee et al. 2017). *Cylindrotheca closterium* and *Porphyridium purpureum* used this method for the optimized extraction of phycobilins (Pagels et al. 2021). Ultrasound-assisted extraction disrupts cyanobacterial cells from cavitation bubble that causes increased pressure. The disruption of the cell wall subsequently causes solvent penetration and results in the enhanced liberation rate of bioactive compounds. UAE boosts the yield of the extraction process with the use of low temperature and lower consumption of solvents (Fernandes et al. 2017).

1.4.3 *High-Pressure Liquid Extraction (PLE)*

The pressurized liquid extraction method for the disruption of cells is one of the potential techniques used for pigment extraction from cyanobacteria. In PLE, biomass is kept at a higher temperature (50–200 °C) with increased pressure in the presence of a solvent. Higher pressure avoids the solvent from boiling as temperature increases, while high temperature affects cell wall stability which permits the solvent penetration into the cell (Lee et al. 2017). Pressurized liquid extraction was reported as an optimal extraction procedure for carotenoids from *Phormidium* spp. (Rodríguez-Meizoso et al. 2008).

1.5 Strategies for Identification and Production of Cyanobacterial Pharmaceuticals and Nutraceuticals

Cyanobacteria are known for their vast portfolio of pharmaceuticals and nutraceuticals, important secondary bioactive metabolites. Although continuous research efforts are being performed to identify and boost the production of bioactive compounds, yet there exists an untapped pool of these compounds which could unfold through strain improvement technologies (Fig. 1.3) like adaptive evolution, omics, and genetic engineering (Amadu et al. 2021).

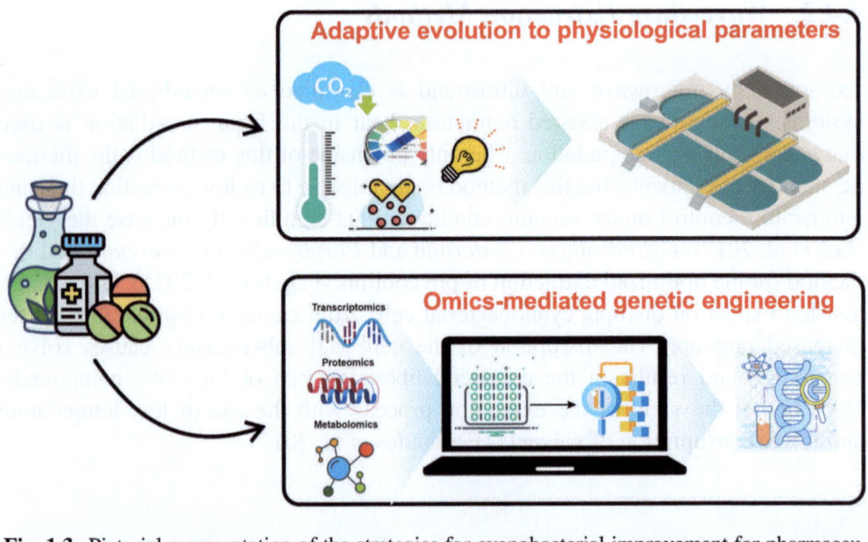

Fig. 1.3 Pictorial representation of the strategies for cyanobacterial improvement for pharmaceutical and nutraceutical purposes

1.5.1 Adaptive Evolution to Physiological Changes

The utilization of indigenous cyanobacterium strains is preferred to produce bioactive compounds as they are already acclimatized to the local environment and do not pose a severe threat to the indigenous biodiversity. However, the commercial robustness of these strains is usually lower as only a few dozen cyanobacterium strains are acclaimed at the commercial level (Bao et al. 2022). Although laboratory exploration of cyanobacterial cultures for secondary metabolite production is more advantageous due to high control on the cell density and the growth dynamics that regulates the biosynthesis and purity of bioactive compounds and permits the extraction of desired compounds in a simpler and reproducible manner (Crnkovic et al. 2018) but, the difficult maintenance of such cultures in real-time settings limits their commercial applicability. Manipulation of environmental factors such as nutrients, temperature, light, etc. is a common strategy to obtain more control over culture conditions to produce enough of the desired compound. Ideal cultivation conditions must induce maximum secondary metabolite diversity, promote growth, and are suitable for several strains. Adaptive evolution of *Nannochloropsis oculata* in response to thermal stress (35 °C) enhances the carotenoid production by 1.23-fold and lipid productivity by 2.24-fold due to upregulation of antioxidant activity as well as rewiring of pathways related to membrane lipid biosynthesis and central carbon metabolism (Arora et al. 2022). Pilot-scale optimization of *Crocosphaera chwakensis* CCY0110 in 120 L photobioreactor at 0.5% CO_2, pH 7.5, and temp. 28 °C under natural sunlight resulted in the production of biomass enriched with 54% of protein, essential amino acids, *n*-3 PUFA, fat-soluble vitamin A and vitamin E, as well as phycocyanin and extracellular polysaccharide cyanoflan (Matinha-Cardoso et al. 2023). Irrespective of cultivation conditions, cyanobacteria tend to accumulate 10–71% of primary metabolites like fatty acids, proteins, lipids, etc., while carotenoid content could range from 0.1 to 14% depending on the species/strain being investigated (Bao et al. 2022). Therefore, strain improvement is the way forward for commercial deployment.

1.5.2 OMICs-Mediated Identification of Gene/Pathway Targets

The definitive knowledge of cellular pathways and molecular targets is essential to develop robust strain(s) for the biosynthesis of bioactive compounds. Omics technologies are used to explore the secondary metabolite production in the stress-subjected cyanobacterium. Twenty-five bioactive compounds with angiotensin-converting enzyme (ACE) inhibitory and antioxidant properties were identified from a pool of 500 peptide sequences by peptidomic analysis of *Scenedesmus obliquus* (Montone et al. 2018). Lipidomics analysis of *Phaeodactylum tricornutum* demonstrated that phenolic stress has a profound effect on the upregulation of key lipogenic pathways as it enhanced the biosynthesis of neutral lipids by 1.54-fold and glycol-lipids by 1.29-fold. It also facilitates the increment of DHA by 2.25-fold and EPA by 1.18-fold. The transcriptomic analysis confirmed the upregulation of genes related to lipogenesis under phenolic stress (Zou et al. 2023). Implementation of integrative technologies is promising to investigate and characterize novel bioactive compounds. Computational mutation discovery, sequencing, and CRISPR-mediated genome editing have been integrated to develop a novel drug discovery approach "DrugTargetSeqR" for the identification of bioactive compound targets (Kasap et al. 2014). Similarly, a multi-omics integrative data analysis platform "TargetMine" has been developed for the robust identification of novel drug targets or bioactive compounds (Chen et al. 2022). OSMAC (one strain multiple products) is a comprehensive approach that uses the cryptic gene cluster through the incorporation of variation in cultivation conditions, omics, and synthetic biology to identify the plethora of bioactive compounds in mimicked natural environment (Maghembe et al. 2020). Currently, OSMAC is mainly applied for the bacterial or fungal system, but it exhibited enormous potential for the omics-based bioprospecting and characterization of metabolic pathways involved in the biosynthesis of cyanobacterial bioactive compounds which could be the future gene targets.

1.5.3 Genetic Modification

A relatively simpler cyanobacterial genome allows the manipulation and regulation of cellular metabolic processes to improve the yield of bioactive compounds (Amadu et al. 2021). In recent years, model cyanobacteria including *Chlamydomonas*, *Synechocystis*, and *Synechococcus* were genetically modified either by the alteration of endogenous genes or through heterologous gene expression to enhance the production of secondary metabolites of pharmaceutical and nutraceutical importance. CRISPR-mediated genome engineering paves the way for efficient strain improvement technologies (Table 1.2).

As the tools for the genetic engineering of cyanobacteria are limited, it is difficult to transfer the genetic information of the target compounds into the host cells. Isolation and expression of cyanobacterial genes need to generate lead compounds for drug development. During the ex vivo production of these metabolites, many housekeeping genes stay silent. Another crucial problem is that several strains of cyanobacteria secrete toxins that have deleterious effects on mammalian cell lines, and this issue needs to be addressed. The success of cyanobacterial therapeutics and nutraceuticals depends on human and environmental safety. More study and research are required to address all these issues.

Table 1.2 Summary of literature related to genetic manipulation of cyanobacteria for enhanced bioactive compound synthesis

Bioactive compound	Genes/enzymes/pathways	Modification	Cyanobacteria	Results	References
Glycogen	Glycogen phosphorylase (glgP), and glycogen-debranching enzyme (glgX)	CRISPR-mediated base editing for the inactivation of glgP and glgX	*Synechococcus elongatus*	95.7% increase in glycogen production (174.7 mg/g) as compared to control	Wang et al. (2023)
Omega-3 fatty acids	*FabG* and *FabZ*	Overexpression of endogenous genes	*Synechococcus elongatus*	Increase in the content of C_{18} fatty acid	Santos-Merino et al. (2022)
Aromatic amino acids	AroG and TyrA enzymes	Heterologous co-expression of feedback inhibition-resistant enzymes	*Synechocystis* sp.	Increase in the production of phenylalanine (580 mg/L) and tyrosine (41 mg/L)	Brey et al. (2020)
Myo-inositol (MI)	MI biosynthesis genes	Elevation of MI biosynthesis genes, downregulation of competing pathways, cofactor enhancement, carbon redirection	*Synechococcus elongatus*	Tenfold improved myo-inositol production with 262.6 mg/L of titer	Sun et al. (2023)
Astaxanthin and β-carotene	β-carotene hydroxylase gene (*crtZ*)	Heterologous expression of *crtZ* with *aadA* gene as the selection marker	*Chlamydomonas reinhardtii*	18% increase in astaxanthin (1.97 mg/g) and 42.2% increase in β-carotene (106 μg/g) production	Huang et al. (2022)
Zeaxanthin	Zeaxanthin epoxidase gene (*ZEP*) and lycopene epsilon cyclase (*LCYE*)	CRISPR-mediated gene knockout	*Chlamydomonas reinhardtii*	60% higher zeaxanthin yield with the final titer of 6.84 mg/L	Song et al. (2020)

1.6 Conclusion and Prospects

Cyanobacteria are effective and unconventional natural sources to combat the issue of prevailing diseases, microbial resistance, environmental sustainability, food scarcity, and malnutrition. Bioactive compounds of cyanobacterial origin are known for their pharmaceutical and nutraceutical potential. Many such bioactive compounds are now in the trial phase for human administration, but their commercial exploration is limited to a few dozen strains. Advances in omics and synthetic biology enable the identification and characterization of novel bioactive compounds and drug targets. Adaptive evolution and genetic engineering approaches are now used to enhance the yield and titer of these valuable compounds.

The commercial feasibility of industrial production is hampered by the constraints related to biomass acquisition, uncertain composition, bioactive profile, the impact of final processing, storage, and consumption practices on the pharmaceutical/nutraceutical potential, associated side effects, consumer acceptance, and legislation requirements. The techno-economic feasibility of the cyanobacterial industry to produce pharmaceuticals and nutraceuticals is only possible by integrating the processing step of multiple bioproducts in a single biorefinery. Thus, considering the biorefinery potential of cyanobacteria, the companies dealing with cyanobacterial products could adopt a synergistic approach where the waste biomass generated after the extraction of pigment or bioactive compounds is supplied to the companies dealing with the production of biofuels, bioplastics, enzymes, etc. It will help in forming a valuable supply chain that will benefit the producers and consumers on the environmental and economic fronts. This harmonization between cyanobacterial industries will ensure the complete valorization of the production process. Rigorous life cycle assessment, techno-economic analysis, and environmental impact analysis of the production pipeline will help in evaluating the sustainability of the process. It will also help in identifying the "hot spots" related to processing, production, and distribution to improve the processing scheme according to market demands.

References

Abrahamsson V, Cunico LP, Andersson N et al (2018) Multicomponent inverse modeling of supercritical fluid extraction of carotenoids, chlorophyll A, ergosterol and lipids from microalgae. J Supercrit Fluid 139:53–61

Agarwal A, Jeevanandham S, Sangam S et al (2022) Exploring the role of carbon-based nanomaterials in microalgae for the sustainable production of bioactive compounds and beyond, vol 7. ACS Omega, p 22061

Al-Awadhi FH, Law BK, Paul VJ, et al. (2017) Grassystatins D–F, potent aspartic protease inhibitors from marine Cyanobacteria as potential antimetastatic agents targeting invasive breast cancer. J Nat Prod 80:2969–2986

Alsenani F, Tupally KR, Chua ET et al (2020) Evaluation of microalgae and cyanobacteria as potential sources of antimicrobial compounds. Saudi Pharm J 28:1834–1841

Amadu AA, KaA D-J, Ameka GK (2021) Industrial applications of cyanobacteria. In: Hozzein WN (ed) Cyanobacteria-recent advances in taxonomy and applications. InTech Open
Amin N, Kannaujiya VK (2021) Metabolic pathways for production of anticancer compounds in cyanobacteria. In: Evolutionary diversity as a source for anticancer molecules. Elsevier
Andler O, Kazmaier U (2021) Total synthesis of apratoxin A and B using Matteson's homologation approach. Organic Biomol Chem 19:4866–4870
Arora N, Lo E, Philippidis GP (2022) A two-prong mutagenesis and adaptive evolution strategy to enhance the temperature tolerance and productivity of *Nannochloropsis oculata*. Bioresour Technol 364:128101
Baig MS, Rajpoot S, Ohishi T et al (2022) Anti-lung cancer properties of cyanobacterial bioactive compounds. Arch Microbiol 204:603
Bao B, Thomas-Hall SR, Schenk PM (2022) Fast-tracking isolation, identification and characterization of new microalgae for nutraceutical and feed applications. Phycologia 2:86
Barba FJ, Grimi N, Vorobiev E (2015) New approaches for the use of non-conventional cell disruption technologies to extract potential food additives and nutraceuticals from microalgae. Food Eng Rev 7:45–62
Bishoyi AK, Sahoo CR, Padhy RN (2022) Recent progression of cyanobacteria and their pharmaceutical utility: an update. J Biomol Struct Dyn:1–34
Brey LF, Włodarczyk AJ, Bang Thøfner JF et al (2020) Metabolic engineering of *Synechocystis* sp. PCC 6803 for the production of aromatic amino acids and derived phenylpropanoids. Metab Eng 57:129–139
Brumley D, Spencer KA, Gunasekera SP et al (2018) Isolation and characterization of anaephenes A–C, alkylphenols from a filamentous cyanobacterium (*Hormoscilla* sp., oscillatoriales). J Nat Prod 81:2716–2721
Bui T-H, Wray V, Nimtz M et al (2014) Balticidins A–D, antifungal hassallidin-like lipopeptides from the Baltic Sea cyanobacterium *Anabaena cylindrica* Bio33. J Nat Prod 77:1287–1296
Business Wire (2022) Global DHA algae oil market 2022-2027
Cabanillas AH, Tena PéRez VC, Maderuelo Corral S et al (2018) Cybastacines A and B: antibiotic sesterterpenes from a *Nostoc* sp. cyanobacterium. J Nat Prod 81:410–413
Carpine R, Sieber S (2021) Antibacterial and antiviral metabolites from cyanobacteria: their application and their impact on human health. Curr Res Biotechnol 3:65–81
Centella MH, Arévalo-Gallegos A, Parra-Saldivar R et al (2017) Marine-derived bioactive compounds for value-added applications in bio-and non-bio sectors. J Clean Prod 168:1559–1565
Chen Y-A, Allendes Osorio RS, Mizuguchi K (2022) TargetMine 2022: a new vision into drug target analysis. Bioinformatics 38:4454–4456
Chilczuk T, Steinborn C, Breinlinger S et al (2020) Hapalindoles from the cyanobacterium *Hapalosiphon* sp. inhibit T cell proliferation. Planta Med 86:96–103
Chirasuwan N, Chaiklahan R, Kittakoop P et al (2009) Anti HSV-1 activity of sulphoquinovosyl diacylglycerol isolated from *Spirulina platensis*. Sci Asia 35:137–141
Crnkovic CM, May DS, Orjala J (2018) The impact of culture conditions on growth and metabolomic profiles of freshwater cyanobacteria. J Appl Phycol 30:375–384
Custom Research (2022). Algae products market- growth, share, opportunities & competitive analysis, 2016-2028
Demay J, Bernard C, Reinhardt A et al (2019) Natural products from cyanobacteria: focus on beneficial activities. Mar Drugs 17:320
Dobretsov S, Teplitski M, Alagely A et al (2010) Malyngolide from the cyanobacterium *Lyngbya majuscula* interferes with quorum sensing circuitry. Environ Microbiol Rep 2:739–744
Fernandes AS, Nogara GP, Menezes CR et al (2017) Identification of chlorophyll molecules with peroxyl radical scavenger capacity in microalgae *Phormidium autumnale* using ultrasound-assisted extraction. Food Res Int 99:1036–1041
Garrison AR, Giomarelli BG, Lear-Rooney CM et al (2014) The cyanobacterial lectin scytovirin displays potent in vitro and in vivo activity against Zaire Ebola virus. Antivir Res 112:1–7

Gautam S, Mannan MAU (2020) The role of algae in nutraceutical and pharmaceutical production. In: Singh J, Meshram V, Gupta M (eds) Bioactive natural products in drug discovery. Springer, pp 665–685

Ghosh T, Bhayani K, Paliwal C et al (2016) Cyanobacterial pigments as natural anti-hyperglycemic agents: an in vitro study. Front Mar Sci 3:146

Giordano D, Coppola D, Russo R et al (2015) Marine microbial secondary metabolites: pathways, evolution and physiological roles. Adv Microb Physiol 66:357–428

Gomes-Dias JS, Teixeira JA, Rocha CMR (2022) Recent advances in the valorization of algae polysaccharides for food and nutraceutical applications: a review on the role of green processing technologies. Food Bioprocess Technol 15:1948–1976

Gómez-Espinoza O, Núñez-Montero K, Díaz LB (2022) Antimicrobial activity of cyanobacteria-derived compounds. In: The pharmacological potential of cyanobacteria. Elsevier

Guedes AC, Amaro HM, Malcata FX (2011) Microalgae as sources of high added-value compounds-a brief review of recent work. Biotechnol Prog 27:597–613

Hachicha R, Elleuch F, Ben Hlima H et al (2022) Biomolecules from microalgae and cyanobacteria: applications and market survey. Appl Sci 12

Hassan S, Meenatchi R, Pachillu K et al (2022) Identification and characterization of the novel bioactive compounds from microalgae and cyanobacteria for pharmaceutical and nutraceutical applications. J Basic Microbiol 62:999–1029

Hayashi T, Hayashi K, Maeda M et al (1996) Calcium Spirulan, an inhibitor of enveloped virus replication, from a blue-green alga *Spirulina platensis*. J Nat Prod 59:83–87

Hernández AJ, Romero A, Gonzalez-Stegmaier R et al (2016) The effects of supplemented diets with a phytopharmaceutical preparation from herbal and macroalgal origin on disease resistance in rainbow trout against *Piscirickettsia salmonis*. Aquaculture 454:109–117

Huang K, Su Z, He M et al (2022) Simultaneous accumulation of astaxanthin and β-carotene in *Chlamydomonas reinhardtii* by the introduction of foreign β-carotene hydroxylase gene in response to high light stress. Biotechnol Lett 44:321–331

Humisto A, Jokela J, Teigen K et al (2019) Characterization of the interaction of the antifungal and cytotoxic cyclic glycolipopeptide hassallidin with sterol-containing lipid membranes. Biochim Biophy Acta Biomembr 1861:1510–1521

Jensen GS, Attridge VL, Beaman JL et al (2015) Antioxidant and anti-inflammatory properties of an aqueous cyanophyta extract derived from *Arthrospira platensis*: contribution to bioactivities by the non-phycocyanin aqueous fraction. J Med Food 18:535–541

Kamble A, Deshmukh R (2021) Phycocyanin market-global opportunity analysis and industry forecast, 2021–2030

Kanekiyo K, Lee J-B, Hayashi K et al (2005) Isolation of an antiviral polysaccharide, nostoflan, from a terrestrial cyanobacterium, *Nostoc flagelliforme*. J Nat Prod 68:1037–1041

Kanekiyo K, Hayashi K, Takenaka H et al (2007) Anti-herpes simplex virus target of an acidic polysaccharide, nostoflan, from the edible blue-green alga *Nostoc flagelliforme*. Biol Pharm Bull 30:1573–1575

Kasap C, Elemento O, Kapoor TM (2014) DrugTargetSeqR: a genomics- and CRISPR-Cas9–based method to analyze drug targets. Nat Chem Biol 10:626–628

Keeffe JR, Gnanapragasam PN, Gillespie SK et al (2011) Designed oligomers of cyanovirin-N show enhanced HIV neutralization. Proc Natl Acad Sci 108:14079–14084

Khan AH, Aziz HA, Khan NA et al (2022) Impact, disease outbreak and the eco-hazards associated with pharmaceutical residues: a critical review. Int J Environ Sci Technol 19:677–688

Kini S, Divyashree M, Mani MK et al (2020) Algae and cyanobacteria as a source of novel bioactive compounds for biomedical applications. In: Singh PK, Kumar A, Singh VK, Shrivastava AK (eds) Advances in cyanobacterial biology. Elsevier

Landrum JT, Bone RA (2001) Lutein, zeaxanthin, and the macular pigment. Arch Biochem Biophys 385:28–40

Laundy M, Gilchrist M, Whitney L (2016) Antimicrobial stewardship. Oxford University Press

Lee SY, Cho JM, Chang YK et al (2017) Cell disruption and lipid extraction for microalgal biorefineries: a review. Bioresour Technol 244:1317–1328

Lee H, Depuydt S, Choi S et al (2021) Chapter 10 - Potential use of nuisance cyanobacteria as a source of anticancer agents. In: Sinha RP, Häder DP (eds) Natural bioactive compounds. Academic Press

Levasseur W, Perré P, Pozzobon V (2020) A review of high value-added molecules production by microalgae in light of the classification. Biotechnol Adv 41:107545

Luan X, Wu Y, Shen Y-W et al (2021) Cytotoxic and antitumor peptides as novel chemotherapeutics. Nat Prod Rep 38:7–17

Luesch H, Moore RE, Paul VJ et al (2001) Isolation of dolastatin 10 from the marine cyanobacterium *Symploca* species VP642 and total stereochemistry and biological evaluation of its analogue symplostatin 1. J Nat Prod 64:907–910

Maghembe R, Damian D, Makaranga A et al (2020) Omics for bioprospecting and drug discovery from bacteria and microalgae. Antibiotics 9

Manochkumar J, Doss CGP, Efferth T et al (2022) Tumor preventive properties of selected marine pigments against colon and breast cancer. Algal Res 61:102594

Mapstone LJ, Leite MN, Purton S et al (2022) Cyanobacteria and microalgae in supporting human habitation on Mars. Biotechnol Adv 59:107946

Marrez DA, Sultan YY (2016) Antifungal activity of the cyanobacterium *Microcystis aeruginosa* against mycotoxigenic fungi. J Appl Pharm Sci 6:191–198

Martins J, Leikoski N, Wahlsten M et al (2018) Sphaerocyclamide, a prenylated cyanobactin from the cyanobacterium *Sphaerospermopsis* sp. LEGE 00249. Sci Rep 8:1–9

Mathur M (2019) Bioactive molecules of *Spirulina*: a food supplement. In: Mérillon JM, Ramawat K (eds) Bioactive molecules in food. Springer

Matinha-Cardoso J, Santos T, Pereira H et al (2023) Pilot scale production of *Crocosphaera chwakensis* CCY0110 and evaluation of its biomass nutritional potential. Algal Res 69:102939

Mazur-Marzec H, Cegłowska M, Konkel R et al (2021) Antiviral cyanometabolites - a review Biomol 11:474

Mcfeeters H, Gilbert MJ, Wood AM et al (2013) Scytovirin engineering improves carbohydrate affinity and HIV-1 entry inhibition. Biochem Physiol S2

Mendes RL, Nobre BP, Cardoso MT et al (2003) Supercritical carbon dioxide extraction of compounds with pharmaceutical importance from microalgae. Inorganica Chim Acta 356: 328–334

Mickymaray S, Alturaiki W (2018) Antifungal efficacy of marine macroalgae against fungal isolates from bronchial asthmatic cases. Molecules 23:3032

Mondal A, Bose S, Banerjee S et al (2020) Marine cyanobacteria and microalgae metabolites - a rich source of potential anticancer drugs. Mar Drugs 18:476

Montaser R, Paul VJ, Luesch H (2011) Pitipeptolides C–F, antimycobacterial cyclodepsipeptides from the marine cyanobacterium *Lyngbya majuscula* from Guam. Phytochemistry 72:2068–2074

Montone CM, Capriotti AL, Cavaliere C et al (2018) Peptidomic strategy for purification and identification of potential ACE-inhibitory and antioxidant peptides in *Tetradesmus obliquus* microalgae. Anal Bioanal Chem 410:3573–3586

Muthukumar J, Chidambaram R, Sukumaran S (2021) Sulfated polysaccharides and its commercial applications in food industries - a review. J Food Sci Technol 58:2453–2466

Nagarajan M, Maruthanayagam V, Sundararaman M (2012) A review of pharmacological and toxicological potentials of marine cyanobacterial metabolites. J Appl Toxicol 32:153–185

Nethravathy MU, Mehar JG, Mudliar SN et al (2019) Recent advances in microalgal bioactives for food, feed, and healthcare products: commercial potential, market space, and sustainability. Compr Rev Food Sci Food Saf 18:1882–1897

Neuhof T, Schmieder P, Preussel K et al (2005) Hassallidin A, a glycosylated lipopeptide with antifungal activity from the cyanobacterium *Hassallia* sp. J Nat Prod 68:695–700

Nguyen AMT, Brettell S, Douanne N et al (2021) Influence of N-methylation and conformation on almiramide anti-leishmanial activity. Molecules 26:3606
Ogawa H, Iwasaki A, Sumimoto S et al (2017) Isolation and total synthesis of hoshinolactam, an antitrypanosomal lactam from a marine cyanobacterium. Org Lett 19:890–893
Pagels F, Pereira RN, Vicente AA et al (2021) Extraction of pigments from microalgae and cyanobacteria-a review on current methodologies. Appl Sci 11:5187
Pandurangan M, Kim DH (2016) Therapeutic potential of cyanobacteria against streptozotocin-induced diabetic rats. 3 Biotech 6:1–7
Pradhan B, Nayak R, Patra S et al (2022) Cyanobacteria and algae-derived bioactive metabolites as antiviral agents: evidence, mode of action, and scope for further expansion; a comprehensive review in light of the sars-cov-2 outbreak. Antioxidants 11:354
Preisitsch M, Harmrolfs K, Pham HT et al (2015) Anti-MRSA-acting carbamidocyclophanes H–L from the Vietnamese cyanobacterium *Nostoc* sp. CAVN2. J Antibiot 68:165–177
Preisitsch M, Niedermeyer TH, Heiden SE, et al. (2016) Cylindrofridins A–C, linear cylindrocyclophane-related alkylresorcinols from the cyanobacterium *Cylindrospermum stagnale*. J Nat Prod 79:106–115
Qamar H, Hussain K, Soni A et al (2021) Cyanobacteria as natural therapeutics and pharmaceutical potential: role in antitumor activity and as nanovectors. Molecules 26:247
Rastogi RP, Sonani RR, Madamwar D (2015) Cyanobacterial sunscreen scytonemin: role in photoprotection and biomedical research. Appl Biochem Biotech 176:1551–1563
Raveh A, Carmeli S (2007) Antimicrobial ambiguines from the cyanobacterium *Fischerella* sp. collected in Israel. J Nat Prod 70:196–201
Rodríguez-Meizoso I, Jaime L, Santoyo S et al (2008) Pressurized fluid extraction of bioactive compounds from *Phormidium* species. J Agric Food Chem 56:3517–3523
Roh JS, Sohn DH (2018) Damage-associated molecular patterns in inflammatory diseases. Immune. Network 18
Rojas V, Rivas L, Cárdenas C et al (2020) Cyanobacteria and eukaryotic microalgae as emerging sources of antibacterial peptides. Molecules 25:5804
Saad MH, El-Fakharany EM, Salem MS et al (2022) The use of cyanobacterial metabolites as natural medical and biotechnological tools. J Biomol Struct Dyn 40:2828–2850
Santos-Merino M, Gutiérrez-Lanza R, Nogales J et al (2022) *Synechococcus elongatus* PCC 7942 as a platform for bioproduction of omega-3 fatty acids. Life 12
Sarada DV, Sreenath Kumar C, Rengasamy R (2011) Purified C-phycocyanin from *Spirulina platensis* (Nordstedt) Geitler: a novel and potent agent against drug resistant bacteria. World J Microbiol Biotechnol 27:779–783
Shahid A, Malik S, Khan AZ et al (2020) Multiproduct algal biorefineries: challenges and opportunities. In: Verma P (ed) Biorefineries: a step towards renewable and clean energy. Singapore, Springer Singapore
Shih C, Teicher BA (2001) Cryptophycins: a novel class of potent antimitotic antitumor depsipeptides. Curr Pharm Des 7:1259–1276
Shishido TK, Humisto A, Jokela J et al (2015) Antifungal compounds from cyanobacteria. Mar Drugs 13:2124–2140
Singh KB, Rajan JP (2022) Therapeutical and nutraceutical roles of cyanobacterial tetrapyrrole chromophore: recent advances and future implications. Front Microbiol 13
Singh SK, Kaur R, Bansal A et al (2020) Biotechnological exploitation of cyanobacteria and microalgae for bioactive compounds. In: Verma ML, Chandel AK (eds) Biotechnological production of bioactive compounds. Elsevier
Sivakumar N, Sundararaman M, Selvakumar G (2011) Efficacy of micro algae and cyanobacteria as a live feed for juveniles of shrimp *Penaeus monodon*. Afr J Biotechnol 10:11594–11599
Song I, Kim J, Baek K et al (2020) The generation of metabolic changes for the production of high-purity zeaxanthin mediated by CRISPR-Cas9 in *Chlamydomonas reinhardtii*. Microb Cell Factories 19:220

Sturdy M, Krunic A, Cho S et al (2010) Eucapsitrione, an anti-*mycobacterium tuberculosis* anthraquinone derivative from the cultured freshwater cyanobacterium *Eucapsis* sp. J Nat Prod 73:1441–1443
Sugumaran A, Pandiyan R, Kandasamy P et al (2022) Marine biome-derived secondary metabolites, a class of promising antineoplastic agents: a systematic review on their classification, mechanism of action and future perspectives. Sci Total Environ 836:155445
Sun T, Li Z, Li S et al (2023) Exploring and validating key factors limiting cyanobacteria-based CO_2 bioconversion: case study to maximize myo-inositol biosynthesis. Chem Eng J 452:139158
Teruya T, Sasaki H, Fukazawa H et al (2009) Bisebromoamide, a potent cytotoxic peptide from the marine cyanobacterium *Lyngbya* sp.: isolation, stereostructure, and biological activity. Org Lett 11:5062–5065
Thuan NH, An TT, Shrestha A et al (2019) Recent advances in exploration and biotechnological production of bioactive compounds in three cyanobacterial genera: *Nostoc*, *Lyngbya*, and *Microcystis*. Front Chem 7:604
Udayan A, Arumugam M, Pandey A (2017) Nutraceuticals from algae and cyanobacteria. In: Rastogi RP, Madamwar D, Pandey A (eds) Algal green chemistry. Elsevier
Verma S, Thapa S, Siddiqui N et al (2022) Cyanobacterial secondary metabolites towards improved commercial significance through multiomics approaches. World J Microbiol Biotechnol 38:100
Vestola J, Shishido TK, Jokela J et al (2014) Hassallidins, antifungal glycolipopeptides, are widespread among cyanobacteria and are the end-product of a nonribosomal pathway. Proc Natl Acad Sci 111:E1909–E1917
Vijayakumar S, Menakha M (2015) Pharmaceutical applications of cyanobacteria - a review. J Acute Med 5:15–23
Vining OB, Medina RA, Mitchell EA et al (2015) Depsipeptide companeramides from a panamanian marine cyanobacterium associated with the coibamide producer. J Nat Prod 78:413–420
Wang S-Y, Li X, Wang S-G et al (2023) Base editing for reprogramming cyanobacterium *Synechococcus elongatus*. Metab Eng 75:91–99
Wilson DW, Nash P, Buttar HS et al (2017) The role of food antioxidants, benefits of functional foods, and influence of feeding habits on the health of the older person: an overview. Antioxidants 6:81
World Health Organization (2020) WHO report on cancer: setting priorities, investing wisely and providing care for all
World Health Organization (2021) Diabetes-fact sheet
Wu C, Chen W, Chen J et al (2015) Preparation of monoPEGylated Cyanovirin-N's derivative and its anti-influenza A virus bioactivity in vitro and in vivo. J Biochem 157:539–548
Xu S, Nijampatnam B, Dutta S et al (2016) Cyanobacterial metabolite calothrixins: recent advances in synthesis and biological evaluation. Mar Drugs 14:17
Xu Q, Zhang Z, Tang M et al (2023) Endogenous production of ω-3 polyunsaturated fatty acids mitigates cisplatin-induced myelosuppression by regulating NRF2-MDM2-p53 signaling pathway. Free Rad Biol Med 201:14–25
Ye B, Jiang P, Zhang T et al (2018) Total synthesis of the highly N-methylated peptides carmabin a and dragomabin. Mar Drugs 16:338
Zou L-G, Zheng D-L, Yao Y-T et al (2023) Polyphenol regulates metabolic and physiological properties of microalgae to accumulate biocompounds. Fuel 352:129085

Chapter 2
Cyanobacterial Pigments: Pharmaceutical and Nutraceutical Applications

Soraya Paz-Montelongo, Cintia Hernández-Sánchez, Fernando Guillén-Pino, Carmen Rubio-Armendáriz, Ángel J. Gutiérrez-Fernández, and Arturo Hardisson

Abstract Cyanobacteria are photosynthetic organisms present in a variety of aquatic and terrestrial environments. In addition to their role in oxygen production and nitrogen fixation, cyanobacteria are also known for producing pigments that have a large variety of applications in the food, pharmaceutical, and cosmetic industries. Some of the cyanobacterial pigments have antioxidant and anti-inflammatory properties and are being investigated for their use in the prevention and treatment of chronic diseases, such as diabetes and Alzheimer's disease. Additionally, some of these pigments have shown antitumoral activity and are being investigated for cancer therapy. The pigments are used as natural colorants in food industry or as active ingredients in skincare products in the cosmetic industry. This chapter delves into the use and properties of these pigments, especially from the food and pharmaceutical point of view. Additionally, possible future applications based on their diverse properties have also been highlighted.

S. Paz-Montelongo (✉) · C. Rubio-Armendáriz · Á. J. Gutiérrez-Fernández · A. Hardisson
Area of Toxicology, Universidad de La Laguna, San Cristóbal de La Laguna, Tenerife, Canary Islands, Spain

Area of Preventive Medicine and Public Health, Universidad de La Laguna, San Cristóbal de La Laguna, Tenerife, Canary Islands, Spain
e-mail: spazmont@ull.edu.es

C. Hernández-Sánchez
Area of Preventive Medicine and Public Health, Universidad de La Laguna, San Cristóbal de La Laguna, Tenerife, Canary Islands, Spain

Grupo interuniversitario de Toxicología Alimentaria y Ambiental, Universidad de La Laguna, San Cristóbal de La Laguna, Tenerife, Canary Islands, Spain

F. Guillén-Pino
Grupo interuniversitario de Toxicología Alimentaria y Ambiental, Universidad de La Laguna, San Cristóbal de La Laguna, Tenerife, Canary Islands, Spain

Area of Legal and Forensic Medicine, Universidad de La Laguna, San Cristóbal de La Laguna, Tenerife, Canary Islands, Spain

M. A. Mehmood et al. (eds.), *Pharmaceutical and Nutraceutical Potential of Cyanobacteria*, https://doi.org/10.1007/978-3-031-45523-0_2

Keywords Cyanobacteria · Pigments · Nutraceuticals · Pharmaceuticals · Natural colorants

2.1 Introduction

Cyanobacteria are photosynthetic organisms found in various aquatic and terrestrial habitats. They are one of the oldest groups of living organisms on Earth and have been around for more than 3.5 billion years. Cyanobacteria are important primary producers in many aquatic and terrestrial ecosystems and play a vital role in the production of oxygen in the atmosphere (Latifi et al. 2009; Ramírez-Mérida et al. 2013; Pathak et al. 2022). In addition, cyanobacteria also produce a wide variety of bioactive compounds, such as pigments, enzymes, and toxins, which have potential applications in the pharmaceutical, nutraceutical, and cosmetic industries (Bhagavathy and Sumathi 2012; Castine et al. 2013; Ramírez-Mérida et al. 2013; Hernández-Reyes et al. 2019).

The use of pigments from cyanobacteria dates to ancient times. Many were the civilizations that used these pigments in their day to day lives. The ancient Egyptians used cyanobacterial pigments for the decoration of temples and statues. In particular, the cyanobacterium *Nostoc* was used to produce blue pigments that were applied to the frescoes in the Luxor Temple. Cyanobacterial pigments have also been found in textiles from Pharaonic times. In the American continent, the ancient Mexicans have also been using pigments from cyanobacteria. In particular, the cyanobacterium *Spirulina* was used to produce green and blue pigments that were used in painting objects and decorating textiles. Additionally, *Spirulina* has been used as a food in various ancient cultures, such as the Aztecs and Mayans. The ancient Chinese also used *Spirulina* as a source of protein and as a traditional medicine. In India, the *Nostoc* cyanobacterium has been used as food and as an Ayurvedic medicine for over a thousand years. *Nostoc* is believed to have beneficial health properties, such as reducing inflammation and improving digestion. On the other hand, in Europe, cyanobacterial pigments were used for dyeing fabrics. The cyanobacterium *Phormidium* was used to produce a red pigment that was used in dyeing wool. Pigments from cyanobacteria were also used in the production of paints and pigments in the Middle Ages.

Due to recent scientific advances, interests in cyanobacterial pigments have increased due to their potential for use in a variety of industrial applications. Phycocyanin, which is a blue pigment and produced by several species of cyanobacteria, has been the subject of numerous studies due to its antioxidant and anti-inflammatory properties. Additionally, phycocyanin has been shown to have antitumor activity and is being investigated for use in cancer therapy. In addition to phycocyanin, cyanobacteria also produce other pigments, such as chlorophyll, carotenoid, and zeaxanthin.

2.2 Classification of Cyanobacteria

Cyanobacteria, also known as cyanophytes or blue-green algae, are classified based on their morphology, physiology, and genetics (Fig. 2.1). Morphology refers to the shape and size of the cyanobacterial cells, while physiology refers to how the cells function and grow. Genetics refers to the genetic information that cyanobacteria possess and how this information is passed from generation to generation.

One of the most common ways to classify cyanobacteria is by their morphology. They can be divided into three main groups: unicellular cyanobacteria, filamentous cyanobacteria, and colonial cyanobacteria. Unicellular cyanobacteria are those that consist of a single cell and do not form colonies. Filamentous cyanobacteria are those that form chains of cells and can have a variety of shapes and sizes. Colonial cyanobacteria are those that consist of cells that have joined together to form a colony.

Another common way to classify cyanobacteria is by their pigmentation. Most cyanobacteria are bluish green in color due to the presence of chlorophyll a and phycocyanin. However, some species of cyanobacteria can also produce other pigments, such as carotenoids, which can give cells an orange or yellow color.

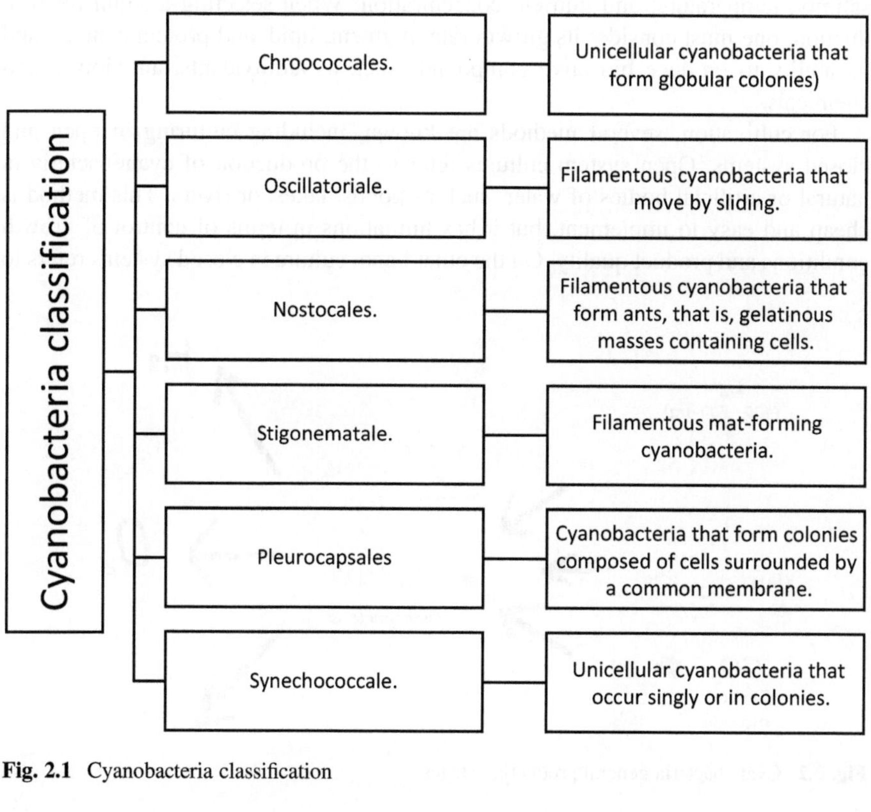

Fig. 2.1 Cyanobacteria classification

Cyanobacteria can also be classified based on their physiology. Some cyanobacteria are capable of nitrogen fixation, which means they can convert gaseous nitrogen into compounds that plants can use to grow. These cyanobacteria are known as nitrogen-fixing cyanobacteria and are important for soil fertility and plant growth. Other cyanobacteria can grow in extreme conditions, such as hot water or highly saline environments.

2.3 Cyanobacteria Cultivation

Cyanobacteria production refers to the cultivation and use of these bacteria for various purposes, such as the production of food, biofuels, chemicals, and medicines (Fig. 2.2) (Rastogi and Sinha 2009; Rajneesh et al. 2017; Singh 2017; Pathak et al. 2018). In this process, it is necessary to consider various factors such as the selection of strains, cultivation methods and their performance, growth conditions, and their subsequent use.

The first stage is the selection of suitable strains. This is a critical step in the production of cyanobacteria and one that will affect the entire production process. Strains have been classified in terms of their characteristics such as light tolerance, salinity, temperature, and nutrient concentration. When selecting a strain for production, one must consider its growth rate, pigment, lipid, and protein content, and its ability to produce bioactive compounds such as antioxidants, antibiotics, and chemicals.

For cultivation, several methods are known, including culturing in open and closed systems. Open system cultures refer to the production of cyanobacteria in natural or artificial bodies of water, such as ponds, lakes, or rivers. This method is cheap and easy to implement, but it has limitations in terms of control of growth conditions and product quality. On the other hand, culture in closed systems refers to

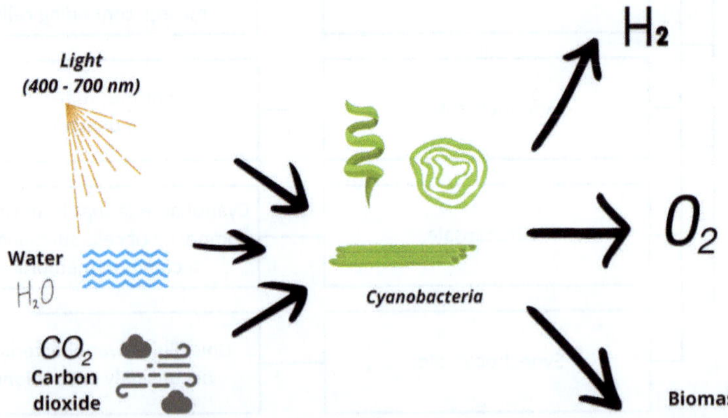

Fig. 2.2 Cyanobacteria general production stages

the production of cyanobacteria in controlled systems such as photobioreactors and culture tanks. These systems allow greater control over growing conditions, such as temperature, lighting, and nutrient concentration, which translates into greater efficiency and product quality.

Then again, considering the growth conditions, cyanobacteria are very versatile and can grow in a wide range of environmental conditions. However, optimal growing conditions vary depending on the strain and the type of culture used. Some important factors to consider are temperature, lighting, nutrient concentration, and acidity of the growing medium:

- **Temperature:** Cyanobacteria can grow in a wide range of temperatures, from around 0 °C to over 40 °C. Optimum growth temperature varies by strain and grow type but is generally in the 20–30 °C range.
- **Light intensity and photoperiod:** Cyanobacteria require light to photosynthesize and grow. The intensity and duration of the necessary light vary according to the strain and the type of culture. The most strains require a light intensity of at least 50–100 μmol/m^2/s and a photoperiod of 12–16 h.
- **Nutrient concentration:** Cyanobacteria require nutrients such as nitrogen, phosphorous, carbon, and other trace elements to grow. The type and amount of nutrients needed vary by strain and type of grow. Excessive use of nutrients can lead to uncontrolled growth and various problems (toxin production, ecosystem alteration, change in the composition of the aquatic environment, or decrease in the amount of dissolved oxygen in water).
- **Acidity:** If the pH of the culture medium is too acidic (pH less than 6), cyanobacteria may experience decreased growth and biomass production. This is because excessive acidity can inhibit the enzyme activity and photosynthesis of cyanobacteria, which can limit their ability to synthesize the nutrients and proteins necessary for their growth. On the other hand, if the pH of the culture medium is too alkaline (pH greater than 9), it can also negatively affect the growth of cyanobacteria. This is because excessive alkalinity can damage cyanobacterial cell membranes and increase the formation of precipitates that can clog nutrient pathways.

In a review carried out by Ramírez-Mérida et al. (2013) (Table 2.1), several methods of obtaining cyanobacteria using photoreactors are listed. Comparing these methods, we found that the flat surface photoreactor methods offer high biomass productivity and are economical and oxygen accumulation is lower (Tredici and Zittelli 1998). However, there are increasing difficulties in controlling parameters such as temperature and wall growth. Likewise, the nutrients required for cyanobacterial culture include water, carbon dioxide, and nitrogenous and phosphorous compounds (Guevara Villalta and Villacreses Reyes 2018).

It should be noted that cyanobacteria can proliferate uncontrollably due to natural causes. For example, blooms of cyanobacteria occur in water bodies such as lakes, marshes, or even oceanic areas (Jang et al. 2003; Paz et al. 2022), as in the recent case of the Macaronesian regions affected by uncontrolled and unusual growth on their coasts (Cordeiro et al. 2020). This unwanted growth causes damage to

Table 2.1 Types of photobioreactors to culture cyanobacteria

Photoreactor type	Advantages	Disadvantages	References
Agitated tank	– Low energy consumption – Less photoinhibition and photooxidation	– Sophisticated materials – Small illumination surface area	Molina et al. (1999), Reyna-Velarde et al. (2010)
airlift system			
Bubbling column			
Flat surface	– High productivity – Economical – Less oxygen accumulation	– Difficult temperature control – Wall growth	Tredici and Zittelli (1998)
Tubular	– Suitable for outdoor cultivation – High productivity – Economical	– Formation of pH gradients – Dissolution of oxygen and CO_2 along the tube	Henrard et al. (2011), Lee et al. (1995)
Hybrid			

ecosystems as well as to the economy of the area itself. Generally, conditions of increased temperature associated with high nutrient content from wastewater or domestic water can lead to the growth of cyanobacteria and the production of cyanotoxins that are harmful to health (El-Shehawy et al. 2011). Cyanotoxins are highly toxic, notably due to their neurotoxic, hepatotoxic, and cytotoxic effects (El-Shehawy et al. 2011; Zilliges et al. 2011; Gehringer and Wannickle 2014).

2.4 The Most Used Cyanobacteria

There are many species of cyanobacteria that are used in different applications, but some of the most common and widely used species are the following:

- ***Arthrospira* (formerly known as *Spirulina*, *Arthrospira platensis*):** It is a filamentous cyanobacterium that is used as a food supplement due to its high content of proteins, vitamins, and minerals.
- ***Nostoc* (*Nostoc commune* and others):** This cyanobacterium is used in the production of biopolymers, which can be used in applications such as the production of biodegradable plastics and biofuels.
- ***Synechococcus* (*Synechococcus elongatus*, *Synechococcus* sp.):** It is a unicellular cyanobacterium that is used in the production of pigments, such as phycocyanin and chlorophyll, which are used in the food industry and in molecular biology research.
- ***Anabaena* (*Anabaena* sp., *Anabaena cylindrica*):** This filamentous cyanobacterium is used in the production of biofuels and as a biofertilizer to improve soil quality and crop production.
- **Nannochloropsis (*Nannochloropsis oceanica, Nannochloropsis gaditana*):** Although not technically a cyanobacterium, but a blue-green microalgae, it is

used in applications like cyanobacteria, such as biofuel production and as a source of omega-3 fatty acids.

It should be noted that the choice of a cyanobacterial species for use in a given application may depend on factors such as its growth rate, its ability to produce the compound of interest, and the ease of culture and manipulation in the laboratory or at an industrial level. Below, we will delve into each of the cyanobacteria mentioned, highlighting their components and main applications.

2.4.1 **Arthrospira *or* Spirulina**

Spirulina is a type of blue-green algae that has been used for centuries as a food source by indigenous peoples in Africa and Central and South America. It is now widely cultivated and consumed around the world for its numerous health benefits and nutritional value (Habib et al. 2008). *Spirulina* growth usually occurs in alkaline and warm environments. It is a single-celled organism that forms long, thin, spiral-shaped filaments. *Spirulina* is blue green in color due to the presence of chlorophyll and phycocyanin, a blue pigment.

This cyanobacterium is a rich source of nutrients as protein, vitamins, minerals, and antioxidants. It is often referred to as a "superfood" due to its high nutritional content (Nagaoka et al. 2005; Gutiérrez-Salmeán et al. 2015). The *Spirulina* is a complete protein source, meaning it contains all nine essential amino acids that the body needs. In fact, it is one of the most protein-dense foods available, with up to 70% protein by weight. About the vitamin content, it includes vitamin B1 (thiamine), vitamin B2 (riboflavin), vitamin B3 (niacin), vitamin B6 (pyridoxine), vitamin B9 (folic acid), vitamin C, and vitamin E (Masten Rutar et al. 2022). For instance, example, one tablespoon of dried *Spirulina* powder (7 g) contains approximately 20–30% of the daily recommended intake of vitamin B1, B2, and B3. For instance,example, one tablespoon of dried *Spirulina* powder (7 g) contains approximately 20–30% of the daily recommended intake of vitamin B1, B2, and B3.

In addition, *Spirulina* contains a variety of minerals or trace elements, including calcium (Ca), iron (Fe), magnesium (Mg), potassium (K), and zinc (Zn). One tablespoon of dried *Spirulina* powder (7 g) contains approximately 11% of the daily recommended intake of calcium, 20% of the daily recommended intake of iron, and 14% of the daily recommended intake of potassium.

Studies conducted by Seghiri et al. (2019) reported a content of protein (76.65 ± 0.15%), carbohydrates (6.46 ± 0.32%), minerals (20.91 ± 0.88%), crude fiber (4.07 ± 1.42%), and lipids (2.45 ± 0.82%) in Moroccan *Spirulina*. According to the results published by Masten Rutar et al. (2022), where the content of minerals was determined in *Spirulina* supplements, these are a good source of calcium (0.15 to 29.5% of RDA, recommended daily intake), phosphorous (3.36–26.7% of RDA), potassium (0.5 to 7.69% of RDA), and selenium (0.01 to 38.6% of RDA) when consumed within recommended amounts.

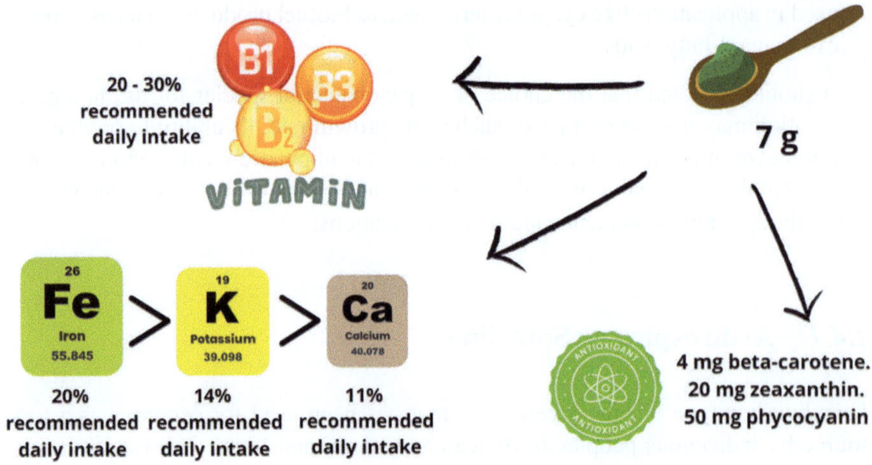

Fig. 2.3 Vitamins, essential elements, and antioxidants from a portion of *Spirulina*

Additionally, this cyanobacterium is rich in antioxidants, which help protect the body from damage caused by free radicals. It contains a range of antioxidants, including beta-carotene, zeaxanthin, and phycocyanin (Fig. 2.3). One tablespoon of dried *Spirulina* powder (7 g) contains approximately 4 mg of beta-carotene, 20 mg of zeaxanthin, and 50 mg of phycocyanin.

Spirulina has also been studied for its potential health benefits, and several studies have found it to be beneficial in a variety of ways. This cyanobacterium may improve immune function, because it contains compounds that have been shown to boost immune function, including phycocyanin and polysaccharides. In addition, it could reduce the inflammation. The nutritional content of *Spirulina* can vary depending on the growing conditions and processing methods used.

Likewise, several studies indicate the usefulness of these cyanobacteria for immunofluorescence assays (Pérez-García et al. 2011). In terms of their industrial use, they can be useful for obtaining biofuel, since, as mentioned, they require less water and make use of solar energy (Chisti 2007).

2.4.2 Nostoc

Nostoc is a genus of cyanobacteria that can be found in a variety of aquatic and terrestrial environments, including soil, freshwater, and marine habitats. This cyanobacterium is part of the diet of Asian populations, although its use has also been highlighted for its beneficial effects (Gao 1998; Fidor et al. 2019). It is a filamentous cyanobacterium that forms colonies of cells called hormogonia. These colonies may be visible to the naked eye. Its characteristic color is bluish-green due to the presence of chlorophyll and phycocyanin (Fidor et al. 2019; Thuan et al. 2019).

Nostoc is a rich source of nutrients, including protein, vitamins, minerals, and antioxidants. It stands out for its protein content, which will reach up to 20% protein by weight (Gao 1998; Liang et al. 2022). It also contains a wide range of vitamins such as vitamin B1 (thiamine), vitamin B2 (riboflavin), vitamin B3 (niacin), vitamin B6 (pyridoxine), vitamin B9 (folic acid), vitamin C, and vitamin E. 100 g of dried *Nostoc* contains approximately 40–50% of the recommended daily intake of vitamin B1, B2, and B3 (Li and Guo 2018; Zhu et al. 2022). However, the required portion is much higher compared to *Spirulina*. Regarding the content of trace elements, Ca, Fe, Mg, K, and Zn stand out, as does *Spirulina* (Gao 1998). Feng et al. (2013) demonstrated that the cyanobacteria *Nostoc* stood out for its Fe and Zn content, especially. On the other hand, this cyanobacterium is also important from the point of view of its antioxidant content, including beta-carotene, zeaxanthin, and phycocyanin. As with *Spirulina*, the nutritional value of *Nostoc* can vary depending on the growing conditions and the processing methods used.

2.4.3 Synechococcus

Synechococcus is a genus of unicellular cyanobacteria that can be found in a variety of aquatic environments, from oceans to fresh and brackish water. They are often found in surface waters, where they photosynthesize and are an important source of primary production in aquatic ecosystems (Olson et al. 1990, 1988; Six et al. 2021; Grébert et al. 2022). *Synechococcus* is a photosynthetic organism and contains chlorophyll and phycocyanin, which give it a characteristic bluish-green color (Olson et al. 1990). This organism is currently of no interest from a food point of view. However, its antioxidant content has been studied. Studies conducted by Lekshmi and Saramma (2018) demonstrated the antioxidant activity of *Synechococcus* sp. Nägeli isolated from Cochin estuary in India.

2.4.4 Anabaena

The genus *Anabaena* includes benthic and planktonic species. It represents solitary filaments and sometimes forms aggregates that can be straight, curved, or rolled. These filamentous cyanobacteria form trichomes, which consist of vegetative cells that participate in fixing nitrogen and photosynthetic growth (Li et al. 2016). The trichomes of these cyanobacteria are not covered by pods but in some species could appear a fine mucilage. The cells show a variety of shapes, from subspherical to cylindrical. The terminal cells are very similar to the rest of the vegetative cells of the filament. In this genus, the presence of gas vesicles is usual.

Anabaena is a common filamentous cyanobacteria bloom worldwide; some species are able to produce life-threatening toxins and for this reason are documented as an important health risk; cyanobacteria due to the production of a variety of toxins like microcystins (MCs), anatoxin-a, anatoxin-a(s), saxitoxin, neosaxitoxin, cylindrospermopsin, and aldo can produce skin irritants (Beltran and Neilan 2000; Al-Tebrineh et al. 2010; Li et al. 2016).

Some species of *Anabaena* are filamentous nitrogen-fixing cyanobacterium that, under controlled laboratory conditions, shows a high growth rate. *Anabaena cylindrica* produces hydrogen and oxygen continuously at the same time that starves nitrogen; for this reason, it is used in the production of biofuels (Tarin et al. 2016). Also, the species *Anabaena cylindrica* is used as biofertilizer to improve soil quality and crop production. *Anabaena* accumulates as part of its biomass phycobiliproteins, photosynthetic antenna pigments (Ojit et al. 2015). In foods and cosmetic industry, phycobiliproteins are used as natural dyes and as fluorescent tracers, presenting many advantages in comparison with other traditional uses.

2.4.5 Nannochloropsis

Technically, *Nannochloropsis* is not categorically a cyanobacterium; actually, it is a blue-green microalga, but it is used in applications similar to cyanobacteria. *Nannochloropsis* sp. includes marine, fresh, and brackish water algal species wide-reaching. Commercial uses of these algae include the production for fish hatcheries, but to provide a sustainable source of this algae, extraction techniques need to be improved to moderate cost before production. It is important to mention that these algae can provide a high-quality protein source, eicosapentaenoic acid, and omega-3 long-chain polyunsaturated fatty acids (Kim et al. 2021). The high concentration of eicosapentaenoic acid in this species has led to its development as vegetarian dietary supplements. In addition, its cosmetic application is under study due to its content of PUFAs, carotenoids, and phenolic compounds (Kim et al. 2021).

Nannochloropsis oceanica is being investigated for biofuel potential resource; it accumulates a higher concentration of lipids, mainly during the stationary phase of its growth. Nearly half of the lipid content of this biofuel is triacylglycerols, which are easily transesterified to biodiesel production (Du Preez et al. 2021). Researchers around the world are hoping to finally realize the potential of these microalgae, to postulate as a sustainable source of biofuels and healthy foods.

2.5 Pigments of Cyanobacteria

As mentioned above, the most prominent pigments from cyanobacteria are chlorophyll, phycocyanin, and phycoerythrin. In the following, the relevant aspects of these pigments will be discussed in more detail.

2.5.1 Chlorophyll

Chlorophyll is the main pigment found in most of the cyanobacteria and is responsible for the green coloring. This green pigment is also found in algae and plants. In terms of its function, it is an essential substance for photosynthesis, the process by which plants and other photosynthetic organisms convert solar energy into chemical energy used to power their metabolic processes (García-Rodríguez and Altamirano-Lozano 2007).

Chemically, chlorophyll is a large, complex molecule containing a porphyrin ring (like hemoglobin) and a central magnesium atom (Fig. 2.4). Its chemical structure is $C_{55}H_{72}MgN_4O_5$. This is a tetrapyrrole derivative (Aronoff 1966). It is the chemical structure of this molecule that gives it the ability to absorb visible light in the 400–700 nm wavelength range, which gives it its characteristic green color (Aronoff 1966).

Chlorophyll is found in leaf cells and is organized into protein complexes called photosystems in the thylakoid membrane of chloroplasts. When light is absorbed by chlorophyll, a flow of electrons is produced and used to generate ATP and NADPH, which are used by the plant to fuel carbohydrate synthesis. The chlorophyll content

Fig. 2.4 Chlorophyll a chemical structure

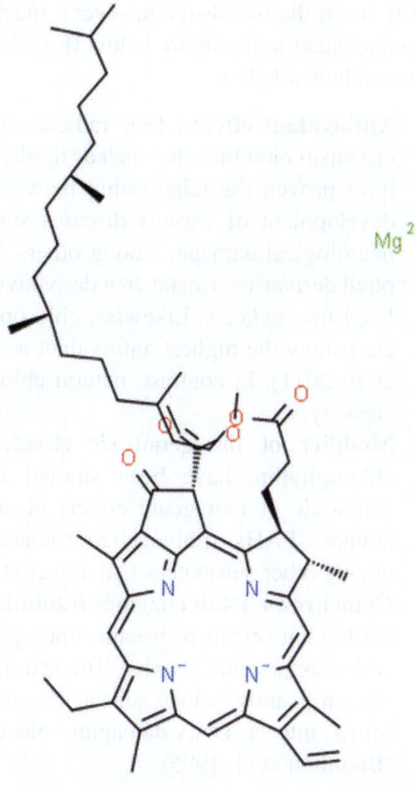

Table 2.2 Chlorophyll a content in different cyanobacteria genera

Cyanobacteria genera	Chlorophyll a (µg/mL)	Chlorophyll b (µg/mL)	References
Anabaena	17.4 ± 1.2	–	Loreto et al. (2003)
Chlorella vulgaris	0.2–1.5		Rinawati et al. (2020)
Nannochloropsis sp.	0.1–0.5		
Porphyridium cruentum	0.5–0.8		
Spirulina platensis	1–3.5		
Chlorella	8.45	4.33	Nwe-Oo et al. (2017)
Nannochloropsis	21.24	9.66	

of cyanobacteria is variable (Table 2.2), which is why some of them are of major interest for pigment production. In addition, pigment content can vary due to various factors such as light, nutrients, and oxygen (López Muñoz et al. 1992; Loreto et al. 2003; Ortega et al. 2004; Moronta et al. 2006; Song et al. 2016).

As for the mechanism of action of chlorophyll, natural chlorophyll derivatives are affected by gastric acidity, which acts by modifying this molecule and producing a metal-free pheophytin derivative. Following degradation in the digestive tract, these derivatives are absorbed by the intestinal cells, passing into the bloodstream. Once they reach the bloodstream, several mechanisms occur (Mishra et al. 2011), which are indicated in the figure below (Fig. 2.5). Some of the most relevant mechanisms are explained below:

- **Antioxidant effects:** Free radicals are the main cause of oxidative damage to organism biomolecules such as lipids, proteins, and nucleic acids. Several studies have proven the relationship between ROS, reactive oxygen species, and the development of various diseases such as cancer, inflammatory diseases, and neurological damage, among others. The antioxidant activity of dietary chlorophyll derivatives (metal-free derivatives) shows a high capacity to scavenge long-lived free radicals. Likewise, chlorophyll derivatives with metals (Mg, Zn, Cu, etc.) show the highest antioxidant activity (Lanfer-Marquez et al. 2005; Mishra et al. 2011). In contrast, natural chlorophyll a and b show a lower antioxidant capacity.
- **Modifier of the genotoxic effect:** Some chlorophyll derivatives, such as chlorophyllin, have been studied for their protective capacity against the genotoxic or mutagenic effects of substances such as aflatoxins, heterocyclic amines, PAHs (polycyclic aromatic hydrocarbons), and alkylating agents, among other substances (Mishra et al. 2011).
- **Cytochrome P450 enzymes inhibition:** Cytochrome P450 enzymes are vital in the human organism because they participate in the elimination of compounds with carcinogenic activity. The activity of dietary chlorophyllin has been shown to significantly reduce substances such as aflatoxin B1 (a substance that transforms into a DNA-damaging electrophilic agent) in experimental animals (Breinholt et al. 1995).

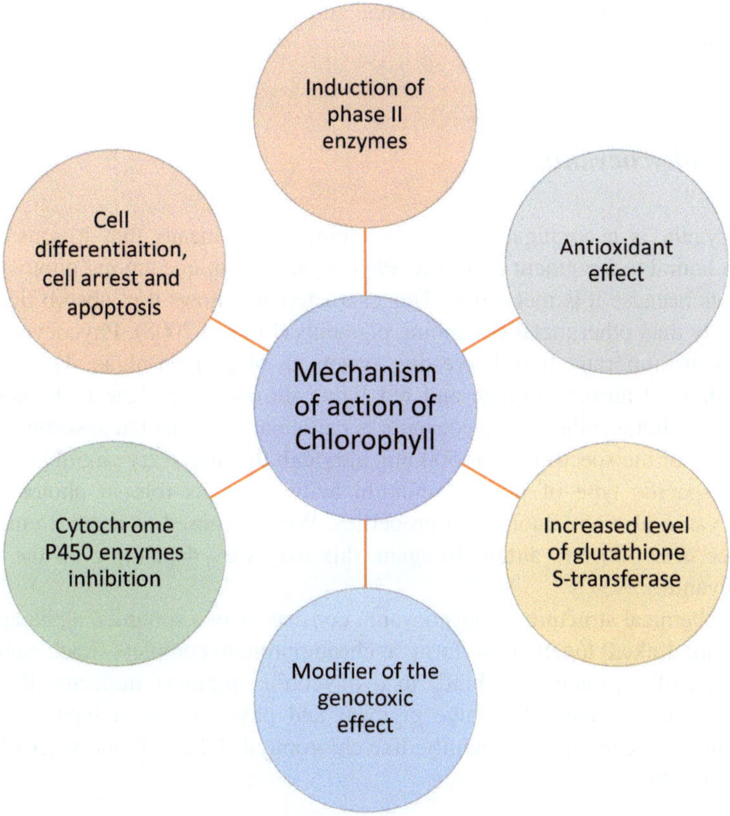

Fig. 2.5 Different modes of action of the chlorophyll

For these reasons, the use of chlorophyll and its derivatives has been extensively studied for pharmaceutical and medical applications. In terms of industrial uses of chlorophyll, one of the most prominent applications is its use as a food coloring, where it is used to give green color to foods such as sweets, ice cream, and drinks. They are also used in starchy products, oils and sauces, condiments, and canned vegetables (Jácome-Pilco et al. 2023). Chlorophyll is used as a food additive and has the international codification INS 140 (E-140) with a color index CI 75810 according to the (EFSA, 2015). Another important application of chlorophyll is in water purification. Chlorophyll is used in wastewater treatment processes to remove organic and inorganic pollutants. Chlorophyll can adsorb pollutants such as heavy metals and pesticides, making them less toxic to aquatic organisms and humans.

On the other hand, research is also focusing on the uses of chlorophyll in the treatment of various pathologies. Chlorophyll has been shown to have anti-inflammatory and antioxidant properties, making it potentially useful in the treatment of diseases such as arthritis and cancer. In fact, a reduction in cancer incidence has been linked to individuals with a diet rich in chlorophyll a (Streit et al. 2015). It has also been shown to have antimicrobial and antiviral properties, making it useful in the treatment of infections.

2.5.2 *Phycocyanin*

Phycocyanin is a conjugated chromoprotein, which means that it consists of a protein bound to a pigment molecule. Phycocyanin is unique among photosynthetic pigments because it is metal-free. This characteristic allows it to absorb light more efficiently than other metal-containing pigments (Eriksen 2008). Phycocyanin functions as an important light-harvesting system in blue-green algae. Together with chlorophyll, it absorbs orange and red light, allowing the algae to harness light energy for photosynthesis. Phycocyanin is particularly efficient at absorbing light in the region of the spectrum near 500 nm, although this may vary slightly depending on the specific type of phycocyanin. In addition to its role in photosynthesis, phycocyanin also has fluorescent properties. When exposed to light, it emits fluorescence near 700 nm, although, again, this may vary depending on the type of phycocyanin.

The chemical structure of phycocyanin consists of two subunits, alpha and beta, which are linked together to form a chromoprotein complex. Each subunit is composed of a protein that binds to a phycobilin pigment molecule (Fig. 2.6). Phycocyanin is a metal-free blue pigment and phycobilin is a type of pigment belonging to the tetrapyrrole family, like chlorophyll (Glazer 1994; MacColl 2004; Sun et al. 2003).

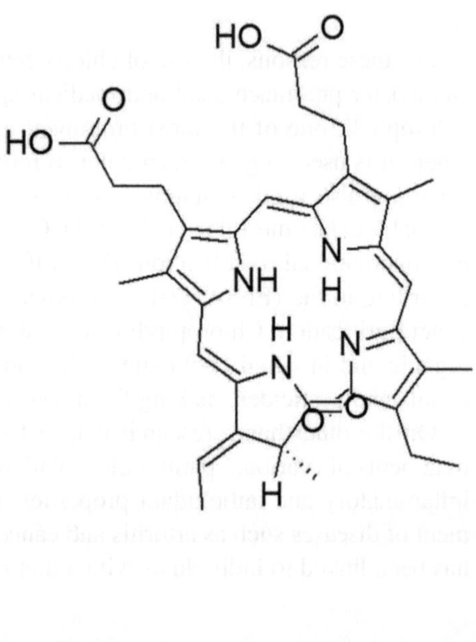

Fig. 2.6 Phycobilin chemical structure

Table 2.3 Health effects of the phycocyanin

Effects	Experiment type	Study subject	Research
Anti-inflammatory, acts to reduce acetic acid-induced colitis, scavenges oxygen radicals	*In vivo*	Rat	González et al. (1999)
Inhibits kainic acid-induced neural damage, scavenges oxygen radicals			Rimbau et al. (1999)
Acts to prevents thioacetamide-induced hepatic encephalopathy, avoids lipid peroxidation			Sathyasaikumar et al. (2007)
Prevents oxalic acid-induced kidney stone formation and the lipid peroxidation			Farooq et al. (2004)
Decrease cardiotoxicity of the drug DOX[a], scavenging of oxygen radicals	*In vitro* (cardiomyocytes)		Kahn et al. (2006)
Anti-inflammatory, blocks nitric oxide synthase expression and reduces nitrite synthesis	*In vitro* (macrophage cell line)	Human	Cherng et al. (2007)
Anticancerogenic, induces apoptosis in human hepatocellular carcinoma cells	*In vitro* (hepatocellular carcinoma cells)		Roy et al. 2007)

[a]DOX (doxorubicin) is a drug used for a wide range of cancers (Carvalho et al. 2009)

The phycobilin molecule binds to the protein through a structure known as a thioether, which is a bond between the sulfhydryl (–SH) group of the protein and the ketone (=O) group of the phycobilin. The three-dimensional structure of phycocyanin is complex and is important for its role in light absorption and energy transfer in the process of photosynthesis. Phycobiliproteins are compounds that, despite not being essential for the cellular functioning of cyanobacteria and algae, play an important role in terms of their capacity to store nitrogen. They are also one of the most abundant proteins found in these organisms.

The degradation of these proteins is selective and occurs when the cells suffer from a nitrogen deficit (Lewitus and Caron 1990; Sloth et al. 2006). It is at this moment that phycobiliproteins are mobilized to avoid this nitrogen deficit. On the other hand, both phycobilin and phycobiliproteins are important as they are often linked together to form gene clusters (Guan et al. 2007). As for the beneficial effects of this pigment, its antioxidant, anti-inflammatory, and even anticarcinogenic effects stand out (Jiang et al. 2017). However, as can be seen in the following table (Table 2.3), these effects have been proven through experimental animals or *in vitro* with human tissues and cells.

Among the effects listed in the Table 2.3, the most important effect is related to cancer prevention. Several authors have studied the application of phycocyanin in the prevention of various types of cancer, with very positive results. For example, studies carried out by Subhashini et al. (2004), which tested the effect of phycocyanin C (C-PC) on the growth and multiplication of the human chronic myeloid leukemia cell line, showed a significant reduction of up to 49% in the proliferation of these cells (treated with 50 μM of C-PC for up to 48 h). Other authors such as Basha et al. (2008) also demonstrated a significant reduction of human hepatoma cell proliferation by inducing cell apoptosis.

However, despite the multiple beneficial effects of this pigment on cancer cell proliferation, among others, it is not currently being used as a drug for clinical use, as the molecular mechanism of phycocyanin is unknown.

However, phycocyanin is currently present in other areas, such as the food industry, where it is used as a dye, although with certain limitations as it is highly sensitive to heat treatment, unless sugars are present, generally fructose. It is commonly used as a food coloring agent in candies, ice cream, and other dairy products (Zeece 2020). It was approved as a food coloring agent by the FDA (Food and Drug Administration) in 2013 (Stanic-Vucinic et al. 2018) and is known as blue *Spirulina* powder.

2.5.3 *Phycoerythrin*

Phycoerythrin is a red pigment of the phycobilin type, holoproteins with linear tetrapyrrole prosthetic groups that are covalently linked to specific cystine residues of the apoprotein, which acts in conjunction with chlorophyll (Ramu Ganesan et al. 2022). This pigment is found in red algae and unicellular algae such as cryptophytes (Rojas and Buitrago 2019; Hamouda and El-Naggar 2021). There are three different types of phycoerythrin present in algae: C-phycoerythrin (found in cyanobacteria), R-phycoerythrin (present in red algae, Rhodophyta), and phycoerythrocyanin (Fig. 2.7) (Morançais et al. 2018).

The presence of phycoerythrin in the various types of cyanobacteria is influenced by several factors, but light conditions, especially light intensity, play a major role in the content of this pigment (Glazer 1984; Hamouda and El-Naggar 2021). C-phycoerythrin, found mainly in cyanobacteria, is a compound whose molecular weight is 225 kDa and has a maximum fluorescent emission at 578 nm, with a single visible absorption peak found between 540 and 570 nm (Hamouda and El-Naggar 2021).

Phycobiliproteins have been studied for their various beneficial effects on health, including the study published by Vargas-Rodríguez et al. (2006), in which they studied the *in vitro* effect of these compounds in the treatment of cervical carcinoma,

Fig. 2.7 Phycoerythrin chemical structure

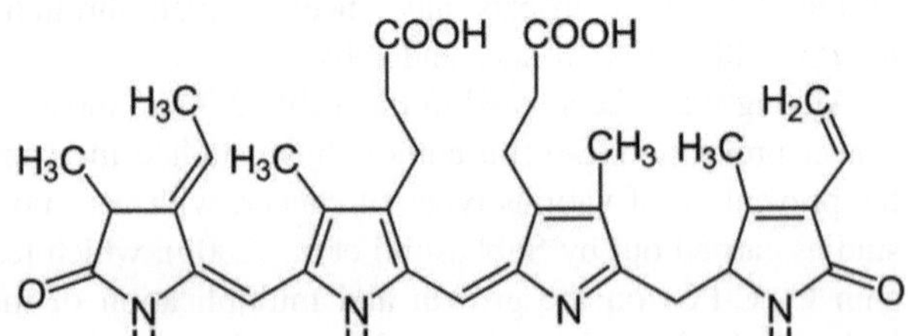

proving their cytotoxic effect on these cells. Another study by Fernández-Duarte et al. (2017) shows the procedure for obtaining pure preparations of B-phycoerythrin and its conjugation with specific antibodies, being a satisfactory process with a high yield (30–51%) and with high levels of purity in phycoerythrin (over 90%), allowing this preparation to be used for diagnostic purposes.

Another beneficial effect of phycoerythrin is its free radical scavenging capacity (Patel et al. 2018; Afreen and Fatma 2018); its antioxidant capacity varies between different species of cyanobacteria (Sonani et al. 2017). In addition, this compound has been shown to have beneficial effects on diabetes (Soni et al. 2009). This red pigment has been extracted and is also used in the food industry as a coloring agent in foods such as ice cream, dairy products, and beverages (Ramu Ganesan et al. 2022). It has also been used in the cosmetic industry (Nguyen et al. 2020).

2.6 Conclusion and Prospects

In this chapter, some emerging prospects of cyanobacterial pigments are highlighted, based on the trend among researchers, pharmaceutical companies, and other interested stakeholders, to develop techniques for the cultivation, extraction, and evaluation of the different compounds applied to fields as diverse as food, health, and even the production of biofuels and other materials. The possibilities for the future of these pigments are therefore based on the following aspects:

1. **Applications in the food industry:** Cyanobacterial pigments have valuable properties for use as natural colorants and are currently used as food additives in some countries or in certain food groups. Their future lies in their implementation in a greater number of foods, trying to minimize or avoid the use of artificial additives that may entail health risks.
2. **Medical and pharmaceutical applications:** The beneficial effects of cyanobacterial pigments have been demonstrated for their role as antioxidants, anti-inflammatory, and even anticarcinogenic. However, some mechanisms are unknown, and this remains a challenge when it comes to using these compounds in therapies or synthesizing drugs based on them.
3. **Industry for obtaining plastics and other sustainable and ecological materials:** These pigments can be used to obtain bioplastics (textiles, packaging, electronic products, etc.), avoiding the excessive use of fossil fuel derivatives. Furthermore, it should be noted that biomass production of cyanobacteria is carbon neutral.
4. **Obtaining biofuels:** The biomass generated by cyanobacteria, together with their rapid and efficient growth, can be used to obtain biofuels.

Therefore, the future of cyanobacterial pigments lies in continuing to focus developing robust and cost-effective methods of their extraction, storage, and transportation to sales counter for the customers for a diverse range of applications.

References

Afreen S, Fatma T (2018) Extraction, purification and characterization of phycoerythrin from Michrochaete and its biological activities. Biocatal Agric Biotechnol 13:84–89

Al-Tebrineh J, Mihali TK, Pomati F, Neilan BA (2010) Detection of Saxitoxin-producing cyanobacteria and *Anabaena circinalis* in environmental water blooms by quantitative PCR. Appl Environ Microbiol 76(23):7836–7842. https://doi.org/10.1128/AEM.00174-10

Aronoff S (1966) Chapter 1. The chlorophylls – an introductory survey. In: Vernon LP, Seely GR (eds) The chlorophylls. Physical, chemical, and biological properties. Academic Press, New York, pp 3–19

Basha OM, Hafez RA, El-Ayouty YM, Mahrous KF, Bareedy MH, Salama AM (2008) C-Phycocyanin inhibits cell proliferation and may induce apoptosis in human HepG2 cells. Egypt J Immunol 15(2):161–167

Beltran EC, Neilan BA (2000) Geographical segregation of the neurotoxin-producing cyanobacterium Anabaena circinalis. Appl Environ Microbiol 66(10):4468–4474

Bhagavathy S, Sumathi P (2012) Evaluation of antigenotoxic effects of carotenoids from green algae Chlorococcum humicola using human lymphocytes. Asian Pac J Trop Biomed 2(2): 109–117

Breinholt, V., Schimerlik, M., Dashwood, R., & Bailey, G. (1995). Mechanisms of chlorophyllin anticarcinogenesis against aflatoxin B1: complex formation with the carcinogen. Chemical research in toxicology, 8(4), 506-514.

Carvalho C, Santos RX, Cardoso S, Correia S, Oliveira PJ, Santos MS, Moreira PI (2009) Doxorubicin: the good, the bad and the ugly effect. Curr Med Chem 16(25):3267–3285

Castine S, Paul N, Magnusson M, Bird M, de Nys R (2013) Algal bioproducts derived from suspended solids in intensive land-based aquaculture. Bioresour Technol 131:113–120

Cherng S-C, Cheng S-N, Tarn A, Chou T-C (2007) Anti-inflammatory activity of c-phycocyanin in lipopolysaccharide-stimulated RAW 264.7 macrophages. Life Sci 81:1431–1435

Chisti Y (2007) Biodiesel from microalgae. Biotechnol Adv 25:294–306

Cordeiro R, Luz R, Vasconcelos V, Fonseca A, Gonçalves V (2020) A critical review of cyanobacteria distribution and cyanotoxins occurrence in Atlantic Ocean islands. Cryptogam Algol 41(9):73–89

Du Preez R, Majzoub ME, Thomas T, Panchal SK, Brown L (2021) Nannochloropsis oceanica as a microalgal food intervention in diet-induced metabolic syndrome in rats. Nutrients 13(11):3991

EFSA (2015). EFSA. Panel on Food Additives and Nutrient Sources added to Food (ANS). "Scientific Opinion on the re-evaluation of chlorophylls (E 140 (i)) as food additives." EFSA Journal 13.5 (2015): 4089.

El-Shehawy R, Gorokhova E, Fernández-Piñas F, del Campo FF (2011) Global warming and hepatotoxin production by cyanobacteria: what can we learn from experiments? Water Res 46:1420–1429

Eriksen NT (2008) Production of phycocyanin—a pigment with applications in biology, biotechnology, foods and medicine. Appl Microbiol Biotechnol 80:1–14

Farooq SM, Asokan D, Kalaiselvi P, Sakthivel R, Varalakshmi P (2004) Prophylactic role of phycocyanin: a study of oxalate mediated renal cell injury. Chem Biol Interact 149:1–7

Feng TT, Chen HB, Shao P, He JZ (2013) The nutritional value and preliminary risk assessment of major and trace elements accumulated in 8 edible seaweeds. In: Advanced materials research, vol 781. Trans Tech Publications, pp 1377–1384

Fernández-Duarte G, Gutiérrez-Calzado E, García-Reyes LM (2017) Obtención de β-Ficoeritrina y su uso en la conjugación de anticuerpos murinos. Rev Cubana Quím 29(2)

Fidor A, Konkel R, Mazur-Marzec H (2019 Sep 29) Bioactive peptides produced by cyanobacteria of the genus Nostoc: a review. Mar Drugs 17(10):561

Gao K (1998 Feb) Chinese studies on the edible blue-green alga, Nostoc flagelliforme: a review. J Appl Phycol 10:37–49

García-Rodríguez MC, Altamirano-Lozano MA (2007) Chlorophyllin as modulator and DNA damage protector: in vivo mouse system experience. Bioquimia 32(1):15–24
Gehringer MM, Wannickle N (2014) Cimate change an regulation of hepatotoxin production in cyanobacteria. FEMS Microbiol Ecol 88:1–25
Glazer AN (1984) Phycobilisome a macromolecular complex optimized for light energy transfer. Biochim Biophys Acta Bioenerg 768(1):29–51
Glazer AN (1994) Phycobiliproteins—a family of valuable, widely used fluorophores. J Appl Phycol 6:105–112
González R, Rodríques S, Romay C, Ancheta O, González A, Armesto J, Remirez D, Merino N (1999) Anti-inflammatory activity of phycocyanin extract in acetic acid-induced colitis in rats. Pharmacol Res 39:55–59
Grébert T, Garczarek L, Daubin V, Humily F, Marie D, Ratin M et al (2022) Diversity and evolution of pigment types in marine Synechococcus cyanobacteria. Genome Biol Evol 14(4): evac035
Guan X, Qin S, Zhao F, Zhang X, Tang X (2007) Phycobilisomes linker family in cyanobacterial genomes: divergence and evolution. Int J Biol Sci 3:434–445
Guevara Villalta TM, Villacreses Reyes MM (2018) Obtención de biomasa empleando la cianobacteria arthrospira sp. con residuos de tilapia. Bachelor's thesis. Universidad de Guayaquil, Facultad de Ingeniería Química, Ecuador
Gutiérrez-Salmeán G, Fabila-Castillo L, Chamorro-Cevallos G (2015) Nutritional and toxicological aspects of spirulina (Arthrospira). Nutr Hosp 32:34–40
Habib MAB, Parvin M, Huntington TC, Hasan MR (2008) A review on culture, production and use of spirulina as food for humans and feeds for domestic animals and fish. Food and Agriculture Organization of the United Nations, Rome, Italy
Hamouda RA, El-Naggar NEA (2021) Cyanobacteria-based microbial cell factories for production of industrial products. In: Microbial cell factories engineering for production of biomolecules. Academic Press, pp 277–302
Henrard AA, de Morais M, Costa J (2011) Vertical tubular photobioreactor for semicontinuous culture of Cyanobium sp. Bioresour Technol 102:4897–4900
Hernández-Reyes BM, Rodríguez-Palacio MC, Castilla-Hernández P, Sánchez-Robles J, Vela-Correa G, Schettino-Bermúdez B (2019) Uso potencial de cianobacterias como biofertilizante para el cultivo de maíz azul en la Ciudad de México. Revista Latinoamericana de Biotecnología Ambiental y Algal 10(1):13–27
Jácome-Pilco CR, Aucatoma Chico KB, Agualongo Sinchipa SA, Callan Chela CR, Montero Silva VD (2023) Biotecnología para la extracción de pigmentos vegetales, para uso industrial. Revista Latinoamericana de Ciencias Sociales y Humanidades 4(1):1475–1488
Jang M-H, Ha K, Joo G-J, Takamura N (2003) Toxin production of cyanobacteria is increased by exposure to zooplankton. Freshw Biol 48:1540–1550
Jiang L, Wang Y, Yin Q, Liu G, Liu H, Huang Y, Li B (2017) Phycocyanin: a potential drug for cancer treatment. J Cancer 8(17):3416
Kahn M, Varadharaj S, Shobha JC, Naidu MU, Parinandi NL, Kutala VK, Kuppusamy P (2006) C-phycocyanin ameliorates doxorubicin-induced oxidative stress and apoptosis in adult rat cardiomyocytes. J Cardiovasc Pharmacol 47:9–20
Kim SY, Kwon YM, Kim KW, Kim JYH (2021) Exploring the potential of Nannochloropsis sp. extract for cosmeceutical applications. Mar Drugs 19(12):690
Lanfer-Marquez, U. M., Barros, R. M., & Sinnecker, P. (2005). Antioxidant activity of chlorophylls and their derivatives. Food research international, 38(8-9), 885-891.
Latifi A, Ruiz M, Zhang C-C (2009) Oxidative stress in cyanobacteria. FEMS Microbiol Rev 33: 258–278
Lee YK, Ding S, Low C, Chang Y, Forday W, Chew P (1995) Design and performance of an α-type tubular photobioreactor for mass cultivation of microalgae. J Appl Phycol 7:47–51
Lekshmi S, Saramma AV (2018) Antioxidant activity of synechococcus sp. Nägeli isolated from cochin estuary, India. Indian J Geo-Mar Sci 47(11):2213–2216

Lewitus AJ, Caron DA (1990) Relative effects of nitrogen or phosphorus depletion and light intensity on the pigmentation, chemical composition, and volume of Pyrenomonas salina (Cryptophyceae). Mar Ecol Prog Ser 61:171–181
Li Z, Guo M (2018) Healthy efficacy of Nostoc commune Vaucher. Oncotarget 9(18):14669
Li X, Dreher TW, Li R (2016) An overview of diversity, occurrence, genetics and toxin production of bloom-forming Dolichospermum (anabaena) species. Harmful Algae 54:54–68
Liang Y, Shu X, Wang W (2022) Biochemical composition, heavy metal content and their geographic variations of the form species Nostoc commune across China. Food Sci Technol 42
López Muñoz I, Abalde J, Herrero C (1992) Crecimiento y contenido en pigmentos de cuatro especies de microalgas marinas cultivadas con diferentes temperaturas e intensidades de luz
Loreto C, Rosales N, Bermúdez J, Morales E (2003) Producción de pigmentos y proteínas de la cianobacteria Anabaena PCC 7120 en relación a la concentración de nitrógeno e irradiancia. Gayana Bot 60(2):83–89
MacColl R (2004) Allophycocyanin and energy transfer. Biochim Biophys Acta 1657:73–81
Masten Rutar J, Jagodic Hudobivnik M, Nečemer M, Vogel Mikuš K, Arčon I, Ogrinc N (2022 Mar 17) Nutritional quality and safety of the spirulina dietary supplements sold on the Slovenian market. Foods 11(6):849
Mishra, A., Kavita, K., & Jha, B. (2011). Characterization of extracellular polymeric substances produced by micro-algae Dunaliella salina. Carbohydrate Polymers, 83(2), 852-857.
Molina E, Acién F, García F, Chisti Y (1999) Photobioreactors: light regime, mass transfer, and scaleup. J Biotechnol 70:231–247
Morançais M, Mouget JL, Dumay J (2018) Proteins and pigments. In: Microalgae in health and disease prevention. Academic Press, pp 145–175
Moronta R, Mora R, Morales E (2006 Jan) Respuesta de la microalga Chlorella sorokiniana al pH, salinidad y temperatura en condiciones axénicas y no axénicas. Rev Fac Agron 23(1):28–43
Nagaoka S, Shimizu K, Kaneko H, Shibayama F, Morikawa K, Kanamaru Y, Otsuka A, Hirahashi T, Kato T (2005) A novel protein C-Phycocyanin plays a crucial role in the hypocholesterolemic action of spirulina platensis concentrate in rats. J Nutr 135:2425–2430
Nguyen HPT, Morançais M, Déléris P, Fleurence J, Nguyen-Le CT, Vo KH, Dumay J (2020) Purification of R-phycoerythrin from a marine macroalga Gracilaria gracilis by anion-exchange chromatography. J Appl Phycol 32:553–561
Nwe Oo, Y.Y.; Su, M.C.; Kyaw, K.T. (2017). Extraction and determination of chlorophyll content from microalgae. Int. J. Adv. Res. Publ. 1, 298–301
Ojit SK, Indrama T, Gunapati O, Avijeet SO, Subhalaxmi SA, Silvia CH et al (2015) The response of phycobiliproteins to light qualities in Anabaena circinalis. J Appl Biol Biotechnol 3(3): 001–006
Olson RJ, Chisholm SW, Zettler ER, Armbrust EV (1988) Analysis of Synechococcus pigment types in the sea using single and dual beam flow cytometry. Deep Sea Res A Oceanogr Res Pap 35(3):425–440
Olson RJ, Chisholm SW, Zettler ER, Armbrust EV (1990) Pigments, size, and distributions of Synechococcus in the North Atlantic and Pacific oceans. Limnol Oceanogr 35(1):45–58
Ortega J, Moronta R, Morales E (2004) Influencia del acetato sobre el crecimiento y contenido de pigmentos de la microalga Chlorella sp. Ciencia 12(1):25–31
Patel SN, Sonani RR, Jakharia K, Bhastana B, Patel HM, Chaubey MG et al (2018) Antioxidant activity and associated structural attributes of Halomicronema phycoerythrin. Int J Biol Macromol 111:359–369
Pathak J, Rajneesh, Maurya PK, Singh SP, Häder D-P, Sinha RP (2018) Cyanobacterial farming for environment friendly sustainable agriculture practices: innovations and perspectives. Front Environ Sci 6:7
Pathak J, Singh PR, Sinha RP, Rastogi RP (2022 Jan 30) Evolution and distribution of cyanobacteria. In: Ecophysiology and biochemistry of cyanobacteria. Springer Nature Singapore, Singapore, pp 1–30

Paz S, Rubio C, Gutiérrez AJ, González-Weller D, Hardisson A (2022) Toxicity of algal foods with respect to human health. In: Mishra A (ed) Algal functional foods and nutraceutical. Bentham Science Publishers, pp 459–479

Pérez-García O, Escalante F, de-Bashan L, Bashan Y (2011) Heterotrophic cultures of microalgae: metabolism and potential products. Water Res 45(1):11–36

Rajneesh SSP, Pathak J, Sinha RP (2017) Cyanobacterial factories for the production of green energy and value-added products: an integrated approach for economic viability. Renew Sust Energ Rev 69:578–595

Ramírez-Mérida LG, Queiroz Zepka L, Jacob-Lopes E (2013) Photobioreactor: tool for mass cultivation of cyanobacteria. Ciencia y Tecnología 6(2):9–19

Ramu Ganesan A, Kannan M, Karthick Rajan D, Pillay AA, Shanmugam M, Sathishkumar P et al (2022) Phycoerythrin: a pink pigment from red sources (rhodophyta) for a greener biorefining approach to food applications. Crit Rev Food Sci Nutr 2022:1–19

Rastogi RP, Sinha RP (2009) Biotechnological and industrial significance of cyanobacterial secondary metabolites. Biotechnol Adv 27:521–539

Reyna-Velarde R, Cristiani-Urbina E, Hernández-Melchor D, Thalasso F, Cañizares-Villanueva R (2010) Hydrodynamic and mass transfer characterization of a flat-panel airlift photobioreactor with high light path. Chem Eng Process Process Intensif 49(1):97–103

Rimbau V, Camins A, Romay C, González R, Pallàs M (1999) Protective effects of C-phycocyanin against kainic acid-induced neuronal damage in rat hippocampus. Neurosci Lett 276:75–78

Rinawati M, Sari LA, Pursetyo KT (2020 Feb 1) Chlorophyll and carotenoids analysis spectrophotometer using method on microalgae. In: IOP conference series: earth and environmental science, vol 441. IOP Publishing, p 012056

Rojas J, Buitrago A (2019) Antioxidant activity of phenolic compounds biosynthesized by plants and its relationship with prevention of neurodegenerative diseases. In: Bioactive compounds. Woodhead Publishing, pp 3–31

Roy KR, Arunasree KM, Reddy NP, Dheeraj B, Reddy GV, Reddanna P (2007) Alteration of mitochondrial membrane potential by Spirulina platensis C-phycocyanin induces apoptosis in the doxorubicin-resistant human hepatocellular-carcinoma cell line HepG2. Biotechnol Appl Biochem 47:159–167

Sathyasaikumar KV, Swapna I, Reddy PVB, Murthy CRK, Roy KR, Gupta D, Senthilkumaran B, Reddanna P (2007) Co-administration of C-phycocyanin ameliorates thioacetamide-induced hepatic encephalopathy in Wistar rats. J Neurol Sci 252:67–75

Seghiri R, Kharbach M, Essamri A (2019 Jul) Functional composition, nutritional properties, and biological activities of Moroccan spirulina microalga. J Food Qual 3:2019

Singh P (2017) Cyanobacterial taxonomy and systematics: a brief review. In: Singh SS (ed) Plants and microbes in ever changing environment. Nova Science, Hauppauge, NY, pp 1–29

Six C, Ratin M, Marie D, Corre E (2021) Marine Synechococcus picocyanobacteria: light utilization across latitudes. Proc Natl Acad Sci 118(38):e2111300118

Sloth JK, Wiebe MG, Eriksen NT (2006) Accumulation of phycocyanin in heterotrophic and mixotrophic cultures of the acidophilic red alga Galdieria sulphuraria. Enzym Microb Technol 38:168–175

Sonani RR, Rastogi RP, Singh NK, Thadani J, Patel PJ, Kumar J et al (2017) Phycoerythrin averts intracellular ROS generation and physiological functional decline in eukaryotes under oxidative stress. Protoplasma 254:849–862

Song D, Xi B, Sun J (2016 Feb) Characterization of the growth, chlorophyll content and lipid accumulation in a marine microalgae Dunaliella tertiolecta under different nitrogen to phosphorus ratios. J Ocean Univ China 15:124–130

Soni B, Visavadiya NP, Madamwar D (2009) Attenuation of diabetic complications by C-phycoerythrin in rats: antioxidant activity of C-phycoerythrin including copper-induced lipoprotein and serum oxidation. Br J Nutr 102(1):102–109

Stanic-Vucinic D, Minic S, Nikolic MR, Velickovic TC (2018) Spirulina phycobiliproteins as food components and complements. In: Microalgal biotechnology. IntechOpen. https://doi.org/10.5772/intechopen.73791

Streit N, Ramírez Mérida LG, Queiroz Zepka L, Jacob-Lopes E, Queiroz M (2015) Producing natural pigments (chlorophyll-a) in agroindustrial bio-refineries. Ciencia y Tecnología 8(2): 27–23

Subhashini J, Mahipal SV, Reddy MC, Reddy MM, Rachamallu A, Reddanna P (2004) Molecular mechanisms in C-Phycocyanin induced apoptosis in human chronic myeloid leukemia cell line-K562. Biochem Pharmacol 68(3):453–462

Sun L, Wang S, Chen L, Gong X (2003) Promising fluorescent probes from phycobiliproteins. IEEE J Sel Top Quantum Electron 9:177–188

Tarin NJ, Ali NM, Chamon AS, Mondol MN, Rahman MM, Aziz A (2016) Optimizing Chlorella vulgaris and Anabaena variabilis growth conditions for use as biofuel feedstock. J Asiat Soc Bangladesh Sci 42(2):191–200

Thuan NH, An TT, Shrestha A, Canh NX, Sohng JK, Dhakal D (2019 Sep 3) Recent advances in exploration and biotechnological production of bioactive compounds in three cyanobacterial genera: Nostoc, Lyngbya, and Microcystis. Front Chem 7:604

Tredici MR, Zittelli G (1998) Efficiency of sunlight utilization: tubular versus flat photobioreactors. Biotechnol Bioeng 57(2):187–197

Vargas-Rodríguez AL, Gallegos ER, Barrita MLV, Orea AC, Ramírez RO (2006) Efecto in vitro de las ficobiliproteínas en el tratamiento del Carcinoma Cervicouterino. Revista del Centro de Investigación Universidad La Salle 7(25):5–12

Zeece M (2020) Food colorants. In: Zeece M (ed) Introduction to the chemistry of food. Academic Press, Cambridge, MA, pp 313–344

Zhu S, Xu J, Adhikari B, Lv W, Chen H (2022) Nostoc sphaeroides cyanobacteria: a review of its nutritional characteristics and processing technologies. Crit Rev Food Sci Nutr 2022:1–17

Zilliges Y, Kehr JC, Meissner S, Ishida K, Mikkat S, Hageman M, Kaplan A, Borner T, Dittmann E (2011) The cyanobacterial hepatotoxin microcystin binds to proteins and increases the fitness of Microcystis under oxidative stress conditions. PLoS One 6:e17615. https://doi.org/10.1371/journal.pone.0017615

Chapter 3
Spirulina as a Food of the Future

Mahwish Amin, Adnan ul Haq, Ayesha Shahid, Raj Boopathy, and Achmad Syafiuddin

Abstract *Spirulina* is an edible, nontoxic, photoautotrophic, multicellular cyanobacterium (blue-green alga) with dynamic metabolic composition. *Spirulina* is rich in proteins (60–70%), carbohydrates (8–25%), lipids (6–20%), essential vitamins, minerals, essential fatty acids, chlorophylls, carotenoids, and phycobiliproteins. It is referred to as the "food for the future, superfood" owing to its higher protein content and the presence of other bioactive compounds. Due to its extraordinary metabolic composition and prolonged history as a food source, it is generally regarded as safe for human consumption. *Spirulina* is known for its antioxidant, anticancer, antidiabetic, and immune booster properties. Therefore, it has been extensively used as a human dietary supplement. It is widely used as a poultry and aquaculture feed additive because it increases the health profile of birds and fish. As *Spirulina* is the food of the future and has been used by NASA as a food supplement in space, its commercial market is continuously flourishing. The outdoor cultivation setup for *Spirulina* has multiple challenges including contamination, culture crash, quality assurance, and a slow harvesting process. A closed photobioreactor (PBR) and wetland-based cultivation system in outdoor setup could be developed to address some of these challenges. The focus of this chapter is to provide the detailed nutritional history of *Spirulina* and its usage as human food and animal feed additives, especially as an aquaculture feed and poultry feed. The potential health benefits and the impact on the growth of animals are discussed in detail. Moreover,

M. Amin · A. ul Haq
Bioenergy Research Center, Department of Bioinformatics and Biotechnology, Government College University Faisalabad, Faisalabad, Pakistan

A. Shahid
US-Pak Center for Advanced Studies in Agriculture and Food Security, University of Agriculture Faisalabad, Faisalabad, Pakistan

R. Boopathy
Department of Biological Sciences, Nicholls State University, Thibodaux, LA, USA

A. Syafiuddin (✉)
Department of Public Health, Universitas Nahdlatul Ulama Surabaya, Surabaya, Indonesia
e-mail: achmadsyafiuddin@unusa.ac.id

M. A. Mehmood et al. (eds.), *Pharmaceutical and Nutraceutical Potential of Cyanobacteria*, https://doi.org/10.1007/978-3-031-45523-0_3

this chapter also covers the challenges associated with the outdoor cultivation of *Spirulina*, its market value, and commercialized products.

Keywords Cyanobacterium · Superfood · Alternative foods · Sustainability · Food security

3.1 Introduction

Spirulina is an edible, nontoxic, aquatic, and photosynthetic cyanobacterium that belongs to the Oscillatoriaceae family (Kavisri et al. 2021). So far, *Spirulina* is known to be the oldest oxygenic cyanobacterium with a history of 3.5 billion years, which has created the oxygenic atmosphere that allowed the evolution of life on Earth. *Spirulina* is a spiral-shaped multicellular organism, measuring between 200 and 500 μm in length and 5 and 10 μm in width (Han et al. 2021). Although it can be found in various habitats but predominantly found in freshwater, seawater, alkaline lakes, and extreme environments where other microorganisms struggle to survive. It can grow in harsh conditions with extreme pH (9.5–10.5) and temperature (30–34 °C) which are inhospitable to most other microorganisms (Touloupakis et al. 2016).

Hernando Cortes, a Spanish explorer, discovered *Spirulina* for the first time in 1519, during his visit to Lake Texcoco of Mexico with his coworkers (Gogna et al. 2022). They observed that the native people (Aztec fishermen) were harvesting the blue-colored things from the pond near the lake by fine nets, which they called "techuitlatl," and made edible cakes after sun-drying (Sasson 1997; Mccarty et al. 2021). In 1940, Pierre Dangeard also discovered *Spirulina* from Lake Chad in Africa and named it "Dihe" (Damessa 2021; Abdulqader et al. 2000). He observed that the Kanembu tribe was harvesting and selling semi-dried *Spirulina* as small cakes for their survival (Ciferri and Tiboni 1985). However, its discovery went unnoticed for almost half a century until Jean Leonard and his colleagues confirmed the findings of Dangeard. Leonard explained that Dihe was composed of dried biomass of *Spirulina* collected from the lakes. The collected *Spirulina* was dried on clean mats, and semi-dried cakes were completely dried in the sunlight (Abdulqader et al. 2000). Leonard was the first to perform a biochemical analysis of *Spirulina* and found that it was rich in protein (almost 50% of dry weight) (Vonshak 1997b). After Leonard's findings, people started growing *Spirulina* commercially to reap its economic benefits.

Zarrouk's detailed biochemical studies of *Spirulina*'s growth requirements in 1966 led to the basis of its commercialization (Wan et al. 2016; Zarrouk 1966). During the early twentieth century, the International Association of Applied Microbiology coined the term "wonderful future food source" for *Spirulina*, and the United Nations declared it as "the best for tomorrow" at the World Food Conference (Reboleira et al. 2019). The French government established the first *Spirulina* production industry, Sosa Texcoco, in 1969 (Koru 2012), and *Spirulina*'s worldwide production is now estimated to be more than 3000 metric tons annually. During the late twentieth century, NASA studied the cultivation processes of *Spirulina* and found that it could be an excellent dietary source for astronauts as it could grow quickly by modifying the growth process and creating an artificial environment (Wright et al. 1993). NASA also reported that the dietary value of only 1 kg of *Spirulina* is equal to 1000 kg of fruits and vegetables.

In 1992, The World Health Organization (WHO) reported that *Spirulina* has no toxic effects on human health and declared it the "best food for the future" to reduce the risk of malnutrition, especially in children. In 2003, The Intergovernmental Institutions (IIMSAM) with United Nations Economic and Social Council (UNECOSOC) studied the impact of *Spirulina* to redress malnutrition in developing countries (Uddin et al. 2018; Mathur 2019). In 2011, the National Institutes of Health (NIH) proposed that *Spirulina* could be used as a dietary food, but NIH requested further research to check the potential benefits of *Spirulina* on human health (Marles et al. 2011). In 2012, the Food and Drug Administration (FDA) declared *Spirulina* a healthy food supplement with no known side effects. The European Food Safety Authority (EFSA) reported that *Spirulina* has excellent potential to control and reduce the blood sugar level in humans (Grosshagauer et al. 2020).

Spirulina is widely known for its considerably balanced amount of all beneficial macro- and micronutrients (Table 3.1). It is highly rich in protein (ten times higher than that of soybean) which generally ranges within 60–65% (dry weight) and up to 70% when excessive nitrogen is available in the growth medium. Additionally, it also contains other metabolites such as carbohydrates, 18 essential amino acids, essential fatty acids (especially eicosapentaenoic acid (EPA), gamma linolenic acid (GLA), and omega-6 fatty acids), vitamins (provitamin A, vitamin E, vitamin B_{12}), and nutritional pigments, namely, phycobilin (phycocyanin, allophycocyanin, phycoerythrin), chlorophyll a, chlorophyll b, and carotenoids along with several minerals (especially iron, magnesium, manganese, phosphorus, potassium, sodium, and zinc) (Michael et al. 2019). In addition to various nutrients, *Spirulina* contains plenty of other secondary metabolites such as phenolic compounds, alkaloids, triterpenoids, flavonoids, glycosides, tannins, terpenoids, and saponins (Lafarga et al. 2021) which are reported to play the role of antioxidants and hypolipidemic and immune booster agents.

3.2 Applications and Market Potential of *Spirulina*

3.2.1 Spirulina *as a Human Food Additive*

Spirulina is one of the leading trends in the functional food industry and gaining worldwide popularity as a food supplement during the last few decades (Lafarga et al. 2020). Nowadays, it is being integrated into many foods like pasta, noodles, baked goods, smoothies, juices, energy bars, and snacks and sold as a nutritive supplement. Currently, most of the organic food supplements in the market contain *Spirulina* biomass and are promoted as "superfood," "food of the future," "rich in protein," and "rich in omega-3 fatty acids" (Lafarga 2019). The recommended daily

Table 3.1 Nutritional value of commercially available *Spirulina* powder (3 g serving)

Nutrients	Quantity per serving (3 g)	Nutrients	Amount per serving (3 g)
Macronutrients			
Calories/energy	8.7 kcal	Carbohydrates	700 mg
Moisture	200 mg	Lipids	200 mg
Protein	1800 mg	Ash	200 mg
Vitamins			
Vitamin A	17.1 IU	Vitamin B2	0.1 mg
Vitamin K	0.0008 mg	Vitamin B3	0.4 mg
Vitamin C	0.3 mg	Vitamin B6	0.02 mg
Vitamin B1	0.07 mg	Vitamin E	0.2 mg
Minerals			
Calcium	3.6 mg	Sodium	30 mg
Iron	0.9 mg	Zinc	0.06 mg
Magnesium	5.9 mg	Copper	0.2 mg
Phosphorus	3.5 mg	Manganese	0.057 mg
Potassium	42 mg	Selenium	0.0002 mg
Amino acids			
Tryptophan	27.9 mg	Valine	105 mg
Threonine	88.1 mg	Arginine	125 mg
Isoleucine	96.3 mg	Histidine	32.4 mg
Leucine	148 mg	Alanine	135 mg
Lysine	90.6 mg	Aspartic acid	174 mg
Methionine	34.5 mg	Glutamic acid	252 mg
Cysteine	19.8 mg	Glycine	92.6 mg
Phenylalanine	83.1 mg	Proline	71.4 mg
Tyrosine	77.4 mg	Serine	89.7 mg

dosage of *Spirulina* ranges between 3 and 10 g (Bobescu et al. 2020), while the maximum intake per day is limited to 30 g for adults. Some of the commercially available *Spirulina*-based food supplements (powder, capsules, and pills) include "Dragon Superfoods *Spirulina* powder (Smart Organic, Germany)," Apollo Hospitals Life *Spirulina* (Apollo Pharmacy, India)," and "Label *Spiruline* La Ferme de Bancel (Label *Spiruline*, France)," respectively.

Spirulina is declared one of the best nutritive healthy human food in the twenty-first century due to its extremely high dietary value by the WHO and the Food and Agriculture Organization (FAO) of the United Nations (Anvar and Nowruzi 2021). The FDA has certified *Spirulina* as "generally recognized as safe" (GRAS)-GRN No. 127 due to its prolonged food source record (Ovando et al. 2018). *Spirulina* has wide applications in human food, animal feed, aquaculture feed, poultry feed, nutraceutical, pharmaceutical, and cosmeceutical.

Spirulina is one of the most prosperous sources of natural proteins. A small dose of *Spirulina* can fulfill the daily protein requirements of the human body. It is enormously used as a nutritive food source, especially as a protein diet, but it also contains other highly valuable metabolites that impart health benefits to the human body. They have easy digestibility of up to 84% due to proteaceous cell walls, lack of cellulose, and the absence of phytates and oxalates (Masten Rutar et al. 2022; Niccolai et al. 2019).

Spirulina is a good source of value-added phycobiliproteins such as C-phycocyanin (C-PC), allophycocyanin (APC), and phycoerythrin (PE). Phycobiliproteins are the photosynthetic accessory pigments that harvest the sunlight and transfer it to chlorophylls during photosynthesis. *Spirulina* produces a good amount of these valuable metabolites when cultivated in low-light conditions. C-PC and APC are bright blue-colored proteins, while PE are red-colored phycobiliproteins. These colored pigments are highly valued and extensively used in the nutraceutical, cosmeceutical, pharmaceutical, textile, and food industries as coloring and nutritive agents. These biomolecules have high health-promoting properties, i.e., antioxidant, antidiabetic, anti-mutative, antiviral, antitumor, and anticancer. C-PC also boosts the immune system to exhibit important health bioactivities such as hepatoprotective, antiplatelet, and neuromodulator activities (Table 3.2). Considering all the properties, these molecules have great potential to be used in functional food formation.

Spirulina also contains other high-value metabolites like minerals, vitamins, and carotenoids that aid in human health. *Spirulina* is decadent in vitamin B, essential in electron transfer, fatty acid synthesis, and DNA repair (Monteverde et al. 2017). Likewise, vitamin B12 is present in minimal concentrations but in the non-active pseudo form that strengthens human immunity. Only 4 g of dried *Spirulina* powder can fulfill the daily body requirements. Carotenoid is another class of photosynthetic accessory pigments found in *Spirulina* as the photoprotective agent. Humans cannot produce carotenoids in their bodies, so they acquire them through alimentation (Park et al. 2018b). A small amount of these molecules in the food may protect the human body from severe disorders such as skin degeneration, aging, cardiovascular disease, cancer, and cataracts that are caused by oxidative damage and inflammatory stress (Rao and Rao 2007).

3.2.2 **Spirulina** *as a Poultry Feed Additive*

Meat production is increasing worldwide for the animal-based protein source because animal-based protein is an important protein source in the human diet that fuels the body's energy, carries oxygen, and helps the body to make antibodies to fight against infectious diseases. Among all the different protein sources, poultry meat is an essential and cheaper protein source compared to other livestock meat. In developing areas, poultry production is getting hype due to its low production cost. Several nutritional strategies have confirmed that the maximum increase in broilers can be achieved by using antibiotics as feed additives as it improves the body weight gain, growth rate, and feed intake and reduces the feed conversion ratio and mortality rate in broilers (Sugiharto et al. 2016). However, the use of health- and growth-

Table 3.2 Bioactivities of different metabolites from *Spirulina*

Bioactive compounds	Bioactivities	Mechanism of action	References
Phycobilins	Antioxidant	Scavenging peroxyl radicals from AAPH	De Amarante et al. (2020)
Phycobilins	Antioxidant	Scavenging ROS and free radicals	Takyar et al. (2019)
Phycobilins	Antioxidant	Scavenging free radicals from DPPH	Chen et al. (2022), Chentir et al. (2018)
Phycobilins	Anticancer	Inhibition of NADPH oxidase activity	Sibiya et al. (2022), Izadi and Fazilati (2018)
Phycobilins	Anticancer	Antiproliferative and cytotoxic effects on cancer cells	Konícková et al. (2014), Soni et al. (2015)
Phycocyanin	Anticancer	Antiproliferative and cytotoxic effects on cancer cells	Hussein et al. (2021), Yu et al. (2019b)
Phycocyanin	Antioxidant	Scavenging free radicals from DPPH	Thangam et al. (2013), Abd El-Baky and El-Baroty (2012)
Phycocyanin	Anti-inflammation	Reduced activity of myeloperoxidase (MPO), inhibits the invasion of inflammatory cells	Gonzálea (1999)
Phycocyanin	Anti-irradiative	Free radicals scavenging effect against damaged cells reduce oxidative stress and DNA damage	Rimbau et al. (2001)
Phycocyanin	Neuroprotective	Increase granulocyte progenitor cells production and sensitivity, decrease neural dysfunction	Liu et al. (2000)
Phycobilins	Anti-inflammation	Downregulate the quantity level of proinflammatory mediation	Minić (2021)
Phycocyanin	Immunomodulation	Increase IgA level, suppress the invasion of allergens, and decrease IgE and IgG1 production to reduce immune responses	Minić (2021)
Phycobilins	Anti-asthma	Inhibit NADPH oxidase	Mccarty et al. (2021)
Phycobilins	Antitumor necrosis	Protect demyelination and axonal loss	Cervantes-Llanos et al. (2018)
Phycobilins	Antidiabetic	Protective effect against human lymphatic endothelial cells (HLEC) apoptosis	Mariey et al. (2012)

(continued)

Table 3.2 (continued)

Bioactive compounds	Bioactivities	Mechanism of action	References
Phycocyanin	Antibacterial	Hamper bacterial growth	Zhu et al. (2018)
Phycocyanin	Anti-obesity	Impede pancreatic lipase activity	Chakdar and Pabbi (2017)
Antidiabetic peptides	Antidiabetic	Impede insulin-signaling molecules, suppress glucose transporter type 4 (GLUT-4) translocation	Mathur (2019)
Glycolipid (H-b2)	Beneficial effects on lipid metabolism	Inhibit pancreatic lipase activity	Han et al. (2006)
γ-linolenic acid (GLA)	Antibacterial, antioxidant, anti-inflammatory, anti-cancer, anti-fibrotic, antiangiogenic	Inhibit β-secretase, and anti-oxidants, express *Ras* and *Bcl-2* oncogenes, induce tumor cell apoptosis, lipid peroxidation, and caspase-3 activation	Youn et al. (2014)
Ascorbic acid	Antioxidant, antiproliferative	Suppress lipid peroxidation, antiproliferative activities, cancer chemopreventive activities	Lu et al. (2006)
Sulfated polysaccharide		DNA damage repair, protection against hypoxia injury, decrease oxidative stress, plasma low-density lipoprotein (LDL) levels	Lorenz (1999)
Sulfolipids	Antiproliferative, antiviral, anticancer	Inhibit DNA polymerase activity, decrease phosphorus demand, inhibit superoxide anion production	Hoseini et al. (2013)
Whole supplement	Anti-inflammatory, anti-oxidant, anticancer, antidiabetic	Increase exercise performance, fat oxidation, and GSH concentration	Bermejo-Bescós et al. (2008)

promoting antibiotics in feed was banned in many countries when antibiotic-resistant bacteria were observed in humans and animals. This ban and withdrawal of in-feed antibiotics in poultry feed led to decreased broiler production and health (Sugiharto et al. 2018; Pourhossein et al. 2015). Finding alternative feed ingredients to replace the antibiotics from poultry feed is essential.

Currently, the integration of natural ingredients in the formulation of poultry feed is widely used as the alternate growth- and health-promoting factors, antibiotics, and other chemicals. Clinical and experimental trials demonstrated that *Spirulina* could also be used as animal and aquaculture feed due to its nutritional profile and potential health benefits (Ahmed et al. 2022; Opoola et al. 2019). *Spirulina* is an excellent in-feed additive that can increase poultry production and reduce production costs. In recent years, *Spirulina*-based poultry feed additives are getting hype because of their probiotics and other beneficial effects on poultry health and production (Joya et al. 2020). Integrating *Spirulina* as a feed additive and new feed formulation protects the environment and natural resources (Madeira et al. 2017). *Spirulina* is more expensive (Holman and Malau-Aduli 2013) than other antibiotics used as in-feed antibiotics to improve the growth of broiler chickens. However, some researchers have confirmed *Spirulina* as a cost-effective in-feed additive due to its enhanced growth and health benefits (Khan et al. 2020).

The inclusion of 10% *Spirulina* in poultry feed has been found to increase the overall growth of poultry. Feeding *Spirulina* to broiler chickens increases the body weight gain, nutrient utilization, and hematological parameters and reduces the feed conversion ratio, aspartate aminotransferase, erythrocyte sedimentation rate, alanine aminotransferase, and mortality rate. In addition, it increases the superoxide dismutase (SOD) and glutathione peroxidase (GPx) enzyme activity, meat quality, eggs, and carcasses in poultry (Park et al. 2018a). *Spirulina* helps fight against viral and bacterial diseases because its bioactive compounds confer antiviral, antibacterial, and immuno booster effects in organisms (Table 3.3).

3.2.3 **Spirulina** *as an Aquaculture Feed Additive*

Global seafood consumption is increasing daily because it is an alternative protein source to animal-based proteins for humans (Xiang et al. 2020). It is predicted by World Bank Report (Bank 2013) that by the year 2030, aquaculture fish production will overtake half of the world's fish requirement. Aquaculture production is expected to reach 102 million tons in 2025 and up to 121 million tons in 2030 (FAO 2016). However, experts predict a decreased growth in aquaculture production because its intensive production requires large quantities of underground freshwater and highly nutritious aquaculture feed. The bulk production of aquaculture food using commercially available fish meal leaves nutrient-rich wastewater that destroys other microbial communities (Wuang et al. 2016) and causes eutrophication of coastal waters.

Fish meal is an essential feed for aquaculture food production because it contains the complete ingredients, especially the protein content (Rosas et al. 2019). However, in the last decade, fish meal production is declining due to the overconsumption of aquaculture food and leading to overpriced fish meal production. Therefore, the total production cost of aquaculture feed accounts for up to 60% of the price of fish meal which would lead to decreased growth in aquaculture feed production. Therefore, an alternative protein source is required that can partially or entirely replace the fish meal.

Spirulina can be used as a functional feed or partially substituted with commercially available fish meal because of its high nutritional profile and amazing biochemical composition (Table 3.4). Although the high cost of *Spirulina* production is a significant problem it can be compensated due to its minimal dosage requirements. *Spirulina* confers improved aquaculture health and growth (Xiang et al. 2020) and

Table 3.3 Impact of different inclusion concentrations of *Spirulina platensis* on poultry health and growth

Spirulina inclusion (%)	Total birds	Experiment duration (days)	Studied characteristics	Findings	References
0, 0.25, 0.5, 0.75, 1.0	800	35	Growth performance, health performance, and breast meat quality	Increase body weight gain, feed conversion ratio, antioxidant enzyme activity, nitrogen digestibility, and cecal *Lactobacillus* population Decrease ammonia gas emission and drip loss after 7 days of storage	Park et al. (2018a)
0, 0.10, 0.15, 0.20	240	364	Egg production, egg weight, feed conversion	High egg production rate, increase in egg yolk percentage and yolk color, decrease cholesterol in egg yolk and plasma, increase fertility and hatchability rate	Mariey et al. (2012)
0, 0.5, 1.0	120	42	Body weight gain, feed conversion ratio	Increase growth rate, carcass quality, feed efficiency, and body weight gain	Bonos et al. (2016)
0, 0.1, 0.15, 0.2	120	42	Growth performance, gut integrity, and immunity	Enhance feed intake and weight gain, decrease feed conversion ratio, and dressing percentage, decrease level of heterophil, and increase level of lymphocytes	Khan et al. (2020)
0, 0.5, 1.0	450	40	Growth performance, serum biochemistry, immunity	Enhance final body weight, decrease feed conversion ratio, increase carcass dressing, carcass percentages, and blood lipid profile	Elbaz et al. (2022)
0, 10	360	37	Growth, carcass health, breast meat quality and health, footpad quality	Decrease foot dermatitis, improve redness, and yellowness, highly desired breast meat quality	Mullenix et al. (2022)
0.03, 0.05, 0.07, 0.09	200	38	Growth performance, immunity, blood chemistry,	Increase body weight, feed intake, growth rate, and immunity, reduce feed conversion	Fathi (2018)

(continued)

Table 3.3 (continued)

Spirulina inclusion (%)	Total birds	Experiment duration (days)	Studied characteristics	Findings	References
			intestinal microflora community	rate, and increase blood parameters Decrease microbial community	
0, 0.025, 0.05, 0.075, 0.1	250	35	Growth, histomorphology, serum biochemistry, immunity, and inflammatory response	Increase growth, weight, serum biochemistry, proteins, albumin, globulins, growth hormone, and thyroxin hormones, decrease FCR, serum cholesterol, and lipoprotein cholesterol	Omar et al. (2022)
0, 0.5, 1.0	120	42	Growth performance, meat quality, and fatty acid profiles	Increase body weight and feed intake, no effect on mortality, breast and thigh meat, and lipid oxidation, enhance meat fatty acid profile with polyunsaturated fatty acids	Bonos et al. (2016)
0, 0.5, 1.0, 2.0	250	45	Antioxidant activity, meat quality, lipid profile, immunity, and growth performance	Decrease stress hormone concentration and lipid parameters, enhance adaptive immunity response and antioxidant status, and increase feed conversion ratio	Mirzaie et al. (2018)
0, 3.0, 6.0	540	35	Growth, antioxidant study, immunity, and intestinal histopathology	Improve growth, calcium-phosphorus metabolism, and lipid metabolism, increase antioxidant activity, and immune profiles in blood, improve gut histology	Abd El-Hady et al. (2022)
0, 1.0	288	35	Growth, hematology, intestinal microbial community, carcass quality	No effect on growth performance; increase cecum weight; decrease hemoglobin, erythrocyte, and hematocrit levels; decrease leukocytes, lymphocytes, and eosinophils	Sugiharto et al. (2018)

(continued)

Table 3.3 (continued)

Spirulina inclusion (%)	Total birds	Experiment duration (days)	Studied characteristics	Findings	References
0.1, 0.2, and 1.0, 2.0 mL aqueous extract	600	35	Growth performance and physiological responses	Decrease body weight and feed conversion ratio, improve carcass quality and dressing percentage, increase lipid profile and adaptive immune response	Elbaz et al. (2022)

compensates for the high economic cost in other areas. Furthermore, the high cost can also be reimbursed by utilizing the aquaculture wastewater for the mass cultivation of *Spirulina* as a feed alternative.

3.2.4 **Spirulina** *as Food for the Future*

The National Aeronautics and Space Administration (NASA) and the European Space Agency (ESA) have both conducted studies on *Spirulina* as a potential food for space travel (Matufi and Choopani 2020) as its properties make it promising food source for astronauts on long missions (Fais et al. 2022). The European Space Agency is investigating the use of *Spirulina* in regenerative technologies for life support in space with the Melissa project (Godia et al. 2002), which involves loading *Spirulina* into a photobioreactor on the International Space Station to measure the amount of oxygen produced and analyze the effects of weightlessness and radiation on the algal cells. *Spirulina*'s high resistance to radiation makes it a promising food source for space travel. In addition, *Spirulina* could help sustain long-duration manned missions and protect astronauts from stress factors (Fais et al. 2022).

3.2.5 Market Potential of **Spirulina**

Worldwide annual production of *Spirulina* is around 10,000 tons (Bhattacharya and Goswami 2020). Europe is at the top in the production of *Spirulina* and conducts *Spirulina*-based projects such as the project entitled Spiral G Project: The first demonstrator of *Spirulina* biorefinery is related to accessing the sustainability and profitability of each *Spirulina* biomass component.

Moreover, multiple companies are working globally on the excessive production of *Spirulina* biomass at a pilot scale. Cyanotech, Earthrise, Cellana, and BlueBioTech are the top companies associated with the cultivation and marketing of *Spirulina*. According to Allied Market Research, the global *Spirulina* market was

Table 3.4 Health effects of different inclusion concentrations of *Spirulina platensis* in aquaculture feed

Spirulina inclusion (%)	Fish	Total fish	Experiment duration (days)	Studied characteristics	Suitable quantity	Results	References
0, 1, 2, 3, 4, 5	Chinese sea bass (*Lateolabrax maculatus*)	800	56	Growth performance, digestive enzyme, hematological parameters, and antioxidant capacity	4% and 5%	Improve body weight gain, specific growth rate, and protease activities, decreased feed conversion ratio (FCR), increased antioxidant activity (SOD, CAT, GSH-Px)	Yu et al. (2019a)
0, 2.5, 5.0, 7.5, 10	Rainbow trout (*Oncorhynchus mykiss*)	216	70	Growth performance and pigmentation	5%	Increased carotenoid concentrations of skin and fillet, highest carotenoid deposition in skin and fillet	Teimouri et al. (2013)
0, 2, 4, 6, 8	Caspian brown trout (*Salmo trutta caspius*)	600	–	Growth performance, body composition, fatty acid profile, amino acid profile, and pigmentation	8%	Increase body weight gain, specific growth rate, protein efficacy, lipid efficacy, decreased feed conversion ratio (FCR), increase high saturated fatty acids, polyunsaturated fatty acids, arginine, lysine profile, luminosity, redness, and yellowness in skin and fillet	Roohani et al. (2019)
0, 1.37, 2.75, 5.5	Oscar fish (*Astronotus ocellatus*)	120	56	Growth, enzyme activity, biochemical parameters, pigmentation	5.5%	Improve growth performance, protease activity, protein content, lipid content, protein, and albumin levels, increase hemoglobin, hematocrit, and erythrocyte, improve lysozyme activity and carotenoid concentration	Mohammadiazarm et al. (2021)

0, 5, 10, 15	Parrot fish (*Oplegnathus fasciatus*)	–	56	Growth performance, biochemical parameters, and immune response	15%	Increase body weight gain, protein efficacy, and feed intake, decrease feed conversion ratio (FCR), increase hematocrit, hemoglobin, and respiratory burst activity, increase muscle protein and decrease lipid concentration, increase dietary polyphenol concentration and antioxidant capacity	Kim et al. (2013)
0, 0.2, 0.3, 0.4, 0.6	Gurami (*Osphronemus goramy*)	45	56	Growth, hematological parameters, body composition, and biochemical parameters	–	Increase growth of body length and weight, decrease feed conversion ratio (FCR), increase RBCs, WBCs, hemoglobin, and hematocrit	Simanjuntak et al. (2018)
0, 0.25, 0.5, 0.75	Nile tilapia (*Oreochromis niloticus*)	80	–	Growth, feed utilization, immunological and biochemical chemistry	0.75%	Enhance body weight gain (BWG), feed conversion ratio (FCR), and protein efficiency ratio	Al-Zayat (2019)
0, 2.5, 5, 7.5	Rainbow trout (*Oncorhynchus mykiss*)	80	84	Fatty acid composition of fish	5%	Improve the quality of FAs, decrease saturated FAs, increase long-chain highly unsaturated FAs	Jafari et al. (2014)

USD 393.6 million and is expected to reach USD 897.61 million by 2027, growing at a compound annual growth rate (CAGR) of 10.5% from 2020 to 2027. In 2019, *Spirulina platensis* had a dominant position in the global *Spirulina* market and would be expected to maintain the highest position than other *Spirulina* species.

According to *Spirulina* Future Market Insight Report in 2019, *Spirulina* powder market size in the food and beverages industry was USD 358 with a 6% CAGR, and it is forecasted that in 2029 the food and beverage industry growth would be 73%. The dietary supplement market in 2025 is expected to be USD 252.1 billion with a 9% CAGR growth from 2021 to 2031. Due to the robust growth rate of the *Spirulina*-based dietary supplement market, it is predicted that by 2031 the market value would be USD 254 billion with a 9.4% CAGR. Geographically, in 2021, the *Spirulina* market size in Western Europe was USD 62,146.9 million, and in 2031, the expected market size in Eastern and Western Europe would be USD 62,146.9 million, in Asia Pacific Region (except Japan), it would be USD 59,789.5 million, and in Japan, it would be USD 22,392.5 million.

In 2019, North America had the leading position as a *Spirulina* supplier and is expected to maintain its dominance with an 11.2% CAGR, while the Asia Pacific region is the biggest and most profitable market for *Spirulina* suppliers.

3.3 Challenges in the Production Pipeline

Multiple challenges are associated with the cultivation and processing of *Spirulina*. The main crucial factors that influence the growth of *Spirulina* are temperature, lighting, inoculum volume, pH, water quality, stirring speed, and overall micronutrient richness, etc.

3.3.1 Challenges Associated with Outdoor Open Pond Cultivation

3.3.1.1 Contamination Risk in Outdoor Cultivation Setup

Open raceway ponds are susceptible to contamination from other microorganisms, which can affect the growth of the *Spirulina* (Fig. 3.1). There are mainly four contamination modes of biological pollutants which are zooplankton, bacteria, fungi, and harmful algae. Zooplanktons are the small aquatic microorganisms of water bodies and mainly include crustaceans, rotifers, aquatic mites, and insects. These are the grazers of microalgae and reduce the biomass production of *Spirulina* during outdoor cultivation (Wang et al. 2013). The bacterial and fungal contaminants are less harmful to the *Spirulina* than zooplanktons because *Spirulina* raises the pH of media (9–10 pH) during its cultivation (Mehar et al. 2019), and this alkaline pH is not optimal for the growth of bacterial and fungal spores. Microalgae

Fig. 3.1 Open pond cultivation system of *Spirulina*. (**a**) Circular open ponds (Source, Nutriphys). (**b**) Raceway ponds (Source, Global Seafood Alliance)

are the most competing contaminants of *Spirulina* culture and the major reason being the unpure inoculum or foreign inclusion of microalgae in the bioreactor. Furthermore, the fate of *Spirulina* culture will be dependent upon the nutrient utilization ability of foreign microalgae as higher utilization will cause the nutrient depletion for the microalgae of interest which results in culture crash. However, elimination of competing microalgae and maintenance of axenic/dominant algal strain are challenging as the contaminating species share physical and biological similarities with the desired strain (Jung et al. 2019). To prevent contamination, the pond must be regularly monitored and maintained to detect any early signs of contamination allowing for prompt action to be taken. Additionally, regular testing and sterilization of equipment can help prevent contamination and ensure the health of the *Spirulina* culture (Grosshagauer et al. 2020).

3.3.1.2 Culture Crash

One challenge associated with the cultivation of *Spirulina* is the risk of a culture crash. Culture crash is among the major risk associated with *Spirulina* cultivation that can occur due to imbalanced culture, causing the algae to stop growing and die off. Several factors including changes in temperature, pH levels, nutrient imbalances, or contamination may cause a culture crash (Yuan et al. 2019).

Controlling temperature is a critical factor in the outdoor cultivation of microalgae, but it could be achieved by (i) choosing the right location for outdoor cultivation that should have access to direct sunlight but also be sheltered from extreme weather conditions such as strong winds, excessive rainfall, or freezing temperatures; (ii) installing shade cloths that can help regulate the temperature of culture by preventing overheating during hot weather or temperature drops during cold weather; (iii) regular mixing of culture can help distribute heat evenly, preventing localized overheating or cooling; and (iv) inclusion of cooling and heating systems such as fans, heaters, or geothermal pumps to achieve optimum temperature.

Nutrient imbalance is another crucial factor for *Spirulina* culture crash as it requires a specific balance of nutrients including nitrogen, phosphorus, and potassium to grow properly. Open ponds can be subjected to nutrient imbalances due to runoff from surrounding fields or other sources of water. So that, to prevent a culture crash, it is important to monitor the *Spirulina* culture closely and make adjustments as needed to maintain the proper balance of nutrients and environmental conditions.

3.3.1.3 Expensive Cultivation Media

In commercial plants, the cost of cultivation media is a challenging factor to maintain the profit of the process. Traditionally, Zarrouk's medium (ZM) is used for *Spirulina* cultivation, but it is known to be relatively expensive as compared to other options due to the high quantity and cost of some of its ingredients such as $MgSO_4$, $NaHCO_3$, and $NaNO_3$. Hence, multiple alternatives (that can maintain the optimal nutrients and conditions conducive to favorable growth and high yield of *Spirulina*) are tried to reduce the cost of cultivation media.

Most of these alternatives are waste based such as cattle waste (Bates et al. 1995), cow dung ash (Jain and Singh 2013), swine effluent (Baker et al. 2021), anaerobically digested pig effluent (Baker et al. 2021), industrial effluent (Zinicovscaia et al. 2019), municipal wastewater (Zhai et al. 2017), and seawater (Leema et al. 2010). The use of industrial residues/by-products is another suitable option due to its appropriate nutrient composition. As a first attempt to use solid media for algal cultivation, sugarcane bagasse (remnant of the sugar industry) with the optimum moisture content of 98.9% was employed for solid-state cultivation of *Spirulina* (Pelizer et al. 2015). Some other agriculture solid waste-based media include molasses-based media (Al Mahrouqi et al. 2022) and cabbage extract-based medium (Akhtar et al. 2014). The alternative nutrients have different impact on the biomass productivity and metabolite content of *Spirulina* than ZM (Table 3.5).

However, in some cases, chemical fertilizers or pesticides are also used in *Spirulina* cultivation. The first example is the use of NPK-10:26:26 fertilizer for *Spirulina* cultivation to check its impact on biomass production and metabolite content. The maximum biomass production and chlorophyll, protein, and lipid concentration were 1.22 gL^{-1}, 8.92 gL^{-1}, 52.35%, and 14.84%, respectively, and this media is 50% more cost-effective than standard media (Kumari et al. 2015). Another study reported that commercial liquid organic fertilizer can replace standard ZM except for nitrate, bicarbonate, and phosphate in culture media. The maximum dry weight and crude protein in the control and experiment groups were comparable such as 1330 mgL^{-1}, 60.17%, and 1256 mgL^{-1}, 59.92%, respectively (Ak 2012). These fertilizers have negative impacts on the environment, including soil and water pollution, if remained untreated. But *Spirulina* cultivation in fertilizer effluent is an eco-friendly and sustainable process.

Table 3.5 Impact of alternative/supplemented media on the biomass productivity and metabolite content of *Spirulina*

Strain	Media composition	Biomass productivity (gL^{-1} day^{-1})	Carbohydrate (%)	Protein (%)	Lipid (%)	Pigment (mgg^{-1})	References
Spirulina sp. LEB 18	Seawater 100%	0.16	22.6	52.5	12.1	–	Bezerra et al. (2020)
Arthrospira platensis UTEX 1926	ZM supplemented with KNO_3 and urea	0.36	–	48.5	26.9	Chl-a: 19.2	Vieira et al. (2012)
Arthrospira (Spirulina) platensis	ZM supplemented with 15.4 mM KNO_3 + 14.1 mM NH_4Cl	0.237	–	49.8	20.5	Chl-a: 21.85	Rodrigues et al. (2010)
Spirulina platensis D9Z	ZM supplemented with buffalo mozzarella cheese whey	0.1793	47.83	44.56	2.56	–	Pereira et al. (2019)
Spirulina sp. LEB18	ZM supplemented with aqua culture wastewater (indoor cultivation)	0.20	69	15	10	Chl-a, 12 μg mL^{-1}; Chl-b, 1 μg mL^{-1}; carotenoids,10 μg mL^{-1}	Cardoso et al. (2020)
Spirulina maxima	Anaerobically digested dairy manure wastewater	0.06	–	–	–	PC: 54.67–58.67	Rahman et al. (2022)

3.3.2 *Challenges Associated with the Processing Pipeline*

3.3.2.1 Quality Assurance

Different analyses to study the chemical, physical, biochemical, and microbiological characteristics of *Spirulina*-based products have paramount importance for food-grade applications. Well-equipped laboratory and trained personnel are essential since it is advisable and economical to carry out these analyses on-site. Occasionally, samples should be sent to independent laboratories for confirmation and standardization. This is necessary to ascertain that the final product meets quality criteria for domestic as well as international markets. Only after each production lot has passed all tests, it should be certified, harvested, and packed for shipping.

3.3.2.2 Harvesting of Biomass

Spirulina grows at an exponential rate, with a doubling time of around 2–3 h under ideal conditions (Jester et al. 2022), and forms a thick layer of cells on the surface of media which is easily harvestable. Under stressed conditions, it grows at a relatively slow rate, with a doubling time of around 24 h. This means that it takes several days or even weeks for a culture to reach harvestable biomass levels.

Large-scale *Spirulina* harvesting has been done for ages through centrifugation, flocculation, gravity sedimentation, and filtration. *Spirulina* is typically harvested by filtration (Patel and Goyal 2013) which involves passing the culture through a fine mesh screen to separate the biomass from the media; it is among the most feasible harvesting method as *Spirulina* like other cyanobacteria produces gas vesicles which help it to float on the surface of media and it makes the harvesting process easy (Kim et al. 2005). There is no doubt that the filtration-based harvesting process of *Spirulina* is easy, but it could still be a slow, labor-intensive, and time-consuming process due to the high concentration of impurities or biomass and varied flotation ability of *Spirulina* according to prevailing conditions (Yuan et al. 2019). Now, automated harvesting systems have been developed to streamline the process and enhance the productivity of culture.

3.3.2.3 Quality Maintenance During Drying and Packaging

Proper and quick drying of *Spirulina* slurry is an essential task to get a high-quality product. There are multiple drying methods such as air-drying, freeze-drying, cast-tape drying, and cast-tape drying under vacuum (Demarco et al. 2022). The air/spray drying method is a more cost-effective method for the pilot-scale drying of *Spirulina* than other methods. So, spray drying is the best and quick choice for drying *Spirulina* at 60 °C. This procedure is safe, as no preservatives, additives, or stabilizers are required, and fortunately, at this temperature, any pigment, enzyme, and heat-sensitive nutrients are not decompose. The most critical point in drying is that the biomass should not be overdried or less-dried (containing a high amount of moisture). In the case of overdrying, some essential components of biomass such as vitamins and pigments (chlorophyll) might be lost from biomass. Moreover, if the moisture content of biomass is more than 8%, then there are very high chances of mold and bacterial growth. After the drying of biomass, another important challenge is its packaging. To maintain the quality of *Spirulina,* the powder coming out of the dryer should immediately be vacuumed away in the packaging room and should be sealed under vacuum in drums with special gas-barrier bags to minimize the oxidation of certain vital pigments like phycocyanin. The *Spirulina's* biomass should ideally have 3–6% moisture content during the time of product packaging/ sealing (Ahsan et al. 2008). The sealed product could be placed in drums for shipment. According to the literature, the *Spirulina*-based product could stay up to 4 years with little changes in the biochemical composition and nutritional properties (Vonshak 1997a).

3.4 Emerging Cultivation-Based Technologies

Emerging cultivation-based technologies for *Spirulina* include wetland farming integrated with CO_2 fixation and closed photobioreactor.

3.4.1 Wetland Farming

Wetlands are natural systems that act as carbon sinks by capturing and storing CO_2 from the atmosphere. Wetlands can be an alternative mode for *Spirulina* cultivation, providing a sustainable and environmentally friendly method for producing this superfood. The *Spirulina* can take advantage of this natural process, using CO_2 as a nutrient and releasing oxygen back into the environment. It can help to reduce greenhouse gas emissions and mitigate climate change (Anderson et al. 2023).

In addition to their environmental benefits, wetlands offer several advantages for *Spirulina* cultivation. They provide a natural source of nutrients, including nitrogen and phosphorus, and can help to convert the organic nutrients to high-value metabolites.

To cultivate *Spirulina* in wetlands, a shallow pond or basin is typically used, with water levels carefully controlled to optimize growth and prevent contamination. The wetland *Spirulina* cultivation strategy is the best approach to producing biomass on a large scale and utilizing nonarable land mainly for the cultivation of microalgae in wastewater (Yehia et al. 2021). After filtration of culture and drying of *Spirulina's* biomass, it can be used for a variety of purposes, including food, supplements, animal feed, and biofuels. In conclusion, the use of wetlands for *Spirulina* cultivation and integrated CO_2 fixation provides a sustainable and environmentally friendly approach to producing this valuable superfood. By utilizing natural processes and promoting ecosystem health, this method offers benefits for human health, the environment, and economy.

3.4.2 Closed Photobioreactors (PBRs)

Closed PBRs offer a controlled environment that eliminates the risk of contamination and allows for precise monitoring of environmental conditions (Fig. 3.2).

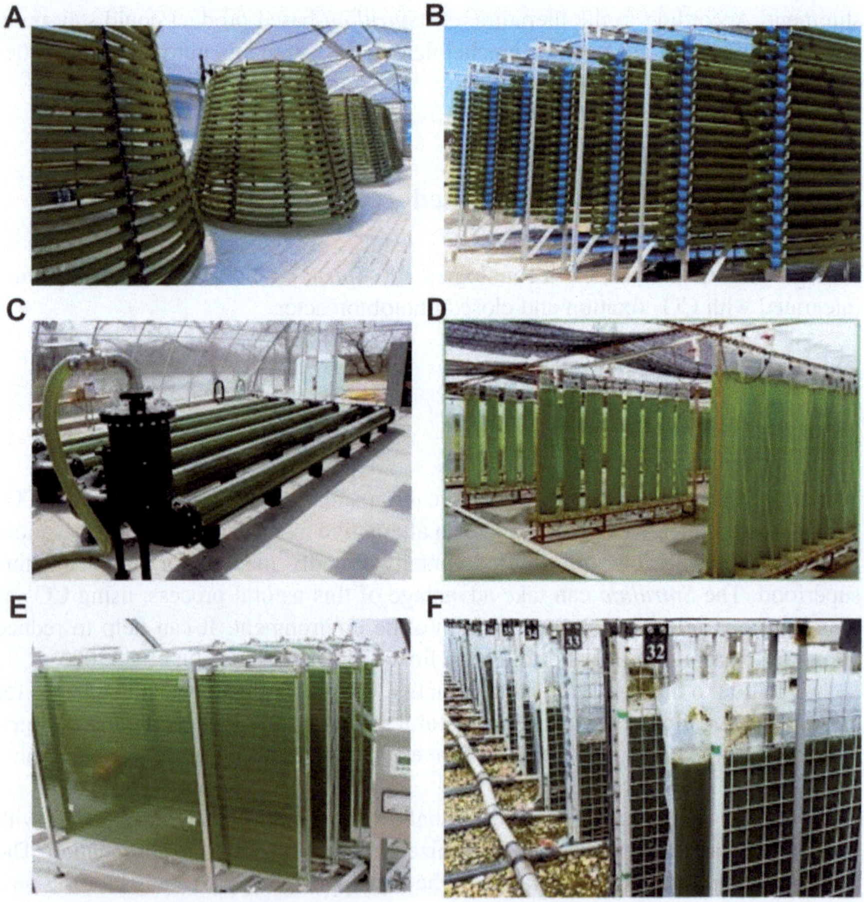

Fig. 3.2 Various configurations of closed photobioreactors (PBRs) for *Spirulina* cultivation. (**a**) Helical PBRs (HPBRs), (**b**) vertical tubular PBRs (VTPBRs), (**c**) horizontal tubular PBRs (HTPBRs), (**d**) annular PBRs (APBRs), (**e**) flat panel PBRs (FPPBRs), (**f**) vertical PBRs (VPBRs) (these images are taken from book chapter of IntechOpen publisher, and as per rules and regulation of publisher, no permission is required)

Additionally, they also capture CO_2 emissions from industrial processes, making them an environmentally friendly option.

Now, LED lighting is also used to optimize growth and the development of automated systems for the production of high-value pigments like chlorophyll and phycobilins. LEDs provide a light output spectrum that is safer and energy efficient for the production of light-sensitive pigments such as phycobilins and chlorophyll (www.thriveagritech.com/algae-growth). Algae strongly absorb and process red and blue light by using chlorophyll a and chlorophyll b, while phycobilins absorb green-red wavelength of light. So, depending on the requirement, the desired pigment can be increased by using the respective LED lights (Frank and Cogdell 2012).

3.5 *Spirulina*-Based Commercialized Products

Spirulina is widely known after NASA successfully used it as a space travel nutritional supplement for astronauts. However, it has multiple health-related applications such as anticancer, antiviral, and antioxidant properties that are being discussed latterly. From a commercial perspective, *Spirulina's* products related to pharmaceuticals, medicine, food, and skin are already commercialized. *Spirulina*-based tablets, capsules, B12 supplements, and phycocyanin are available commercially and have been used for health problems such as skin degeneration, aging, cardiovascular disease, cancer, and cataracts that are caused by oxidative damage and inflammatory stress (Rao and Rao 2007).

Spirulina has also been the top trend in the food industry, produced and consumed as a food (Jung et al. 2019). It has been used in reforming multiple traditional food products such as cookies (Şahin 2020), pasta (Grahl et al. 2020), ice cream (Tiepo et al. 2021), and juices (El-Beltagi et al. 2020). In the case of skin treatment, multiple *Spirulina*-based extracts are commercialized such as Bio-Botanica (USA) that commercializes a liquid blend of *Spirulina platensis* extract in glycerin and water. This extract offers skin conditioning benefits, and it is used in a variety of personal care formulations (Ragusa et al. 2021). Spiruline AP® is a water-soluble blue algae extract, formulated by SEPPIC (France, headquarter). This brand claims that its extract has excellent antiradical, anti-inflammatory, photoprotective, and cells renewal effects. It may be used as an active ingredient for antiaging skin products (Ragusa et al. 2021). It is likely that, in the coming years, the diversity, quality, and topical applications of *Spirulina* will rapidly increase. Table 3.6 summarized the already commercialized products of *Spirulina*.

Overall, *Spirulina* has various value chains which can be monetized. By leveraging the unique properties of *Spirulina* and its various applications, procedures can create a range of valuable products and services.

Table 3.6 Commercialized products of *Spirulina*

Application	Product	Brand	Price ($)	URL
Nutraceuticals	*Spirulina* 500mg capsules	Nicholas Nutraceuticals	8.37	www.amazon.in
	Spirulina tablets	SUNOVA	4.55	
	Organic *Spirulina*	Parry Nutraceuticals	–	www.parrynutraceuticals.com
	Spirulina capsules	Ancient Nutra	7.08	www.ancientnutra.com
	Spirulina tablets	Bionova	5.67	www.bionovastore.com
	Spirulina B12	Pnk Impex Trade	1.82	www.indiamart.com
Cosmetics	*Spirulina* facial cream	Pnk Impex Trade	1.03	www.indiamart.com
	Spirulina shampoo		1.21	
	Spirulina soap		0.61	
	Spirulina hair oil		1.15	
	Spirulina mask	Aromata Group	–	www.aromatagroup.net
	Spirulina lip balm			
	KELLY powder mask	PuroBio Cosmetics (Itlay)	349.04	www.ecco-verde.com/purobio-cosmetics
	Sukin Super Greens detoxifying clay masque 100 mL	SUKIN SKINCARE—(Australia)	5.95	www.cosmeticcapital.com
	Sukin Super Greens detoxifying clay scrub 125 mL		5.95	
	Sukin Super Greens detoxifying cleansing oil 125 mL		6.95	
Human food and beverages	*Spirulina* cereal	Beatus Natura	58.50	https://beatusnatura.com
	Spirulina smoothie, *Spirulina* energy bar, *Spirulina* shake and juice, *Spirulina* spaghetti, bright blue bread	DIC Color & Comfort	–	www.dic-global.com
	CrunchLina *Spirulina* Snack Clusters	prepared FOODS	–	www.preparedfoods.com

(continued)

Table 3.6 (continued)

Application	Product	Brand	Price ($)	URL
	Spirulina and Zitrone dm Bio 40g	dm Bio	–	world.openfoodfacts.org
	dm Bio Superfood Riegel *Spirulina* Zitrone 40 Gram			naschkater.com
	Extra-dark chocolate and *Spirulina*, milk chocolate, and *Spirulina*, extra-dark chocolate, hemp seeds and *Spirulina*	The Algae™ Factory	–	thealgaefactory.com
	Weleet *Spirulina* and Millet Digestive Healthy Cookies	Weleet	3.63	www.jiomart.com
Animal Feed	AQUATIC ARTS *SPIRULINA* FLAKES PREMIUM FRESHWATER FOOD	–	13.99	aquaticarts.com/products/*Spirulina*-flakes
	Kojima Supreme Color Enhancing Diet *Spirulina* fish food 1 kg	–	9.24	www.daraz.pk
	Freeze-dried *Spirulina* Brine shrimps Aquarium fish food 50g		1.22	
	Hikari *Spirulina* Brine shrimp	–	3.99	kingkoigoldfish.com
	Stiefel *Spirulina* pellet for horses	–	73.23	www.equusvitalis.com
	Spirulina powder organic for pets	SHARRETS NUTRITIONS	28.45	www.sharrets.com/products/*Spirulina*-powder-organic-for-pets
	MIDA	HealthyBio	2.43–60.63	healthybio.in/animal/
	MIDACAL		3–67.30	

3.6 Conclusions and Prospects

Spirulina is a filamentous, spiral-shaped, photoautotrophic, edible, nontoxic blue-green alga. It has many habitats but is mainly found in freshwater, seawater, alkaline lakes, and extreme environments, where other microorganisms would rarely survive. *Spirulina* is used as an excellent nutritional food in the human diet, due to its extremely high nutritional profile. It is considered the most nutritious diet on the planet and the richest source of protein (60–70%), carbohydrates (8–14%), lipids (10–20%), essential vitamins, essential fatty acids, photosynthetic pigments, amino acids, minerals, and phycobiliproteins. In addition, *Spirulina* is well known for its health benefits because it contains antioxidant, anticancer, immune-modulatory, antidiabetic, and immune-protective agents. Therefore, it is widely used as human food supplements, poultry, and aquaculture feed. The literature showed that when 02–08% *Spirulina* was mixed with the poultry and aquaculture feed, then it showed positive impacts on the growth and health of chicks and fish. The extensive literature on *Spirulina*'s food and feed applications has demonstrated beyond doubt that *Spirulina* can be effectively applied for pilot-scale outdoor cultivation to produce protein-rich biomass. However, outdoor cultivation of *Spirulina* has multiple challenges, but those challenges can be met by using closed PBR (Fig. 3.3). Recently, the *Spirulina* market is continuously flourishing, according to Allied Market Research; the global *Spirulina* market was USD 393.6 million in 2020 and is expected to reach USD 897.61 million by 2027, growing at a compound annual growth rate (CAGR) of 10.5% from 2020 to 2027. There are multiple commercialized products of *Spirulina* related to human food, animal feed, pharmaceutical and nutraceutical, and cosmetics. Some of the products are *Spirulina* cereal (Beatus Natura), *Spirulina* spaghetti and bright blue bread (DIC Color & Comfort), *Spirulina* powder organic for pets (Sharrets Nutritions), *Spirulina* B12 (Pnk Impex Trade), and *Spirulina* lip balm and mask (Aromata Group).

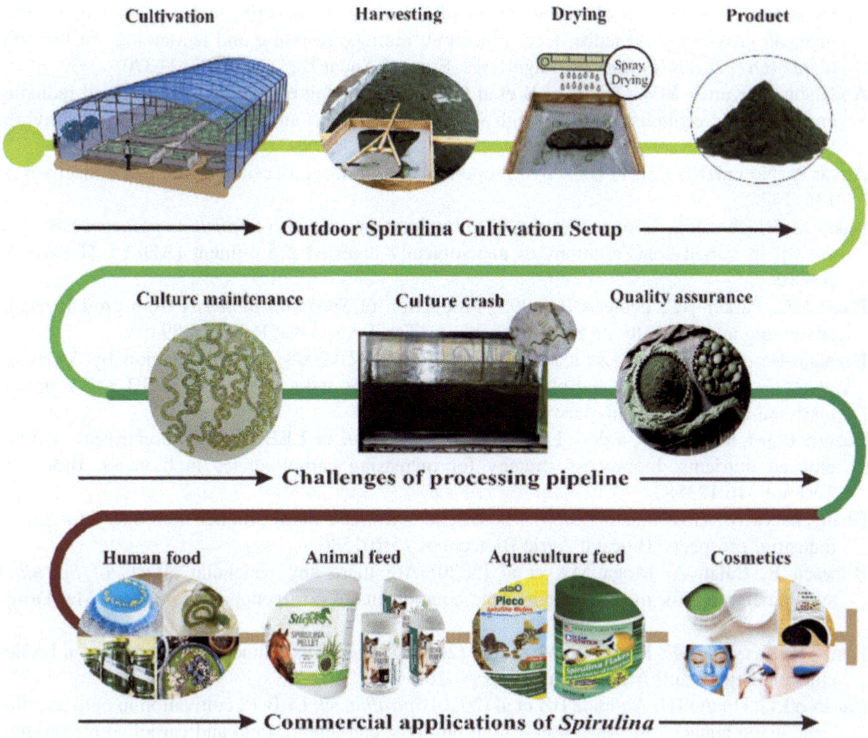

Fig. 3.3 A schematic representation of the cultivation setup, challenges, prospects, and applications of *Spirulina* farming

References

Abd El-Baky HH, El-Baroty GS (2012) Characterization and bioactivity of phycocyanin isolated from *Spirulina maxima* grown under salt stress. Food Funct 3:381–388

Abd El-Hady AM, Elghalid OA, Elnaggar AS et al (2022) Growth performance and physiological status evaluation of *Spirulina platensis* algae supplementation in broiler chicken diet. Livest Sci 263:105009

Abdulqader G, Barsanti L, Tredici MR (2000) Harvest of *Arthrospira platensis* from Lake Kossorom (Chad) and its household usage among the Kanembu. J Appl Phycol 12:493–498

Ahmed MM, Abdel El Deim MA, Orabi S et al (2022) Enhancement effect of *Spirulina platensis* extract on broiler chicks' growth performance and immunity. J Curr Vet Res 4:156–167

Ahsan M, Habib B, Parvin M, et al (2008) A review on culture, production and use of *Spirulina* as food for humans and feeds for domestic animals

Ak İ (2012) Effect of an organic fertilizer on growth of blue-green alga *Spirulina platensis*. Aquac Int 20:413–422

Akhtar N, Ahmed M, Hossain A et al (2014) Growth response of *Spirulina platensis* (nordst) geitler, in cabbage extract and antibacterial activities in different culture media. Bangladesh J Sci Ind Res 49:205–210

Al Mahrouqi H, Dobretsov S, Avilés A et al (2022) *Spirulina* optimization using cane molasses as the cost-effective alternative of sodium bicarbonate. Biol Bull 49:1–9

Al-Zayat MA (2019) Effect of various levels of *Spirulina (Arthrospira platensis)* as feed supplement on growth performance, feed utilization, immune response and hematology of the Nile tilapia (*Oreochromis niloticus)* fingerlings. Egypt J Aquat Biol Fish 23:361–370

Anderson A, Kumar MV, Gavurová B et al (2023) Optimizing engine performance and reducing emissions of greenhouse gases through *Spirulina* microalgae and nano-additive blends. Environ Res 231:115958

Anvar A, Nowruzi B (2021) Bioactive properties of *spirulina*: a review. Microbial Bioactives 4: 134–142

Baker M, Blackman S, Cooper E et al (2021) Exploratory analysis of *Spirulina platensis* LB 2340 growth in varied concentrations of anaerobically digested pig effluent (ADPE). Heliyon 7: e08065

Bates MC, Parker NC, Fedler CB (1995) Production of *Spirulina platensis* from growth media containing anaerobically digested cattle waste. Tex J Agric Nat Resour 8:89–98

Bermejo-Bescós P, Piñero-Estrada E, Del Fresno ÁMV (2008) Neuroprotection by *Spirulina platensis* protean extract and phycocyanin against iron-induced toxicity in SH-SY5Y neuroblastoma cells. Toxicol In Vitro 22:1496–1502

Bezerra PQM, Moraes L, Cardoso LG et al (2020) *Spirulina* sp. LEB 18 cultivation in seawater and reduced nutrients: bioprocess strategy for increasing carbohydrates in biomass. Bioresour Technol 316:123883

Bhattacharya M, Goswami S (2020) Microalgae – a green multi-product biorefinery for future industrial prospects. Biocatal Agric Biotechnol 25:101580

Bobescu E, Bălan A, Moga MA et al (2020) Are there any beneficial effects of *Spirulina* supplementation for metabolic syndrome components in postmenopausal women? Mar Drugs 18:651

Bonos E, Kasapidou E, Kargopoulos A et al (2016) *Spirulina* as a functional ingredient in broiler chicken diets. South Afr J Anim Sci 46:94–102

Cardoso LG, Duarte JH, Andrade BB et al (2020) *Spirulina* sp. LEB 18 cultivation in outdoor pilot scale using aquaculture wastewater: high biomass, carotenoid, lipid and carbohydrate production. Aquaculture 525:735272

Cervantes-Llanos M, Lagumersindez-Denis N, Marín-Prida J et al (2018) Beneficial effects of oral administration of C-phycocyanin and phycocyanobilin in rodent models of experimental autoimmune encephalomyelitis. Life Sci 194:130–138

Chakdar H, Pabbi S (2017) Algal pigments for human health and cosmeceuticals. In: Algal green chemistry. Elsevier

Chen Q, Li S, Xiong H et al (2022) Effect of different extraction methods on physicochemical characteristics and antioxidant activity of C-phycocyanin from dry biomass of *Arthrospira platensis*. Foods 11:1296

Chentir I, Hamdi M, Li S et al (2018) Stability, bio-functionality and bio-activity of crude phycocyanin from a two-phase cultured Saharian *Arthrospira* sp. strain. Algal Res 35:395–406

Ciferri O, Tiboni O (1985) The biochemistry and industrial potential of *Spirulina*. Annu Rev Microbiol 39:503–526

Damessa F (2021) Nutritional and functional values of microalgae (*Spirulina*) naturally found in East Africa. NM-AIST

De Amarante MCA, Braga ARC, Sala L et al (2020) Colour stability and antioxidant activity of C-phycocyanin-added ice creams after in vitro digestion. Food Res Int 137:109602

Demarco M, Moraes JOD, Ferrari MC et al (2022) Production of *Spirulina (Arthrospira platensis)* powder by innovative and traditional drying techniques. J Food Process Eng 45:e13919

Elbaz AM, Ahmed AM, Abdel-Maqsoud A et al (2022) Potential ameliorative role of *Spirulina platensis* in powdered or extract forms against cyclic heat stress in broiler chickens. Environ Sci Pollut Res 29:1–11

El-Beltagi HS, Dhawi F, Ashoush IS et al (2020) Antioxidant, anti-cancer and ameliorative activities of *Spirulina platensis* and pomegranate juice against hepatic damage induced by CCl4. Not Bot Horti Agrobot Cluj Napoca 48:1941

Fais G, Manca A, Bolognesi F et al (2022) Wide range applications of *Spirulina*: from earth to space missions. Mar Drugs 20:299

FAO Fisheries (2016) Aquaculture Department. Port State Measures

Fathi MA (2018) Effect of dietary supplementation of algae meal (*Spirulina platensis)* as growth promoter on performance of broiler chickens. Egypt Poult Sci J 38:375–389

Frank HA, Cogdell RJ (2012) 8.6 light capture in photosynthesis. In: Egelman EH (ed) Comprehensive biophysics. Elsevier, Amsterdam

Godia F, Albiol J, Montesinos J et al (2002) MELISSA: a loop of interconnected bioreactors to develop life support in space. J Biotechnol 99:319–330

Gogna S, Kaur J, Sharma K et al (2022) *Spirulina* - an edible cyanobacterium with potential therapeutic health benefits and toxicological consequences. J Am Nutr Assoc 42:559–572

Gonzálea R (1999) Ca¡diodinámíca. República Dominicana

Grahl S, Strack M, Mensching A et al (2020) Alternative protein sources in Western diets: food product development and consumer acceptance of *spirulina*-filled pasta. Food Qual Prefer 84: 103933

Grosshagauer S, Kraemer K, Somoza V (2020) The true value of *Spirulina*. J Agric Food Chem 68: 4109–4115

Han L-K, Li D-X, Xiang L et al (2006) Isolation of pancreatic lipase activity-inhibitory component of *Spirulina platensis* and it reduce postprandial triacylglycerolemia. Yakugaku Zasshi 126:43–49

Han P, Li J, Zhong H et al (2021) Anti-oxidation properties and therapeutic potentials of *Spirulina*. Algal Res 55:102240

Holman B, Malau-Aduli A (2013) *Spirulina* as a livestock supplement and animal feed. J Anim Physiol Anim Nutr 97:615–623

Hoseini SM, Khosravi-Darani K, Mozafari MR (2013) Nutritional and medical applications of *Spirulina* microalgae. Mini Rev Med Chem 13:1231–1237

Hussein NA, Ebied S, Saleh M (2021) Evaluation of the anticancer effect of violacein, phycocyanin and phycocyanobilin on apoptotic genes expression and glycan profiles in breast cancer cells. Int J Cancer Biomed Res 5:81–97

Izadi M, Fazilati M (2018) Extraction and purification of phycocyanin from *Spirulina platensis* and evaluating its antioxidant and anti-inflammatory activity. Asian J Green Chem 2:364–379

Jafari SMA, Rabbani M, Emtyazjoo M et al (2014) Effect of dietary *Spirulina platensis* on fatty acid composition of rainbow trout (*Oncorhynchus mykiss*) fillet. Aquac Int 22:1307–1315

Jain S, Singh SG (2013) Low cost medium formulation using cow dung ash for the cultivation of cyanobacterium: *Spirulina (Arthrospira) platensis*. Emir J Food Agric 2013:682–691

Jester BW, Zhao H, Gewe M et al (2022) Development of *Spirulina* for the manufacture and oral delivery of protein therapeutics. Nat Biotechnol 40:956–964

Joya M, Ashayerizadeh O, Dastar B (2020) Effects of *Spirulina (Arthrospira) platensis* and *Bacillus subtilis* PB6 on growth performance, intestinal microbiota and morphology, and serum parameters in broiler chickens. Anim Prod Sci 61:390–398

Jung F, Krüger-Genge A, Waldeck P et al (2019) *Spirulina platensis*, a super food? J Cell Biotechnol 5:43–54

Kavisri M, Abraham M, Prabakaran G et al (2021) Phytochemistry, bioactive potential and chemical characterization of metabolites from marine microalgae (*Spirulina platensis*) biomass. Biomass Convers Biorefin 13:1–8

Khan S, Mobashar M, Mahsood FK et al (2020) *Spirulina* inclusion levels in a broiler ration: evaluation of growth performance, gut integrity, and immunity. Trop Anim Health Prod 52: 3233–3240

Kim S-G, Choi A, Ahn C-Y et al (2005) Harvesting of *Spirulina platensis* by cellular flotation and growth stage determination. Lett Appl Microbiol 40:190–194

Kim S-S, Rahimnejad S, Kim K-W et al (2013) Partial replacement of fish meal with *Spirulina pacifica* in diets for parrot fish (*Oplegnathus fasciatus*). Turk J Fish Aquat Sci 13

Koníčková R, Vanková K, Vaníková J et al (2014) Anti-cancer effects of blue-green alga *Spirulina platensis*, a natural source of bilirubin-like tetrapyrrolic compounds. Ann Hepatol 13:273–283
Koru E (2012) Earth food *Spirulina (Arthrospira)*: production and quality standards. In: Food additive. IntechOpen, pp 191–202
Kumari A, Pathak AK, Guria C (2015) Cost-effective cultivation of *Spirulina platensis* using NPK fertilizer. Agric Res 4:261–271
Lafarga T (2019) Effect of microalgal biomass incorporation into foods: nutritional and sensorial attributes of the end products. Algal Res 41:101566
Lafarga T, Fernández-Sevilla JM, González-López C et al (2020) *Spirulina* for the food and functional food industries. Food Res Int 137:109356
Lafarga T, Sánchez-Zurano A, Villaró S et al (2021) Industrial production of *Spirulina* as a protein source for bioactive peptide generation. Trends Food Sci Technol 116:176–185
Leema JM, Kirubagaran R, Vinithkumar N et al (2010) High value pigment production from *Arthrospira (Spirulina) platensis* cultured in seawater. Bioresour Technol 101:9221–9227
Liu Y, Xu L, Cheng N et al (2000) Inhibitory effect of phycocyanin from *Spirulina platensis* on the growth of human leukemia K562 cells. J Appl Phycol 12:125–130
Lorenz R (1999) A review of *Spirulina* and *Haematococcus* algae meal as a carotenoid and vitamin supplement for poultry. Spirulina Pac Tech Bull 53:1–14
Lu H-K, Hsieh C-C, Hsu J-J et al (2006) Preventive effects of *Spirulina platensis* on skeletal muscle damage under exercise-induced oxidative stress. Eur J Appl Physiol 98:220–226
Madeira MS, Cardoso C, Lopes PA et al (2017) Microalgae as feed ingredients for livestock production and meat quality: a review. Livest Sci 205:111–121
Mariey Y, Samak H, Ibrahem M (2012) Effect of using *Spirulina platensis* algae as afeed additive for poultry diets: 1-productive and reproductive performances of local laying hens. Egypt Poul Sci J 32:201–215
Marles RJ, Barrett ML, Barnes J et al (2011) United States pharmacopeia safety evaluation of *Spirulina*. Crit Rev Food Sci Nutr 51:593–604
Masten Rutar J, Jagodic Hudobivnik M, Nečemer M et al (2022) Nutritional quality and safety of the *Spirulina* dietary supplements sold on the Slovenian market. Foods 11:849
Mathur M (2019) Bioactive molecules of: a food supplement. In: Bioactive molecules in food. Springer
Matufi F, Choopani A (2020) *Spirulina*, food of past, present and future. Health Biotechnol Biopharma 3:1–20
Mccarty MF, Dinicolantonio JJ, Lerner A (2021) Review–nutraceuticals can target asthmatic bronchoconstriction: NADPH oxidase-dependent oxidative stress, RhoA and calcium dynamics. J Asthma Allergy 14:685
Mehar J, Shekh A, Sarada R et al (2019) Automation of pilot-scale open raceway pond: a case study of CO2-fed pH control on *Spirulina* biomass, protein and phycocyanin production. J CO_2 Util 33:384–393
Michael A, Kyewalyanga MS, Lugomela CV (2019) Biomass and nutritive value of *Spirulina (Arthrospira fusiformis)* cultivated in a cost-effective medium. Ann Microbiol 69:1387–1395
Minić S (2021) Bioactive properties of *Spirulina*-derived phycobiliproteins and phycobilins. Biol Serb 43
Mirzaie S, Zirak-Khattab F, Hosseini SA et al (2018) Effects of dietary *Spirulina* on antioxidant status, lipid profile, immune response and performance characteristics of broiler chickens reared under high ambient temperature. Asian Australas J Anim Sci 31:556
Mohammadiazarm H, Maniat M, Ghorbanijezeh K et al (2021) Effects of *spirulina* powder (*Spirulina platensis*) as a dietary additive on Oscar fish, Astronotus ocellatus: assessing growth performance, body composition, digestive enzyme activity, immune-biochemical parameters, blood indices and total pigmentation. Aquac Nutr 27:252–260
Monteverde D, Gómez-Consarnau L, Suffridge C et al (2017) Life's utilization of B vitamins on early earth. Geobiology 15:3–18

Mullenix G, Maynard C, Owens C et al (2022) *Spirulina platensis* meal inclusion effects on broilers fed a reduced protein diet. J Appl Poult Res 31:100199

Niccolai A, Venturi M, Galli V et al (2019) Development of new microalgae-based sourdough "crostini": functional effects of *Arthrospira platensis (Spirulina)* addition. Sci Rep 9:1–12

Omar AE, Al-Khalaifah HS, Osman A et al (2022) Modulating the growth, antioxidant activity, and immunoexpression of proinflammatory cytokines and apoptotic proteins in broiler chickens by adding dietary *Spirulina platensis* phycocyanin. Antioxidants 11:991

Opoola E, Makinde O, Lawal A (2019) Effect of *Spirulina platensis* supplementation on performance, haematological and serum biochemical profiles of broiler chickens reared under tropical environment. Niger J Anim Sci 21:352–360

Ovando CA, Carvalho JCD, De Melo V, Pereira G et al (2018) Functional properties and health benefits of bioactive peptides derived from *spirulina:* a review. Food Rev Intl 34:34–51

Park J, Lee S, Kim I (2018a) Effect of dietary *Spirulina (Arthrospira) platensis* on the growth performance, antioxidant enzyme activity, nutrient digestibility, cecal microflora, excreta noxious gas emission, and breast meat quality of broiler chickens. Poult Sci 97:2451–2459

Park WS, Kim H-J, Li M et al (2018b) Two classes of pigments, carotenoids and c-phycocyanin, in *Spirulina* powder and their antioxidant activities. Molecules 23:2065

Patel S, Goyal A (2013) Current and prospective insights on food and pharmaceutical applications of *spirulina*. Curr Trends Biotechnol Pharm 7:696–707

Pelizer LH, De Carvalho JCM, De Oliveira MI (2015) Protein production by *Arthrospira (Spirulina) platensis* in solid state cultivation using sugarcane bagasse as support. Biotechnol Rep 5: 70–76

Pereira MIB, Chagas BME, Sassi R et al (2019) Mixotrophic cultivation of *Spirulina platensis* in dairy wastewater: effects on the production of biomass, biochemical composition and antioxidant capacity. PLoS One 14:e0224294

Pourhossein Z, AaA Q, Seidavi A et al (2015) Effect of different levels of dietary sweet orange (*Citrus sinensis*) peel extract on humoral immune system responses in broiler chickens. Anim Sci J 86:105–110

Ragusa I, Nardone GN, Zanatta S et al (2021) *Spirulina* for Skin care: a bright blue future. Cosmetics 8:7

Rahman D, Hidhayati N, Apriastini M (2022) Utilization of anaerobically digested dairy manure wastewater for *Spirulina maxima* cultivation. In: IOP conference series: earth and environmental science. IOP Publishing, p 012022

Rao AV, Rao LG (2007) Carotenoids and human health. Pharmacol Res 55:207–216

Reboleira J, Freitas R, Pinteus S et al (2019) Spirulina. In: Nonvitamin and nonmineral nutritional supplements. Elsevier

Rimbau V, Camins A, Pubill D et al (2001) C-phycocyanin protects cerebellar granule cells from low potassium/serum deprivation-induced apoptosis. Naunyn Schmiedeberg's Arch Pharmacol 364:96–104

Rodrigues MS, Ferreira LS, Converti A et al (2010) Fed-batch cultivation of *Arthrospira (Spirulina) platensis*: potassium nitrate and ammonium chloride as simultaneous nitrogen sources. Bioresour Technol 101:4491–4498

Roohani AM, Abedian Kenari A, Fallahi Kapoorchali M et al (2019) Effect of *Spirulina platensis* as a complementary ingredient to reduce dietary fish meal on the growth performance, whole-body composition, fatty acid and amino acid profiles, and pigmentation of Caspian brown trout (*Salmo trutta caspius*) juveniles. Aquac Nutr 25:633–645

Rosas VT, Poersch LH, Romano LA et al (2019) Feasibility of the use of *spirulina* in aquaculture diets. Rev Aquac 11:1367–1378

Şahin OI (2020) Functional and sensorial properties of cookies enriched with *Spirulina* and *Dunaliella* biomass. J Food Sci Technol 57:3639–3646

Sasson A (1997) Microalgal biotechnologies: recent developments and prospects for developing countries. In: Asia-Pacific marine biotechnology conference and 3. Asia-Pacific conference on algal biotechnology, Phuket, Thailand, 7–10 May 1997

Sibiya T, Ghazi T, Chuturgoon A (2022) The potential of *Spirulina platensis* to ameliorate the adverse effects of highly active antiretroviral therapy (HAART). Nutrients 14:3076

Simanjuntak SBI, Indarmawan I, Wibowo ES (2018) Impact of fed containing different levels of diets supplementation *Spirulina platensis* on growth, haematological, body composition and biochemical parameters, of Gurami (*Osphronemus gouramy*). Turk J Fish Aquat Sci 18:681–690

Soni A, Dubey M, Verma M et al (2015) Revisiting the role of phycocyanin in current clinical practice. Int J Pharm Sci Res 6:4588–4600

Sugiharto S, Yudiarti T, Isroli I (2016) Haematological and biochemical parameters of broilers fed cassava pulp fermented with filamentous fungi isolated from the Indonesian fermented dried cassava. Livest Res Rural Dev 28

Sugiharto S, Yudiarti T, Isroli I et al (2018) Effect of feeding duration of *Spirulina platensis* on growth performance, haematological parameters, intestinal microbial population and carcass traits of broiler chicks. South Afr J Anim Sci 48:98–107

Takyar MBT, Khajavi SH, Safari R (2019) Evaluation of antioxidant properties of *Chlorella vulgaris* and *Spirulina platensis* and their application in order to extend the shelf life of rainbow trout (*Oncorhynchus mykiss*) fillets during refrigerated storage. LWT 100:244–249

Teimouri M, Amirkolaie AK, Yeganeh S (2013) Effect of *Spirulina platensis* meal as a feed supplement on growth performance and pigmentation of rainbow trout (*Oncorhynchus mykiss*). World J Fish Mar Sci 5:194–202

Thangam R, Suresh V, Princy WA et al (2013) C-Phycocyanin from *Oscillatoria tenuis* exhibited an antioxidant and in vitro antiproliferative activity through induction of apoptosis and G0/G1 cell cycle arrest. Food Chem 140:262–272

Tiepo CBV, Gottardo FM, Mortari LM et al (2021) Addition of *Spirulina platensis* in handmade ice cream: Phisicochemical and sensory effects Adição de *Spirulina platensis* em sorvete caseiro: Efeitos físico-químicos e sensoriais. Braz J Dev 7:88106–88123

Touloupakis E, Cicchi B, Benavides AMS et al (2016) Effect of high pH on growth of *Synechocystis* sp. PCC 6803 cultures and their contamination by golden algae (*Poterioochromonas* sp.). Appl Microbiol Biotechnol Appl Biochem 100:1333–1341

Uddin A, Mahbuba S, Rahul S et al (2018) Super food *Spirulina (Spirulina platensis*): prospect and scopes in Bangladesh. Int J Bus Soc Sci Res 6:51–55

Vieira DCM, Matsudo MC, Sato S et al (2012) Simultaneous use of urea and potassium nitrate for *Arthrospira (Spirulina) platensis* cultivation. Biotechnol J 7:649–655

Vonshak A (1997a) Spirulina platensis (Arthrospira). Taylor & Francis, London

Vonshak A (1997b) Spirulina platensis arthrospira: physiology, cell-biology and biotechnology, CRC Press, Boca Raton, FL

Wan D, Wu Q, Kuča K (2016) Spirulina. In: Nutraceuticals. Elsevier

Wang H, Zhang W, Chen L et al (2013) The contamination and control of biological pollutants in mass cultivation of microalgae. Bioresour Technol 128:745–750

World Bank (2013) FISH TO 2030 prospects for fisheries and aquaculture World Bank Report Number 83177-GLB. Washington, DC

Wright J, Clift J, Dumais B, et al (1993) Design of components for the NASA OCEAN project

Wuang SC, Khin MC, Chua PQD et al (2016) Use of *Spirulina* biomass produced from treatment of aquaculture wastewater as agricultural fertilizers. Algal Res 15:59–64

Xiang W, Zhang X, Chen J et al (2020) Biochar technology in wastewater treatment: a critical review. Chemosphere 252:126539

Yehia S, El-Saadi A, Galal MM (2021) Application of algae to free surface wetlands for effluent reuse. Water Environ J 35:748–758

Youn K, Lee J, Yun E-Y et al (2014) Biological evaluation and in silico docking study of γ-linolenic acid as a potential BACE1 inhibitor. J Funct Foods 10:187–191

Yu J, Hu H, Wu X et al (2019a) Continuous cultivation of *Arthrospira platensis* for phycocyanin production in large-scale outdoor raceway ponds using microfiltered culture medium. Bioresour Technol 287:121420

Yu W, Yang Y, Chen Z et al (2019b) Dietary effect of *Spirulina platensis* on growth performance, digestive enzymes, haematological indices and antioxidant capacity of Chinese sea bass (Lateolabrax maculatus). South China Fish Sci 15:57–67

Yuan D, Yao M, Wang L et al (2019) Effect of recycling the culture medium on biodiversity and population dynamics of bio-contaminants in *Spirulina platensis* mass culture systems. Algal Res 44:101718

Zarrouk C (1966) Contribution a l'etude d'une Cyanophycee. Influence de Divers Facteurs Physiques et Chimiques sur la croissance et la photosynthese de *Spirulina mixima*. Thesis University of Paris, France

Zhai J, Li X, Li W et al (2017) Optimization of biomass production and nutrients removal by *Spirulina platensis* from municipal wastewater. Ecol Eng 108:83–92

Zhu C, Zhai X, Wang J et al (2018) Large-scale cultivation of *Spirulina* in a floating horizontal photobioreactor without aeration or an agitation device. Appl Microbiol Biotechnol Appl Biochem 102:8979–8987

Zinicovscaia I, Safonov A, Ostalkevich S et al (2019) Metal ions removal from different type of industrial effluents using *Spirulina platensis* biomass. Int J Phytoremediation 21:1442–1448

Chapter 4
Potential of Cyanobacterial Biomass as an Animal Feed

Muhammad Usman, Iqra Akbar, Sana Malik, Liya Deng, Md Asraful Alam, and Xu Jingliang

Abstract Increasing demand of poultry, livestock, and their derived products puts additional pressure on the agriculture sector. Major components of conventional animal meal stuff are corn and soybean, which also increase the competition for human food. Therefore, the usage of such kind of components in animal feed is unsustainable and requires a viable alternative. Selected cyanobacteria and microalgae could be suitable sources of animal feed due to their promising metabolites such as lipids, carbohydrates, proteins, pigments, and their derived products. Other secondary metabolites astaxanthin, lutein, β-carotene, chlorophyll, phycobilins, polyunsaturated fatty acids, and β-1,3-glucan have also shown a great potential in poultry, livestock, and aquaculture field. The replacement of corn, soybean, and other foods with microalgae could be an economically reliable alternative due to their lower production cost and multiproduct biorefinery. Although microalgae are very promising, there are certain challenges that need to be considered such as contamination risks in open pond cultivation, nutrient availability, poor penetration of light, and optimum temperature.

Keywords Cyanobacteria · Animal feed · Food security · Sustainability · Carbon neutrality

M. Usman · L. Deng · M. A. Alam · X. Jingliang (✉)
School of Chemical Engineering, Zhengzhou University, Zhengzhou, China
e-mail: alam@zzu.edu.cn; xujl@zzu.edu.cn

I. Akbar
Bioenergy Research Centre, Department of Bioinformatics and Biotechnology, Government College University Faisalabad, Faisalabad, Pakistan

S. Malik
Climate Change Cluster, University of Technology Sydney, Ultimo, NSW, Australia

M. A. Mehmood et al. (eds.), *Pharmaceutical and Nutraceutical Potential of Cyanobacteria*, https://doi.org/10.1007/978-3-031-45523-0_4

4.1 Introduction

According to the FAO (Food and Agriculture Organization) of the United Nations, the overall demand for products generated from animals will be more than double by the year 2050. The impact of this rising meat consumption is expected to be significant on livestock farming due to the widespread use of corn and soybeans as conventional feedstuffs (Food and Agriculture Organization of United Nations 2022). It appears that the existing corn and soybean allocations are unsustainable, necessitating the development of viable alternatives to keep the food, feed, and biofuel sectors functioning in harmony. Degradation of the land, insufficient water supplies, and severe climate change are also significant obstacles to animal farming. New, sustainable raw materials and increased efficiency in the use of existing resources will be crucial in ensuring the long-term viability of the livestock production process.

Microalgae are a vital aquatic resource. There are certain heterotrophic microalgal species, although most of the microalgae are autotrophs. Although microalgae are genetically a very diverse group of organisms genetically, with a vast diversity of physicochemical features, the great interest gaining phototrophic species include *Haematococcus*, *Chlorella, Dunaliella*, and *Arthrospira* (previously known as *Spirulina*, blue-green algae). Moreover, heterotrophic marine species such as *Crypthecodinium*, *Schizochytrium*, and *Ulkenia* are utilized for the generation of n-3 fatty acids in the marine environment. Microalgal photosynthesis has been shown to have a potential source of valuable chemicals or energy. This has led to advanced research into how this biomass can be used for food, feed, and biofuels (Chew et al. 2017). Moreover, microalgae can be grown for their protein, carbohydrate, lipid, micronutrient, food additive, cosmetic, and animal feed constituents and supplements (Sathasivam et al. 2019). Microalgal biomass has the potential to enrich feeds with a wide variety of nutrients, including vitamins, essential amino acids, polysaccharides, monounsaturated fatty acids (MUFA), and polyunsaturated fatty acids (PUFA), as well as minerals and colors (including carotenoids and chlorophylls) (Sathasivam et al. 2019). Feed supplements made from microalgal biomass have been the subject of numerous nutritional and toxicological studies (Becker 2013). There are many different examples, *Arthrospira*, of domesticated animals kept as pets (such as dogs, cats, and beautiful birds, horses, and cattle). *Chlorella*, *Isochrysis*, *Pavlova*, and *Phaeodactylum* are some of the most essential microalgae species for fish farming, especially for feeding fish larvae. Microalgae can be used as a protein supplement in chicken feed and to improve the yolk and skin color of the birds (Table 4.1). Due to the abundance of microalgae species, careful biochemical analysis is required to choose the best microalgae for use in modern animal diets (Batista et al. 2013). This chapter is focused on the microalgal metabolites, especially astaxanthin, lutein, β-carotene, chlorophyll, phycobilins, polyunsaturated fatty acid, and β-1,3-glucan, and the role of microalgae and cyanobacteria in livestock feed, poultry feed, aquaculture feed, shrimp feed, and their products. Additionally,

Table 4.1 Metabolites of different microalgae used in animal feed

Microalgae	Proteins (% dwt)	Carbohydrate (% dwt)	Lipids (% dwt)	Phycocyanin (% dwt)	β-carotene (% dwt)	Lutein (% dwt)	Astaxanthin (% dwt)
Spirulina	53.21	16.4	8.86	9.27	0.14	–	–
Chlorella	46.53	15.84	15.82	–	0.014	0.4	0.203
Dunaliella	40.46	20.44	15.51	–	0.102	–	0.083
Chlamydomonas	48	17	21	–	–	–	20.85
Scenedesmus	43.66	25.39	16.02	–	0.07	0.54	0.15
Haematococcus	30.87	37.93	23.07	–	0.05	–	3.07
Porphyridium	33.55	40.87	9.02	1.2	–	–	–
Isochrysis	41	14.46	17.72	–	–	1.8	–
Nannochloropsis	29.7	22.84	20.39	–	0.048	–	0.64
Schizochytrium	21	22.5	43.05	–	–	–	1.25

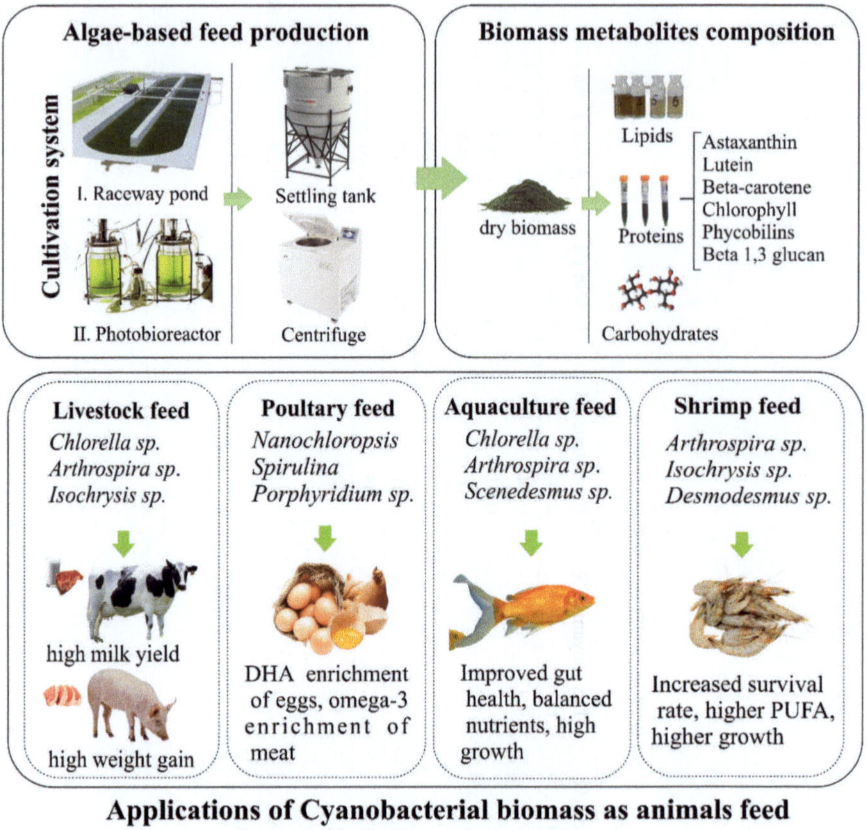

Fig. 4.1 Application of cyanobacterial biomass as animal feed

the economic applicability for the usage of microalgae for animal feed is also discussed critically. Diverse applications of cyanobacteria are shown in Fig. 4.1.

4.2 Cyanobacterial and Microalgal Metabolites

4.2.1 Lipids

Triacylglycerols (TAGs) are the predominant energy storage form in microalgal lipids. TAGs are made up of three fatty acid chains connected to a glycerol backbone; the exact fatty acid content varies between microalgal species. Microalgal lipids have shown promise as a source of high-value products in addition to their prospective use as biofuel. For instance, microalgae-based astaxanthin, a pigment with antioxidant characteristics, is used as a dietary supplement and in cosmetics. Microalgae lipids may also be used in the synthesis of omega-3 dietary supplements, animal feed, and even food products such as algae-based vegan meat analogues (Dineshbabu et al. 2019). To augment the quality and quantity of the fatty acids in animal feed, microalgal lipids are added as a nutritional supplement. The high concentration of omega-3 fatty acids in microalgal lipids has shown positive effects on animal health by lowering inflammatory responses and boosting the immune system (Adarme-Vega et al. 2012). As a rich source of energy and essential fatty acids, microalgal lipids are used as a feed element for aquaculture species like fish and shrimp. Microalgal lipids can also be used to enhance the flavor and texture of seafood, making it more attractive to consumers. Because of their nutritional value and possible benefits for animal health and performance, microalgal lipids have become an essential component in animal feed (Nagarajan et al. 2021).

4.2.2 Proteins

Microalgae, especially green algae and cyanobacteria, are high in protein, with a protein concentration ranging from 18 to 46% (Tibbetts et al. 2015). The cyanobacterium *Anabaena cylindrica* has been found to contain quite enough 69% protein (Tarento et al. 2018). Essential amino acids can also be abundantly observed in microalgal proteins. Microalgae are comparable to other protein sources including eggs, meat, milk, and soybeans in terms of the necessary amino acid content (tryptophan, methionine, phenylalanine, leucine, isoleucine, lysine, histidine, valine, and threonine). When it comes to protein quality, digestibility, and absorption, microalgal proteins are up to snuff for use in animal feed (Becker 2013). The overall free amino acid concentration of commercial *Spirulina* powders was found to be between 11.49 and 56.14 mg per 100 g, with essential amino acids accounting for between 2.06 and 31.72 mg per 100 g. Two types of *Spirulina* powder were found to be deficient in tryptophan and lysine, while *Chlorella pyrenoidosa* powder was deficient in isoleucine. The most abundant amino acids in *Chlorella vulgaris* SAG 211–19 were alanine (10.7%) and glutamic acid (10.3%), accounting for 38% of the total protein content. Instead of feeding soybean meal as a protein source, Finnish Ayrshire dairy cows were fed a combination of three different types of microalgae: *S. platensis*, *C. vulgaris*, and *N. gaditana* (Lamminen et al. 2019). It was noted that the microalgal meal used to replace soybean meal had no effect on the cows' growth or feed consumption. Milk fat went from 41 g/kg to 45 g/kg when microalgal meal was added, while milk PUFA (ALA and EPA) went from 3.44 g/kg to 4.85 g/kg. Although microalgal meal consumption decreased urinary nitrogen loss, its unappealing flavor was noted. Residual biomass after oil extraction is also good sources of protein. Although both microalgal diets increased weight gaining efficiency by 13–40% in broiler hens, there was no difference in performance or intestinal health between full fat and residual biomass of *Scenedesmus* sp. with 41% and 39% crude protein content, respectively (Sun et al. 2021).

4.2.3 Carbohydrates

Carbohydrates can also be abundantly encountered in microalgae. Starch, sugars, cellulose, and other polysaccharides are found in the cytosol and chloroplast of microalgae. Carbohydrates have multiple roles in microalgae, including providing energy reserves and supporting the cell walls. Microalgae can store large amounts of carbohydrates, up to 50 wt/wt% dry weight, because of their high photoconversion efficiency. *Porphyridium cruentum* (40–57 wt/wt%), *Prymnesium parvum* (25–33 wt/wt%), *Scenedesmus quadricauda* (21–52 wt/wt%), and others are among the algae with the highest carbohydrate content (Van Krimpen et al. 2013). Nonetheless, algal species differ in their glucose metabolism and composition. Consequently, it is important to select algae species that produce a high quantity of carbohydrates and have a sugar composition that is safe for human and animal consumption. Abiotic factors including light availability, salt stress, temperature, nutrients, and so on are all environmental elements that can be used to influence algae's production and carbohydrate content. Sugar content in microalgae is also influenced by a number of other factors, including the types of carbon sources used and the metabolic process (Buono et al. 2014). Light is essential for the photosynthetic process and so has a major impact on algae growth and biomass composition when grown in a controlled environment. In most cases, the light intensity used for algal cultivation is between 200 and 400 mol photons m^2s^{-1}. Carbohydrate buildup in microalgae can be manipulated via the metabolic pathway by using methods like nutrition limitation (Ismail et al. 2020).

4.2.4 Astaxanthin

One oxidized form of β-carotene is astaxanthin, also known as 3,3′-dihydroxy-β,β′-carotene-4,4′-dione. It was initially found in coastal environment, where it has since spread (Brendler and Williamson 2019). High amounts of this carotenoid pigment are present in the tissues of salmonids and the exoskeletons of many crustaceans (including shrimp, crabs, lobsters, and crayfish). In addition to being abundant in the tissues of salmonids and the exoskeletons of many crustaceans (including shrimp, crabs, lobsters, and crayfish), this carotenoid pigment is also widely distributed among marine bacteria and microalgae (Patel et al. 2022). As aquatic animals can only produce a little amount of astaxanthin biochemically, they must rely on diets rich in pigment to get the proper hue. Microalgae, which are found in the primary production stage of the food chain, are responsible for the majority of the biosynthesis of the carotenoid pigment astaxanthin (Lu et al. 2021). Astaxanthin is transmitted to higher trophic levels when fish and other aquatic animals eat the crustaceans, zooplankton, or insects that accumulated it by eating the microalgae. Astaxanthin can be biosynthesized by a variety of species, some of which have already been discovered. Included in this category are marine *Agrobacterium aurantiacum*, *Chlorococcum* sp., *Chlorella zofingiensis*, and *Haematococcus pluvialis*. The highest concentration of astaxanthin (>30 g of astaxanthin 1.0 kg dry biomass) was discovered in *Haematococcus pluvialis*, one of the exploited sources (Lu et al. 2021). Being a natural and modest supply of astaxanthin, this discovery has garnered a lot of attention. This carotenoid pigment has become widely recognized for its importance as a component of aquaculture feed. Iti s what gives seafood like salmon, trout, ornamental fish, shrimp, lobsters, and crayfish their appealing pinkish-red hue, which in turn increases their marketability (Lim et al. 2018). The demand for pigment astaxanthin has increased dramatically due to the aquaculture industry's continued growth. Furthermore, astaxanthin has been demonstrated to have beneficial effects when used as a dietary supplement in animals. Several aquatic species rely on astaxanthin for both growth and reproduction. Evidence is mounting that astaxanthin has a notable impact on reproductive success, egg production, and egg quality in aquatic species. Carotenoids and astaxanthin are pigments essential to health, but higher animals cannot make them on their own. Carotenoids, and especially astaxanthin, are important nutrients with a wide range of physiological functions (Elbahnaswy and Elshopakey 2023). Many studies have shown that increasing dietary astaxanthin has a beneficial effect on reproductive success and brood stock performance in a variety of marine species. The fertility of the goldfish species *Carassius auratus* was dramatically increased after being fed a diet containing 150 mg of astaxanthin per kg of feed each day for 150 days (Tizkar et al. 2015). Another related study found that feeding *P. monodon* brood stock a diet containing 500 mg of astaxanthin per kilogram of food considerably increased their reproductive efficiency, as evaluated by the quantity of spermatozoa in male shrimp and the amount of eggs in gravid females (Paibulkichakul et al. 2008). More than 60% of the cost of running an aquaculture hatchery is attributable to feed, but this varies widely depending on output scale and growing methods. Feeds containing nutritious components that enhance the growth and survival of the species being farmed are therefore required to lower production costs. Studies on a wide range of aquatic species have been done to examine the potential advantages of astaxanthin as a crucial nutritional addition required for optimum growth and survival (Lim et al. 2018).

4.2.5 Lutein

Cyanobacteria and microalgae may be a prospective biomass feedstock for pharmaceutical industries and animal feed due to their high protein, lipid, carbohydrate, pigment, vitamin, and antioxidant content (Udayan et al. 2017). Cyanobacteria and microalgae and the products that can be made from them are frequently used as feed components or supplements in aquaculture and poultry production (Świątkiewicz et al. 2015). Due to their availability and reduced cost compared to synthetic analogues, natural carotenoid pigments are seeing an increase in demand in the aquaculture sector. Carotenoids are natural lipid-soluble pigments that receive a lot of attention while studying fish. In contrast, fish must obtain carotenoids from their diet because they are unable to produce them. One of these naturally occurring carotenoid compounds is the pigment lutein, which is sourced from microalgae and cyanobacteria. Even though astaxanthin is one of the most important carotenoid pigments for fish, lutein pigment has shown comparable efficacy in promoting fish growth, survival, and coloration (Liu et al. 2021). Including carotenoids in a fish's diet can boost not just the skin color but also the market value of ornamental fish by making them look more vibrant. Several animal species may use integumentary carotenoids for photoprotection, camouflage, and signaling (e.g., breeding color) purposes. Research on the effects of carotenoids on fish growth has produced contradictory findings, with some studies showing a positive influence on fish growth and others finding no effect in anyway (Rashidian et al. 2021). Potent antioxidant carotenoids like lutein may protect the nutritional stores and embryos of brood stock from oxidation. It is observed that they play a role as a precursor for vitamin A and as pigment reserves for the formation of chromatophores and eyespots in embryos and larvae (Maoka 2011). During sexual maturation, lutein gathers in the hepatopancreas. During vitellogenesis, they are transported in the hemolymph as carotenoglycolipoproteins and ultimately accumulate in the eggs as a part of the lipovitellin protein. Carotenoids increase egg quality by protecting the eggs from harmful UV light and other environmental pro-oxidants (Maoka 2011). Research suggests that feeding Oscar fish (*Astronotus ocellatus*), Caspian brown trout (*Salmo trutta* caspius), and red cherry shrimp with *A. platensis* may cause these fish and shrimp to grow faster. African catfish fed with *A. platensis* have higher total body protein and flesh lipid levels (*Clarias gariepinus*). *A. platensis* not only increased the size of ornamental fish, but it also made their skin darker. The nutritional, biofuel, and medical industries may benefit from using cyanobacteria and microalgae as a biomass feedstock due to their high protein, fat, carbohydrate, pigment, vitamin, and antioxidant content. Cyanobacteria and the products that can be made from them are frequently used as feed additives or supplements in aquaculture and poultry production. Carotenoids cannot do their functions without the lipids to which they are connected. Fish, on the other hand, must obtain carotenoids from their diet because they are unable to synthesis them. One of these naturally occurring carotenoid compounds is the pigment lutein, which is sourced from microalgae and cyanobacteria. Fishes' growth, survival, and pigmentation were all positively affected by a combination of astaxanthin and lutein pigment; moreover, astaxanthin is one of the most necessary carotenoids (Ansari et al. 2021). The cellular mechanism of proteins, pigments, and TAG biosynthesis is shown in Fig. 4.2.

4.2.6 β-Carotene

Plants get their color from a compound called beta-carotene. Carrot is the Latin origin of the name beta-carotene (Rajkumar et al. 2023). It is responsible for the vibrant tones of yellow and orange in fruits and vegetables. Foods like margarine are

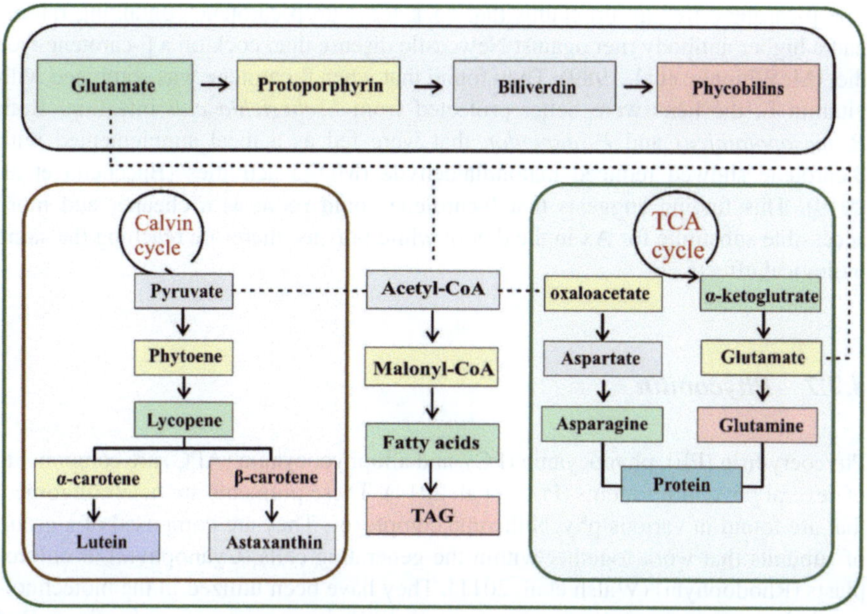

Fig. 4.2 Metabolism of pigments, TAGs, and protein biosynthesis in microalgae

colored by β-carotene. β-carotene is a precursor to vitamin A in the body (retinol). β-carotene, with its two β-ionone rings and eight isoprene units, is the most common dietary provitamin A carotenoid (Anand et al. 2022). Vitamin A is essential for normal eye development and function, a healthy immune system, and glowing skin and mucous membranes. Large amounts of vitamin A can be dangerous, but your body only converts as much as it needs from β-carotene. This means that β-carotene can be used without worry as a vitamin A supplement. Carotenoids can be obtained from microalgae. Fucoxanthin, violaxanthin, neoxanthin, and lutein are all considered provitamin A, and they can be produced in high numbers by *Spirulina*, *Chlorella*, *Dunaliella*, and *Haematococcus* (Foong et al. 2021). Hepatoprotection, anti-inflammation, anti-obesity, anticancer, and immunomodulation are just few of the biological actions associated with the raw extract and pure compound of β-carotene from microalgae (Marcelino et al. 2020; Riccio and Lauritano 2019; Ambati et al. 2019). Extracts of the microalgae *Spirulina* and *Dunaliella* β-carotene have been found to decrease transaminase activity in CCl_4-induced liver injury in Wistar rats. Supplementation of *Spirulina* biomass increases antioxidant enzymes in the liver and serum of Wistar rats, demonstrating its antioxidant properties (Ambati et al. 2019).

Bone, skin, and mucosal health, as well as eye development, relies on vitamin A (Nazrana et al. 2020). This demonstrates that β-carotene has a synergistic impact in poultry. β-carotene is believed to have a beneficial effect on the immune system and productivity of poultry that is susceptible to diseases and the effects of mycotoxins. It is also evident that β-carotene in eggs will improve people's health (Çalışlar 2019).

The β-carotene free group claims that cock fed with β-carotene-containing rations had a higher antibody titer against Newcastle disease than cock on a β-carotene-free diet (McWhinney et al. 1989). They found that when β-carotene was combined with vitamin E, the hens were better protected from *Escherichia coli* infection. Both *P. mesopotamicus* and *P. monodon* that were fed as a meal supplemented with β-carotene showed reduced malondialdehyde (MDA) activities (Bacchetta et al. 2019). This finding suggests that β-carotene could serve as a cheaper and more accessible substitute for Ax in the diet of white prawns, therefore reaching the same biological effects.

4.2.7 Phycobilin

Phycoerythrin (PE), phycocyanin (PC), and allophycocyanin (APC) are components of several phycobiliproteins (Price et al. 2014). These pigments are heterooligomers that are found in various phycobilisome complexes. They are composed of a group of subunits that work together within the generating cells (Cyanophyta) or chloroplasts (Rhodophyta) (Walsh et al. 2011). They have been utilized in the biotechnology industry as natural dyes and as nutraceuticals in a variety of applications. Red algae, cyanobacteria, some cryptomonads, and glaucophytes all have phycobilins (phycoerythrin and phycocyanin) in the stroma of their chloroplasts. Phycocyanin, a blue pigment, is found largely in cyanobacteria, while phycoerythrin, a red pigment, is the norm in red algae. The link between phycobilins and phycobiliproteins sets them apart from all other photosynthetic stain pigments (water-soluble proteins). Its primary role is to transfer the captured light energy to chlorophylls, where it can be used in the process of photosynthesis. Phycobilin fluorescence at a specific wavelength is commonly employed as chemical tags. The binding of phycobiliproteins to antibodies is visualized by phycobilin fluorescence, which is employed in immunofluorescence techniques. These phycobiliproteins, which originate from algae, fetch the highest prices on the market. Because of their strong coloring effects, phycobilins are used in the cosmetics and food industries in addition to their usage as chemical identifiers. Fucan, phycocyanin, allophycocyanin, diacylglycerol, etc. are just few of the additional bioactive chemicals found in marine algae (Soleimani 2022).

These are average values from different papers (Andrade et al. 2018; Salmeán et al. 2015; Koyande et al. 2019; Kumar et al. 2014; Li et al. 2019; Matos et al. 2017; Pagels et al. 2019; Shah et al. 2016; Molino et al. 2018; Shi et al. 2018).

4.2.8 Polyunsaturated Fatty Acids (PUFAs)

Fatty acids (FAs) with two or more double bonds in their acyl chain are called polyunsaturated fatty acids (PUFAs). Short-chain polyunsaturated fatty acids

(SCPUFAs) have a shorter carbon chain, typically consisting of 16 or 18 carbon atoms, while long-chain PUFAs (LCPUFAs) have a longer carbon chain, typically consisting of more than 18 carbon atoms. It is generally known that humans benefit from consuming PUFAs, which have been linked to increased metabolic rates, normalized blood pressure and glucose levels, and protection against a wide variety of disorders, including some forms of cancer (Kapoor et al. 2021). Moreover, EPA and DHA help protect chronic inflammatory illnesses, and there is some evidence that they can reduce the risk of obesity in animals and humans (Simopoulos 2016). The precursors of AA and DHA, linoleic acid (LA-C18:2 omega-6), and alpha-linolenic acid (ALA-C18:3 omega-3) are PUFAs that humans and other mammals cannot or only poorly produce. Consequently, it is essential to take in these substances directly from outside sources. Foods originating from fish, mollusks, and crustaceans (meat, milk, and eggs), fungi, bacteria, some plants, and microalga are common examples. One strategy for reducing fish meal consumption in aquaculture is to switch to oils derived from plants. Fish oil production is seen to be bad for the environment because of the rapid loss of natural fish stocks in response to rising demand. There are currently some items on the market that are derived from microalgae. Regulation of growing conditions and/or genetic alterations targeted at changing algal metabolism can be used to fine-tune algal biochemical composition and, thus, the yields of their high-value lipids. Microalgae's lipid metabolism can be managed by adjusting environmental conditions during growth. These include heat, light, CO_2 concentration, and nutrition availability. Increasing growth rates and, by extension, biomass and lipid productivities are possible through optimization of such factors. Tuning physical and nutritional conditions to alter microalgal biochemical composition results in parallel alterations in the biochemical composition of microalgae-fed organisms at both the larval and adult stages.

Initiation of host defense mechanisms in response to pathogen is thought to be the responsibility of beta-1,3-glucan. Apart from oil, beta-1,3-glucan is the most valuable component of *Chlorella* sp. that generates annual sales of more than US$ 38 billion (Dolganyuk et al. 2023). Zymosan was the original name given to beta-1,3-glucan in Norway during the 1940s. Beta-1,3-glucan, an immunostimulant, can improve both the general and targeted immune responses as well as resistance to bacterial infection in healthy fish (Sahoo and Mukherjee 2001). Beta-1,3-glucan is commonly found in *Chlorella* spp. Shrimp phenol oxidase activity was shown to be increased by pattern recognition proteins, which included beta-1,3-glucan binding protein (LGBP) and lipopolysaccharide.

4.3 Potential of Cyanobacteria as Feed

4.3.1 Livestock

Microalgae have a wide range of nutritional value for cattle production. In the first place, it is dependent on the species of the microalgae and its chemical makeup (such as protein, vitamins, antioxidants, lipids, polysaccharides, and minerals), and in the second place, it is dependent on how well the animal has adapted to the component. Microalgae have been shown to have a significant nutritional value in aquaculture, as well as in the diets of cows, pigs, sheep, chickens, and several other domestic animals and pets (Dolganyuk et al. 2023). *Chlorella* sp., *Isochrysis* sp., *Arthrospira* sp., and *Porphyridium* sp. account for as much as 30% of the world's current algal production sold for animal feed applications (Becker 2007). Animals' immunological responses, antiviral and antibacterial action, gastrointestinal function, disease resistance, and probiotic colonization can all be enhanced by including even a little quantity of microalgae biomass in their meal. The aggregate effects of these factors boost the reproductive success, feed conversion ratio, and body weight of the animals (Saadaoui et al. 2021). The meat quality of pigs, lambs, and broilers is known to improve when they are fed microalgae because of the vitamins, minerals, and polyunsaturated fatty acids (PUFA), especially EPA and DHA. Poultry is the most obvious choice as the domestic animal for whom microalgae will be used as feed, largely because the incorporation of microalgae into poultry diets presents the most promising prospect for their eventual commercial application (Kalia and Lei 2022). Nonetheless, ruminants are ideal animal models for feeding microalgae because they can break down the cell wall of raw microalgae. As a result, microalgae have been employed in ruminant diets in recent years to enhance their products, particularly meat (Kholif and Olafadehan 2021).

4.3.2 Microalgae for Meat Quality

Since humans and livestock are unable to produce omega-3 fatty acids on their own, these facts are considered essential and must be received through diet. It is well established that PUFAs like ALA, EPA, and DHA are beneficial to health (Sharma et al. 2020). As a consequence of this, it has been proposed that consuming foods that contain considerable quantities of omega-3 PUFAs can have significant positive effects on one's health (Saini et al. 2021). Although relatively low amounts of these polyunsaturated fatty acids were discovered in the meat of ruminant animals, the ruminant animal's diet is based on forage or cereals that are considered to be rich in linoleic acid (LA, C18:2 n-6) and ALA (C18:3 n-3) (Roque-Jiménez et al. 2021). Marine microalgae, algae-like microorganisms, and fish have been added to standard animal feed as well as other methods to increase PUFA content in meat. *A. platensis* or *Chlorella* sp. to a pig's diet has been found to improve the lipid profile of the meat. Recently, it was shown that increasing the percentage of *Schizochytrium* sp. microalgae in a pig's diet from 0% to 5% or 7% increased the amount of omega-3 fatty acids in the pig's flesh and caused alterations in the pigs' phenotype, appearance, and function (Martins et al. 2021). The addition of microalgae in the diets of pigs in amounts up to 33%, which is the greatest level that has been reported so far, was proven to have no adverse effects on the animals. Broiler hens' body weight gain, immunological features, and *Lactobacillus* bacteria production in their intestinal microflora are all enhanced when fresh liquid algae (1% of the meal) are added (Abdelnour et al. 2019). Finally, *Arthrospira* and *Chlorella* defatted biomass obtained during biofuel production was found to improve the quality of poultry meat (Rajagopal et al. 2021).

4.3.3 Microalgae for Milk Production and Quality

There is growing evidence to support the addition of microalgae to the diet to increase the production of milk high in healthy fatty acids (Lamminen et al. 2019). However, the digestive system of the animal (ruminant vs. nonruminant) and the metabolic capacities of the animal greatly influence the impact of microalgal metabolites on lactation and the transfer of nutrients to the milk (Safafar et al. 2015). The enzymes in the small intestine and the rumen bacteria in ruminants influence the digestion and absorption of fatty acids from the intestine. Intestinal and microbial enzymes work together to convert unsaturated fatty acids into short-chain saturated fatty acids, which may then be absorbed by the body. This process alters the molecular composition of the substances that are eventually incorporated into the animal's tissue. In contrast, the fatty acids in the diets of nonruminant animals are mostly unaltered, making them readily available for absorption by the small intestine and subsequent tissue incorporation. Because it has been proven that the quality of milk is directly impacted by the variety and quantity of fatty acids that are ingested by cattle, it is imperative that bio-hydrogenation in the rumen be avoided. Recent evidence suggests that microalgae are a viable protein feed alternative to soybean meal for dairy cattle (Wu et al. 2019). Because of this, ruminant animals should be fed coated microalgae biomass to preserve the feedstock's nutritional value (Lamminen 2021). This mechanism will facilitate the enhanced absorption of omega-3 fatty acids by the small intestine, leading to their transportation to the mammary glands. In addition, it has been demonstrated that the abundance of LA, DHA, and vaccenic acids in milk fat can be increased with the use of mixed algal feedstocks (Zhu et al. 2022). It has been demonstrated that supplementation with microalgae can enhance the level of DHA in milk by up to four times (Altomonte et al. 2018). *Schizochytrium* and *Nannochloropsis* are the most often employed microalgae strains for increasing milk quality in terms of usable fatty acids (Lopes et al. 2017). Furthermore, *Nannochloropsis* sp. has a greater level of EPA as well as other PUFAs when compared to strains such as *A. platensis* (cyanobacterium) and *C. vulgaris*, which have EPA contents comparable to fish oil (Zittelli et al. 1999). According to these studies, microalgae from different genera have diverse metabolic profiles and will have different effects on animals when utilized as feedstocks. Despite increasing amounts of LA, EPA, and DHA in milk enhanced with omega-3 FAs, the milk's oxidative stability remains unaltered (Demets et al. 2022). Prostaglandin secretion can be decreased through the consumption of omega-3 FAs during lactation, leading to an increase in fertility and a higher rate of embryo survival. In addition, giving cattle a diet that contains microalgae at inclusion rates of 5–10% increases the amount of certain minerals, such as iron, iodine, potassium, and zinc that can be found in the milk and flesh of the animals. Moreover, the amount of urea in the milk produced by cows that were fed protected microalgae had a lower concentration (2.98 mg/dL as opposed to 3.22–1.27 mg/dL; $P = 0.01$) (Saadaoui et al. 2021).

4.3.4 Poultry

Due to the increasing demand for protein-rich food for humans, there is currently a scarcity of novel feed resources that can provide livestock with a dependable and secure supply of nutrients. Multiple feeding experiments have demonstrated that residual biomass derived from biofuel production can be effectively integrated into the diets of chickens. This integration has resulted in favorable impacts on the birds' health, performance, and the quality of their meat and eggs (Armylisas et al. 2023). Eggs enriched with health-promoting long-chain n-3 polyunsaturated fatty acids (LCPUFAs n-3) are of particular importance for the chicken industry, and recent research shows that microalgal biomass can be used effectively to produce such eggs (Miles et al. 2021). Eggs can be enriched with LCPUFAs n-3 by feeding the layer linseed or fish oil; however, the latter technique has limitations due to the strong demand for marine products and the possibility of contamination with heavy metals. Therefore, incorporating certain species of microalgae such as into poultry nutrition could be of interest not only as a source of nutrients but also as an alternative method of enriching eggs with LCPUFAs n-3 (Nagarajan et al. 2021).

4.3.5 Microalgae Feed Supplement for Egg Production

Much research has been conducted on the manipulation of polyunsaturated fatty acids (PUFAs) in the diet, which has resulted in the manufacture of specialized egg yolks that are high in PUFAs (Woods and Fearon 2009). These egg yolks have been designed to match specific nutritional criteria that are deemed desirable. The utilization of microalgae as a substitute feedstock for crucial omega-3 fatty acids can enhance the nutritional quality of eggs meant for human consumption. Several omega-3 fatty acids exhibit noteworthy anti-inflammatory properties that are vital for growth. The addition of *N. gaditana*-supplemented diets in the feeding regimen of laying hens led to the accrual of long-chain omega-3 fatty acids, namely, eicosapentaenoic acid (EPA), docosapentaenoic acid (DPA), and docosahexaenoic acid (DHA), in the egg yolk (Bruneel et al. 2013). Eggs' omega-3 fatty acid content varies with the age and breed of the laying hens, as well as the digestibility of the microalgae they eat (Feng et al. 2020). Compared to eggs generated from conventional feed, those produced from hens fed with algal-blended feedstocks enriched with omega-3 FAs contained more beneficial fatty acids. The yolk's phospholipids are almost entirely composed of these long-chain fatty acids. The incorporation of microalgal biomass into feedstocks resulted in notable modification of the omega-3 fatty acid levels in eggs, even with a small quantity. Eggs laid by chickens given algae-based feed had a DHA content six times higher than control eggs. Egg yolks from hens fed algal feedstocks like *Porphyridium* sp. showed lower levels of cholesterol and lower ratios of n-6: n-3 fatty acids, while the chickens themselves consumed 10% less grain. Microalgal inclusion as low as 1.5% to 10% is claimed to be favorable for broilers, with further increases resulting in negative effects on growth and egg quality (Moran et al. 2019) (Table 4.2). Carotenoids gave the yolks of the eggs a distinct orange to red color. Studies have demonstrated that both naturally occurring and artificially produced carotenoids can significantly increase egg weight and enhance feed conversion in poultry. The oxidative stability of yolk lipids was found to be greatly improved when *Chlorella* was added to the feedstocks (Englmaierová et al. 2013). This is because the naturally occurring lutein was integrated more efficiently. Eggshell thickness and other desirable physical qualities were also enhanced by carotenoid-rich diets.

4.3.6 Suitability of Algal Biomass for Aquafeed

Algae production is the most efficient and cost-effective means of generating aquafeed, which can reduce overall aquaculture input costs by as much as 40% (Bhattacharjya et al. 2021). In the aquaculture food chain, microalgae play a crucial role due to their consumption by zooplankton. It has been reported that even small amounts of algae are sufficient for the fish diet to improve growth rate, feed utilization efficiency, intestinal microbiota, physiological activity, disease resistance, protein storage capacity over the winter, and control of the stress response. Algae are widely considered to be the main ingredient for fish in aquafeed. Microalgae's nutritional worth is determined by their cell size, shape, digestibility, biochemical content, and the needs of the animals they are fed to. Many species of microalgae have different nutritional profiles, with some being high in carbs and others in protein, vitamins, or lipids. Algae and microalgae are projected to be in high demand as fish food as the demand for fish production rises, creating substantial opportunities for the algae and aquaculture industries. The nutritional value, culture conditions (temperature, pH, and light), and environmental factors all affect the amount of protein in microalgal biomass, which is the primary metabolite. Fish meal with a high protein content that includes all the essential amino acids is highly recommended; however, fish meal is typically made with plant (crop) protein, which lacks some of these nutrients. Microalgae have been shown to contain all essential amino acids, which is important because fish meal requires an acceptable amount of amino acid (lysine, methionine, etc.) (Chrapusta et al. 2017). However, different algae have different amino acid profiles; marine algae, for example, have less amino acids with sulfur than freshwater microalgae (Niccolai et al. 2019). Certain

Table 4.2 Role of microalgae in animal feed

Microalgae	Animals	Weight (kg)	Age	Level in diet (% dry matter)	Time duration	Main findings	References
Arthrospira platensis	Dairy cows	37.6	6 weeks	1.18%	90 days	Increase of body condition by 8.5–11% Increase of body weight (10% dosage) Increase of backfat thickness (20% dosage)	Tremblay et al. (2007)
Arthrospira platensis	Lambs	46.5	6 weeks	0.01%	5 weeks	Increase of ADG, ADFI, and final body weight	EL-Sabagh et al. (2014)
Schizochytrium sp.	Cows	22.7		3.97%	6 weeks	Decrease of feed intake	Franklin et al. (1999)
Schizochytrium sp.	Lambs	34.8	3 months	1.92%	6 weeks	No effect on ADG, FCR, and carcass weight	Hopkins et al. (2014)
Schizochytrium sp.	Weaned male lambs	15.3	8 weeks	2%	6 weeks	Decrease ADG and feed intake	Díaz et al. (2017)
Arthrospira maxima	Piglets	20.9		20%	6 weeks	No effect on ADG, ADFI, and FCR	Saeid et al. (2013)
Arthrospira platensis	Pig	30.6		0.20%	5 weeks	Increase of ADG and FCR No effect on backfat thickness	Šimkus et al. (2013)
Arthrospira platensis & *Chlorella vulgaris*	Weaned piglets	9.1	28 days	1%	14 days	No effect on ADG, ADFI, and FCR	Furbeyre et al. (2017)
Chlorella spp.	Female pigs	30	3 months	0.00%	4 weeks	No effect on ADG, body weight, hot carcass weight, lean muscle thickness, and backfat thickness	Baňoch et al. (2013)
Chlorella vulgaris	Growing pigs	26.6		0.20%	6 weeks	Increase of ADG (0.1% dosage) No effect on ADFI, G:F, and body weight	Yan et al. (2012)
Schizochytrium sp.	Barrows	118		0.25%	8 weeks	No effect on ADG, FCR, and backfat thickness	Sardi et al. (2006)
Arthrospira sp.	Male chicken	678	21 days	8%	16 days	No effect on ADG	Toyomizu et al. (2001)

Arthrospira platensis	Hens	1.5	63 weeks	2.50%	4 weeks	No effect on ADFI and FCR	Zahroojian et al. (2013)
Arthrospira platensis	Broiler chicken		30 weeks	21%	21 days	No effect on ADG, ADFI, and FCR	Evans et al. (2015)
Arthrospira platensis	Broiler chicken		1 days	1.50%	36 days	Increase of ADG (1% dosage) Decrease of FCR (1% dosage)	Shanmugapriya et al. (2015)
Arthrospira platensis	Chicks		49 g	1%	42 days	No effect on ADG and FCR	Bonos et al. (2016)
Chlorella sp.	Broiler chicken		1 days	0.00%	42 days	Increase of ADG No effect on F:G	Dlouhá et al. (2008)
Chlorella vulgaris	Broiler chicken		1 days	0.21%	42 days	Decrease of FCR	Rezvani et al. (2012)
Chlorella sp.	Male broiler		1 days	1%	28 days	Increase of ADG No effect on ADFI and FCR	Kang et al. (2013)
Chlorella sp.	Laying hens		25 days	1.25%	39 weeks	Decrease of FCR No effect on feed intake	Englmaierová et al. (2013)
Chlorella sp.	Male Pekin ducks		1 days	0.20%	42 days	Increase of ADG and feed intake No effect on G:F	Oh et al. (2015)
Porphyridium sp.	Chickens		30 weeks	10%	10 days	No effect on body weight Decrease of ADFI	Ginzberg et al. (2000)
Schizochytrium JB5	Broiler chicks	0.04		0.20%	35 days	No effect on ADG, ADFI, and FCR	
Schizochytrium sp.	Broilers		21 days	7.40%	35 days	No effect on FCR and carcass yield Increase of ADG (7.4% dosage) and ADFI (7.4% dosage)	Ribeiro et al. (2013), Bonos et al. (2016)

microalgae, such as *S. platensis* and *C. vulgaris*, have a high protein content, making them a valuable food source for aquatic life (Bhattacharjya et al. 2021).

4.4 Challenges in Producing Microalgae/Cyanobacteria Biomass as Feedstock

There are several major problems that can arise during the cultivation of microalgae, including the following:

4.4.1 *Contamination Risks*

Microalgae and cyanobacteria are vulnerable to contamination by bacteria, fungi, and other microbes, which can outcompete the algae for nutrients and light if they are present in sufficient numbers (Pleissner et al. 2020). The presence of contamination can result in reduced growth rate as well as lower yields. The environment or tainted equipment can introduce bacteria into the culture, which can then use those nutrients for their own growth at the expense of the microalgae, or they can emit toxic substances that kill the microalgae (Saravanan et al. 2021). Microalgae may encounter competition with surface fungi for space, light, and nutrients. Some types of fungi are also capable of producing poisonous chemicals, which can be harmful to the microalgae. In addition, protozoans are single-celled organisms that are capable of feeding on the microalgae, which can lower the yield as well as the purity of the culture (Wang et al. 2022). In addition, harmful substances such as herbicides or pesticides could make their way into the culture via contaminated water or air, causing the microalgae to become harmed or perhaps extinct. It is imperative that during the cultivation process, aseptic conditions be maintained, that the culture be routinely monitored for signs of contamination, and that appropriate actions be taken to either contain or eradicate the organisms that are causing the contamination. Only then can contamination be avoided.

4.4.2 *Light Availability*

Light is crucial for the cultivation of microalgae because it is the primary energy source for photosynthesis, the process by which microalgae convert carbon dioxide and water into organic compounds and oxygen (Kumar et al. 2021). Microalgae use pigments, chlorophyll, to absorb light energy and then convert it into chemical energy that fuels their growth and reproduction. Microalgae require adequate amounts of light for photosynthesis and growth. However, excessive light can also damage the cells and decrease growth rates. The intensity, duration, and spectral quality of light can all influence the growth and biomass yield of microalgae (López Muñoz and Bernard 2021). For example, higher light intensity can increase the rate of photosynthesis and biomass production but can also lead to photoinhibition, a condition where excess light damages the photosynthetic machinery. On the other hand, lower light intensity can result in slower growth rates and lower biomass yields. Light can also regulate the biochemical composition of microalgae, including their protein, lipid, and pigment content (Xiao et al. 2020). For example, some microalgal species produce more lipids under conditions of high light and nutrient limitation, which can be harvested for biodiesel production. Other species may produce more pigments under low light conditions, which can be used for food and nutraceutical applications.

4.4.3 Nutrient Availability

Microalgae require various nutrients, including carbon, nitrogen, phosphorus, sulfur, and micronutrients, to support their growth and reproduction. The availability of these nutrients in the environment can significantly impact the growth rate and overall productivity of microalgae (Leong et al. 2022). In general, when nutrients are abundant, microalgae will grow and reproduce quickly, resulting in high biomass and productivity (Yaakob et al. 2021). However, when nutrients are limited, microalgae will experience reduced growth rates, decreased biomass, and lower productivity. For example, nitrogen is an essential nutrient for microalgae, and its availability can limit growth. Nitrogen-deficient microalgae may exhibit reduced photosynthesis, decreased cell division, and reduced biomass production (Nagappan and Kumar 2021). Similarly, phosphorus limitation can affect microalgae growth by reducing energy production, inhibiting photosynthesis, and decreasing cell division. Carbon is the primary building block for all organic matter and is therefore essential for the growth of microalgae. The availability of carbon can therefore limit the growth of microalgae, and increasing the supply of CO_2 can enhance their growth rate. Phosphorus is a key component of nucleic acids, ATP, and cell membranes and is therefore essential for microalgal growth (Yaakob et al. 2021). Microalgae can obtain phosphorus from inorganic sources, such as phosphate (PO43-), or organic sources, such as phytic acid. Like carbon and nitrogen, the availability of phosphorus can also limit microalgal growth, and increasing the supply of phosphate can enhance their growth rate. If these nutrients are limited, the microalgae may experience stunted growth, reduced photosynthetic efficiency, and even death. This can lead to a decrease in the yield of valuable products such as biofuels, nutraceuticals, and pharmaceuticals, which can have negative economic and environmental impacts. Moreover, in some cases, nutrient deficiency in microalgae can trigger changes in their metabolic pathways, leading to the production of unwanted compounds or toxins, which can make the microalgae unsuitable for certain applications.

4.4.4 Temperature

Microalgae are sensitive to temperature fluctuations, and optimal growth conditions can vary depending on the species. High temperatures can cause cell damage or even death, while low temperatures can slow down growth rates (Wang et al. 2022). The rate of photosynthesis in microalgae is affected by temperature. Optimum temperature ensures that the photosynthetic efficiency of the algae is maximized, resulting in increased biomass production (Wang et al. 2022). Optimum temperature provides the necessary thermal energy for metabolic processes in microalgae, leading to increased biosynthesis of proteins, carbohydrates, lipids, and other metabolites. Moreover, temperature affects the uptake of nutrients by microalgae. Optimum temperature enhances nutrient uptake, leading to increased growth rates and biomass productivity.

4.5 Economic Feasibility of Microalgae Production and Market Values

Due to their impressive productivity rates, microalgal biomass has been suggested as a sustainable option for generating energy and various other commodities. Microalgal biomass is being considered as a sustainable source for the creation of energy and various other products. The utilization of wastewater is a viable option for cultivation, rendering the need for agricultural land unnecessary (Safafar et al. 2015). However, the cost of producing microalgal biomass is comparatively almost equal to other feedstocks such as wheat (which is priced at only $0.035/kg). To enhance the feasibility of farming microalgae for biomass and valuable compounds on a commercial scale, it is imperative to enhance the cultivation and harvesting systems as well as improve the microalgal processing technologies. Almost 5 kt (thousand tons) per year of biomass is produced from microalgae at a cost of $ 25,000 per ton. The cost of recovering biomass can add further 20% to 30% to overall production costs (Das et al. 2019). Biomass is difficult to harvest because microalgal cells are so small (between 3 and 60 μm in diameter). Energy requirements vary depending on whether centrifugation, filtration, or gravity sedimentation is used to dewater and concentrate the microalgal biomass. The utilization of chemical flocculants may have an impact on the lipid extraction and/or the quality of the final product during biomass processing. However, the harvesting procedure must be developed empirically for each strain, considering factors like the intended use. It was calculated that annual production of dry algal biomass was 19,000 tons, worth approximately $5.7 billion (Jacob-Lopes et al. 2019). While there are only a limited number of microalgae strains that are commercially cultivated, *Arthrospira* has the highest annual output at 3000 tons, followed by *Chlorella* at 2000 tons, *Dunaliella* at 1200 tons, *Aphanizomenon* at 500–600 tons, and *Haematococcus pluvialis* at 300 tons (Batista et al. 2013). Processed biomass has various applications such as nutraceuticals, medicines, animal feed, aquaculture, human food, coloring compounds, and antioxidants. The global algae market is projected to attain a value of $1.143 billion, with a compound annual growth rate of 7.39% from 2016 to 2024. Microalgal compounds, such as omega-FAs, antioxidants, and coloring agents, which have been purified create far more income than the full, unprocessed biomass. For instance, natural functional components like lutein are produced by microalgae like *Chlorella* and *Scenedesmus*. Carotenoids saw a similar increase in value, from $1.24 billion in 2016 to an anticipated $ 1.53 billion by 2021 (Ambati et al. 2019). Despite the current disparities between production capabilities and market requirements, microalgae exhibit significant potential for meeting global market demands. Microalgae are now an uncompetitive feed choice due to their high production cost (Peng et al. 2018). However, it is possible that this scenario may undergo a transformation in the upcoming years due to technological progress and diverse policy measures, such as incentives and carbon pricing.

4.6 Conclusion and Prospects

The nutritional benefits can be substantially provided by using selected cyanobacteria and microalgae as feed or as a feed supplement. Animals raised on microalgae/cyanobacteria will have higher concentrations of beneficial nutrients, antioxidants, disease-fighting agents, and minerals, all of which will benefit humans who eat them. Microalgae and its derivatives will be commercially feasible, profitable, and widely available after growing and harvesting methods have been perfected. *Spirulina* sp. for protein replenishment, *Haematococcus* sp. for carotenoid improvement, and so on are just two examples of microalgal species that could be employed in feed production. Certain types of microalgae feed require processing before being given to animals. In addition to being more nutritious, microalgae feed also grows more quickly, requires less space, requires no arable land, and may be cultivated in salt water. It also prevents the needless waste of food and agricultural crops, such as grains used in aquaculture and animal feed. Microalgae will play an important role in feeding the world's growing population during the next few years. Further detailed studies involving milk and meat producing animals will be required to assure the potential applications of cyanobacteria and microalgae biomass to become part of the animal feed across the world. Efficient cultivation systems for large-scale production of the biomass to fulfill the demands should be developed alongside the impact of large-scale cultivation on the environment and biodiversity.

References

Abdelnour S, Abd El-Hack M, Arif M et al (2019) The application of the microalgae *Chlorella* spp. as a supplement in broiler feed. Worlds Poult Sci J 75:305–318

Adarme-Vega TC, Lim DK, Timmins M et al (2012) Microalgal biofactories: a promising approach towards sustainable omega-3 fatty acid production. Microb Cell Factories 11:1–10

Altomonte I, Salari F, Licitra R et al (2018) Use of microalgae in ruminant nutrition and implications on milk quality – a review. Livest Sci 214:25–35

Ambati RR, Gogisetty D, Aswathanarayana RG et al (2019) Industrial potential of carotenoid pigments from microalgae: current trends and future prospects. Crit Rev Food Sci Nutr 59: 1880–1902

Anand R, Mohan L, Bharadvaja N (2022) Disease prevention and treatment using β-carotene: the ultimate provitamin A. Rev Bras 32:491–501

Andrade L, Andrade C, Dias M et al (2018) *Chlorella* and *spirulina* microalgae as sources of functional foods, nutraceuticals, and food supplements: an overview. Food Process Technol. 6: 45–58

Ansari FA, Guldhe A, Gupta SK et al (2021) Improving the feasibility of aquaculture feed by using microalgae. Environ Sci Pollut Res 28:43234–43257

Armylisas AHN, Hoong SS, Tuan Ismail TNM (2023) Characterization of crude glycerol and glycerol pitch from palm-based residual biomass. Biomass Convers Biorefin 2023:1–13

Bacchetta C, Rossi AS, Cian RE et al (2019) Dietary β-carotene improves growth performance and antioxidant status of juvenile Piaractus mesopotamicus. Aquac Nutr 25:761–769

Baňoch T, Svoboda M, Kuta J et al (2013) The effect of iodine from iodine-enriched alga *Chlorella* spp. on the pork iodine content and meat quality in finisher pigs. Acta Vet Brno 81:339–346

Batista AP, Gouveia L, Bandarra NM et al (2013) Comparison of microalgal biomass profiles as novel functional ingredient for food products. Algal Res 2:164–173

Becker EW (2007) Micro-algae as a source of protein. Biotechnol Adv 25:207–210

Becker EW (2013) Microalgae for human and animal nutrition. In: Handbook of microalgal culture: applied phycology and biotechnology. Springer, pp 461–503

Bhattacharjya R, Singh PK, Mishra B et al (2021) Phycoprospecting the nutraceutical potential of Isochrysis sp as a source of aquafeed and other high-value products. Aquac Res 52:2988–2995

Bonos E, Kasapidou E, Kargopoulos A et al (2016) *Spirulina* as a functional ingredient in broiler chicken diets. South Afr J Anim Sci 46:94–102

Brendler T, Williamson EM (2019) Astaxanthin: how much is too much? A safety review. Phytother Res 33:3090–3111

Bruneel C, Lemahieu C, Fraeye I et al (2013) Impact of microalgal feed supplementation on omega-3 fatty acid enrichment of hen eggs. J Funct Foods 5:897–904

Buono S, Langellotti AL, Martello A et al (2014) Functional ingredients from microalgae. Food Funct 5:1669–1685

Çalışlar S (2019) The important of beta carotene on poultry nutrition. Selcuk J Agric Food Sci 33: 252–259

Chew KW, Yap JY, Show PL et al (2017) Microalgae biorefinery: high value products perspectives. Bioresour Technol 229:53–62

Chrapusta E, Kaminski A, Duchnik K et al (2017) Mycosporine-like amino acids: potential health and beauty ingredients. Mar Drugs 15(10):326

Das P, Thaher M, Khan S et al (2019) The effect of culture salinity on the harvesting of microalgae biomass using pilot-scale tangential-flow-filter membrane. Bioresour Technol 293:122057

Demets R, Bonneux L, Dejonghe C et al (2022) Photo-oxidative stability of aqueous model systems enriched with Omega-3 long-chain polyunsaturated fatty acid-rich microalgae as compared to autoxidative stability. J Agric Food Chem 70:5691–5700

Díaz M, Pérez C, Sánchez C et al (2017) Feeding microalgae increases omega 3 fatty acids of fat deposits and muscles in light lambs. J Food Compos Anal 56:115–123

Dineshbabu G, Goswami G, Kumar R et al (2019) Microalgae–nutritious, sustainable aqua-and animal feed source. J Funct Foods 62:103545

Dlouhá G, Sevcikova S, Dokoupilova A et al (2008) Effect of dietary selenium sources on growth performance, breast muscle selenium, glutathione peroxidase activity and oxidative stability in broilers. Czeh J Anim Sci 53:265

Dolganyuk V, Sukhikh S, Kalashnikova O et al (2023) Food proteins: potential resources. Sustainability 15:5863

El-Sabagh MR, Abd Eldaim MA, Mahboub D et al (2014) Effects of *Spirulina platensis* algae on growth performance, antioxidative status and blood metabolites in fattening lambs. J Agric Sci 6:92

Elbahnaswy S, Elshopakey GE (2023) Recent progress in practical applications of a potential carotenoid astaxanthin in aquaculture industry: a review. Fish Physiol Biochem 2023:1–30

Englmaierová M, Skrivan M, Bubancová I (2013) A comparison of lutein, spray-dried *Chlorella*, and synthetic carotenoids effects on yolk colour, oxidative stability, and reproductive performance of laying hens. Czeh J Anim Sci 58:412–419

Evans A, Smith D, Moritz J (2015) Effects of algae incorporation into broiler starter diet formulations on nutrient digestibility and 3 to 21 d bird performance. J Appl Poult Res 24:206–214

Feng J, Long S, Zhang H-J et al (2020) Comparative effects of dietary microalgae oil and fish oil on fatty acid composition and sensory quality of table eggs. Poult Sci 99:1734–1743

Food and Agriculture Organization of United Nations (2022). https://www.fao.org/home/en/

Foong LC, Loh CWL, Ng HS et al (2021) Recent development in the production strategies of microbial carotenoids. World J Microbiol Biotechnol 37:1–11

Franklin ST, Martin KR, Baer RJ et al (1999) Dietary marine algae (Schizochytrium sp.) increases concentrations of conjugated linoleic, docosahexaenoic and transvaccenic acids in milk of dairy cows. J Nutr 129:2048–2054

Furbeyre H, Van Milgen J, Mener T et al (2017) Effects of dietary supplementation with freshwater microalgae on growth performance, nutrient digestibility and gut health in weaned piglets. Animal 11:183–192

Ginzberg A, Cohen M, Sod-Moriah UA et al (2000) Chickens fed with biomass of the red microalga *Porphyridium* sp. have reduced blood cholesterol level and modified fatty acid composition in egg yolk. J Appl Phycol 12:325–330

Hopkins D, Clayton E, Lamb T et al (2014) The impact of supplementing lambs with algae on growth, meat traits and oxidative status. Meat Sci 98:135–141

Ismail MM, Ismail GA, El-Sheekh MM (2020) Potential assessment of some micro- and macroalgal species for bioethanol and biodiesel production. Energy Sources A Recovery Util Environ Eff. https://doi.org/10.1080/15567036.2020.1758853

Jacob-Lopes E, Maroneze MM, Deprá MC et al (2019) Bioactive food compounds from microalgae: an innovative framework on industrial biorefineries. Curr Opin Food Sci 25:1–7

Kalia S, Lei XG (2022) Dietary microalgae on poultry meat and eggs: explained versus unexplained effects. Curr Opin Biotechnol 75:102689

Kang H, Salim H, Akter N et al (2013) Effect of various forms of dietary *Chlorella* supplementation on growth performance, immune characteristics, and intestinal microflora population of broiler chickens. J Appl Poult Res 22:100–108

Kapoor B, Kapoor D, Gautam S et al (2021) Dietary polyunsaturated fatty acids (PUFAs): uses and potential health benefits. Curr Nutr Rep 10:232–242

Kholif AE, Olafadehan OA (2021) Microalgae in ruminant nutrition: a review of the chemical composition and nutritive value. Ann Anim Sci 21:789–806

Koyande AK, Chew KW, Rambabu K et al (2019) Microalgae: a potential alternative to health supplementation for humans. Food Sci Human Wellness 8:16–24

Kumar K, Dasgupta CN, Das D (2014) Cell growth kinetics of *Chlorella sorokiniana* and nutritional values of its biomass. Bioresour Technol 167:358–366

Kumar V, Sharma N, Jaiswal KK et al (2021) Microalgae with a truncated light-harvesting antenna to maximize photosynthetic efficiency and biomass productivity: recent advances and current challenges. Process Biochem 104:83–91

Lamminen M (2021) Nutritional value of microalgae for ruminants and implications from microalgae production. CABI Rev 2021

Lamminen M, Halmemies-Beauchet-Filleau A, Kokkonen T et al (2019) Different microalgae species as a substitutive protein feed for soya bean meal in grass silage based dairy cow diets. Anim Feed Sci Technol 247:112–126

Leong WH, Kiatkittipong W, Lam MK et al (2022) Dual nutrient heterogeneity modes in a continuous flow photobioreactor for optimum nitrogen assimilation to produce microalgal biodiesel. Renew Energy 184:443–451

Li T, Xu J, Wu H et al (2019) Growth and biochemical composition of *Porphyridium purpureum* SCS-02 under different nitrogen concentrations. Mar Drugs 17:124

Lim KC, Yusoff FM, Shariff M et al (2018) Astaxanthin as feed supplement in aquatic animals. Rev Aquac 10:738–773

Liu C, Li Y, Chen Z et al (2021) Effects of dietary whole and defatted *Arthrospira platensis* (cyanobacterium) on growth, body composition and pigmentation of the yellow catfish Pelteobagrus fulvidraco. J Appl Phycol 33:2251–2259

Lopes PA, Bandarra NM, Martins SV et al (2017) Markers of neuroprotection of combined EPA and DHA provided by fish oil are higher than those of EPA (Nannochloropsis) and DHA (Schizochytrium) from microalgae oils in Wistar rats. Nutr Metab 14:1–17

López Muñoz I, Bernard O (2021) Modeling the influence of temperature, light intensity and oxygen concentration on microalgal growth rate. Processes 9:496

Lu Q, Li H, Zou Y et al (2021) Astaxanthin as a microalgal metabolite for aquaculture: a review on the synthetic mechanisms, production techniques, and practical application. Algal Res 54: 102178

Maoka T (2011) Carotenoids in marine animals. Mar Drugs 9:278–293

Marcelino G, Machate DJ, Freitas KDC et al (2020) β-Carotene: preventive role for type 2 diabetes mellitus and obesity: a review. Molecules 25:5803

Martins CF, Ribeiro DM, Costa M et al (2021) Using microalgae as a sustainable feed resource to enhance quality and nutritional value of pork and poultry meat. Foods 10:2933

Matos J, Cardoso C, Bandarra NM et al (2017) Microalgae as healthy ingredients for functional food: a review. Food Funct 8:2672–2685

Mcwhinney S, Bailey C, Panigrahy B (1989) Immunoenhancing effect of β-carotene in chicks. Poult Sci 68:94

Miles EA, Childs CE, Calder PC (2021) Long-chain polyunsaturated fatty acids (LCPUFAs) and the developing immune system: a narrative review. Nutrients 13:247

Molino A, Iovine A, Casella P et al (2018) Microalgae characterization for consolidated and new application in human food, animal feed and nutraceuticals. Int J Environ Res Public Health 15: 2436

Moran C, Morlacchini M, Keegan J et al (2019) Increasing the omega-3 content of hen's eggs through dietary supplementation with Aurantiochytrium limacinum microalgae: effect of inclusion rate on the temporal pattern of docosahexaenoic acid enrichment, efficiency of transfer, and egg characteristics. J Appl Poult Res 28:329–338

Nagappan S, Kumar G (2021) Investigation of four microalgae in nitrogen deficient synthetic wastewater for biorefinery based biofuel production. Environ Technol Innov 23:101572

Nagarajan D, Varjani S, Lee D-J et al (2021) Sustainable aquaculture and animal feed from microalgae–nutritive value and techno-functional components. Renew Sust Energ Rev 150: 111549

Nazrana N, Jain T, Verma S (2020) Role of nutrition in maintaining normal eyesight - a review. Int J Med Biomed Stud 4:194–198

Niccolai A, Zittelli GC, Rodolfi L et al (2019) Microalgae of interest as food source: biochemical composition and digestibility. Algal Res 42:101617

Oh ST, Zheng L, Kwon H et al (2015) Effects of dietary fermented Chlorella vulgaris (CBT®) on growth performance, relative organ weights, cecal microflora, tibia bone characteristics, and meat qualities in Pekin ducks. Asian Australas J Anim Sci 28:95

Pagels F, Guedes AC, Amaro HM et al (2019) Phycobiliproteins from cyanobacteria: chemistry and biotechnological applications. Biotechnol Adv 37:422–443

Paibulkichakul C, Piyatiratitivorakul S, Sorgeloos P et al (2008) Improved maturation of pond-reared, black tiger shrimp (Penaeus monodon) using fish oil and astaxanthin feed supplements. Aquaculture 282:83–89

Patel AK, Tambat VS, Chen C-W et al (2022) Recent advancements in astaxanthin production from microalgae: a review. Bioresour Technol 364:128030

Peng K, Li J, Jiao K et al (2018) The bioeconomy of microalgal biofuels. In: Energy from microalgae. Springer, New York, pp 157–169

Pleissner D, Lindner AV, Ambati RR (2020) Techniques to control microbial contaminants in nonsterile microalgae cultivation. Appl Biochem Biotechnol 192:1376–1385

Price S, Toal S, Anandan S (2014) The TrpA protein of Trichodesmium erythraeum IMS101 is a non-fibril-forming collagen and a component of the outer sheath. Microbiology 160:2148–2156

Rajagopal R, Mousavi SE, Goyette B et al (2021) Coupling of microalgae cultivation with anaerobic digestion of poultry wastes: toward sustainable value added bioproducts. Bioengineering 8:57

Rajkumar M, Sivakumar J, Venu S et al (2023) β-Carotene analysis and histological investigation on different egg stages of anomuran crab, Emerita Asiatica (H. Milne Edwards, 1837). Asian J Sci Appl Technol. 12(1):1–11

Rashidian G, Rainis S, Prokić MD et al (2021) Effects of different levels of carotenoids and light sources on swordtail fish (*Xiphophorus helleri*) growth, survival rate and reproductive parameters. Nat Prod Res 35:3675–3686

Rezvani M, Zaghari M, Moravej H (2012) A survey on *Chlorella vulgaris* effect's on performance and cellular immunity in broilers. Int J Agric Sci Res 3:9–15

Ribeiro T, Lordelo M, Alves S et al (2013) Direct supplementation of diet is the most efficient way of enriching broiler meat with n-3 long-chain polyunsaturated fatty acids. Br Poult Sci 54:753–765

Riccio G, Lauritano C (2019) Microalgae with immunomodulatory activities. Mar Drugs 18:2

Roque-Jiménez JA, Rosa-Velázquez M, Pinos-Rodríguez JM et al (2021) Role of long chain fatty acids in developmental programming in ruminants. Animals 11:762

Saadaoui I, Rasheed R, Aguilar A et al (2021) Microalgal-based feed: promising alternative feedstocks for livestock and poultry production. J Anim Sci Biotechnol 12:76

Saeid A, Chojnacka K, Korczyński M et al (2013) Biomass of *Spirulina maxima* enriched by biosorption process as a new feed supplement for swine. J Appl Phycol 25:667–675

Safafar H, Van Wagenen J, Møller P et al (2015) Carotenoids, phenolic compounds and tocopherols contribute to the antioxidative properties of some microalgae species grown on industrial wastewater. Mar Drugs 13:7339–7356

Sahoo P, Mukherjee S (2001) Effect of dietary β-1,3 glucan on immune responses and disease resistance of healthy and aflatoxin B1-induced immunocompromised rohu (Labeo rohita Hamilton). Fish Shellfish Immunol 11:683–695

Saini RK, Prasad P, Sreedhar RV et al (2021) Omega-3 polyunsaturated fatty acids (PUFAs): emerging plant and microbial sources, oxidative stability, bioavailability, and health benefits—a review. Antioxidants 10:1627

Salmeán GG, Castillo LHF, Chamorro-Cevallos G (2015) Nutritional and toxicological aspects of *Spirulina* (*Arthrospira*). Nutr Hosp 32:34–40

Saravanan A, Kumar PS, Varjani S et al (2021) A review on algal-bacterial symbiotic system for effective treatment of wastewater. Chemosphere 271:129540

Sardi L, Martelli G, Lambertini L et al (2006) Effects of a dietary supplement of DHA-rich marine algae on Italian heavy pig production parameters. Livest Sci 103:95–103

Sathasivam R, Radhakrishnan R, Hashem A et al (2019) Microalgae metabolites: a rich source for food and medicine. Saudi J Biol Sci 26:709–722

Shah MMR, Liang Y, Cheng JJ et al (2016) Astaxanthin-producing green microalga *Haematococcus pluvialis*: From single cell to high value commercial products. Front Plant Sci 7:531

Shanmugapriya B, Babu SS, Hariharan T et al (2015) Dietary administration of *Spirulina platensis* as probiotics on growth performance and histopathology in broiler chicks. Int J Recent Sci Res 6:2650–2653

Sharma J, Sarmah P, Bishnoi NR (2020) Market perspective of EPA and DHA production from microalgae. In: Nutraceutical fatty acids from oleaginous microalgae: a human health perspective. Wiley, pp 281–297

Shi H, Luo X, Wu R et al (2018) Production of eicosapentaenoic acid by application of a delta-6 desaturase with the highest ALA catalytic activity in algae. Microb Cell Factories 17:1–15

Šimkus A, Šimkienė A, Černauskienė J et al (2013) The effect of blue algae *Spirulina platensis* on pig growth performance and carcass and meat quality. Veterinarija ir zootechnika 61:70–74

Simopoulos AP (2016) An increase in the omega-6/omega-3 fatty acid ratio increases the risk for obesity. Nutrients 8:128

Soleimani S (2022) Antioxidant compounds with antioxidant properties in seaweed and their application in various industries. J Plant Process Funct 2022:77–98

Sun T, Wang K, Wyman B et al (2021) Supplemental dietary full-fatted and defatted *Desmodesmus* sp. exerted similar effects on growth performance, gut health, and excreta hydrothermal liquefaction of broiler chicks. Algal Res 54:102205

Świątkiewicz S, Arczewska-Włosek A, Józefiak D (2015) Application of microalgae biomass in poultry nutrition. World Poult Sci J 71:663–672

Tarento TD, Mcclure DD, Vasiljevski E et al (2018) Microalgae as a source of vitamin K1. Algal Res 36:77–87

Tibbetts SM, Milley JE, Lall SP (2015) Chemical composition and nutritional properties of freshwater and marine microalgal biomass cultured in photobioreactors. J Appl Phycol 27: 1109–1119

Tizkar B, Kazemi R, Alipour A et al (2015) Effects of dietary supplementation with astaxanthin and β-carotene on the semen quality of goldfish (Carassius auratus). Theriogenology 84:1111–1117

Toyomizu M, Sato K, Taroda H et al (2001) Effects of dietary *Spirulina* on meat colour in muscle of broiler chickens. Br Poult Sci 42:197–202

Tremblay R, Cartier S, Miner P et al (2007) Effect of *Rhodomonas salina* addition to a standard hatchery diet during the early ontogeny of the scallop Pecten maximus. Aquaculture 262:410–418

Udayan A, Arumugam M, Pandey A (2017) Nutraceuticals from algae and cyanobacteria. In: Algal green chemistry. Elsevier

Van Krimpen M, Bikker P, Van Der Meer I et al (2013) Cultivation, processing and nutritional aspects for pigs and poultry of European protein sources as alternatives for imported soybean products. Wageningen UR Livestock Research

Walsh PJ, Smith S, Fleming L et al (2011) Oceans and human health: risks and remedies from the seas. Academic Press

Wang K, Khoo KS, Leong HY et al (2022) How does the internet of things (IoT) help in microalgae biorefinery? Biotechnol Adv 54:107819

Woods VB, Fearon AM (2009) Dietary sources of unsaturated fatty acids for animals and their transfer into meat, milk and eggs: a review. Livest Sci 126:1–20

Wu Y, Li L, Wen Z et al (2019) Dual functions of eicosapentaenoic acid-rich microalgae: enrichment of yolk with n-3 polyunsaturated fatty acids and partial replacement for soybean meal in diet of laying hens. Poult Sci 98:350–357

Xiao Y, Guo J, Zhu H et al (2020) Inhibition of glucose assimilation in *Auxenochlorella protothecoides* by light. Biotechnol Biofuels 13:1–13

Yaakob MA, Mohamed RMSR, Al-Gheethi A et al (2021) Influence of nitrogen and phosphorus on microalgal growth, biomass, lipid, and fatty acid production: an overview. Cells 10:393

Yan L, Lim S, Kim I (2012) Effect of fermented *Chlorella* supplementation on growth performance, nutrient digestibility, blood characteristics, fecal microbial and fecal noxious gas content in growing pigs. Asian Australas J Anim Sci 25:1742

Zahroojian N, Moravej H, Shivazad M (2013) Effects of dietary marine algae (*Spirulina platensis*) on egg quality and production performance of laying hens. J Agric Sci Technol 15:1353–1360

Zhu H, Wang X, Zhang W et al (2022) Dietary Schizochytrium microalgae affect the fatty acid profile of goat Milk: quantification of docosahexaenoic acid (DHA) and its distribution at Sn-2 position. Foods 11:2087

Zittelli GC, Lavista F, Bastianini A et al (1999) Production of eicosapentaenoic acid by *Nannochloropsis* sp. cultures in outdoor tubular photobioreactors. In: Progress in industrial microbiology. Elsevier

Chapter 5
Cost-Effective Cultivation of Cyanobacteria for Biotechnological Applications

Muhammad Nabeel Haider, Fatima Tahir, Syed Ghulam Musharraf, Farhat Jabeen, and Sana Malik

Abstract Cyanobacteria are photosynthetic organisms with extraordinary potential to be utilized for various biotechnological applications toward establishing a green and sustainable future. Despite enormous research efforts, the commercial-scale production of cyanobacteria and the economic sustainability of the upstream process are still under question. Devising cost-effective cultivation strategies offering higher biomass productivity with lower water and carbon footprints is critically important for the successful commercialization of cyanobacteria-based products. Different designs for algae photobioreactor (PBR) have evolved from the conventional open pond algae cultivation system. The first part of this chapter comprehensively covers the configuration of different cultivation systems and their comparison in terms of cost, energy, biomass productivity, and the utilization of wastewater as a low-cost cultivation media in addition to phycoremediation which provides a win-win opportunity to produce higher amounts of low-cost cyanobacterial biomass with concomitant treatment/recycling of wastewater. Additionally, different cost-effective cultivation systems, namely, biofilm-based cyanobacterial cultivation systems and heterotrophic cultivation systems, are also discussed. It is also important to investigate the impact of different biotic and abiotic factors on the growth and metabolite composition of cyanobacteria. In the second part of the chapter, various biotechnological applications of cyanobacterium biomass including biofuels, biofertilizers, nutraceuticals and pharmaceuticals, and bioplastics. Despite the extensive research

M. N. Haider · F. Tahir
Bioenergy Research Center, Department of Bioinformatics and Biotechnology, Government College University Faisalabad, Faisalabad, Pakistan

S. G. Musharraf
H.E.J. Research Institute of Chemistry, International Center for Chemical and Biological Sciences, University of Karachi, Karachi, Pakistan

F. Jabeen
Department of Zoology, Government College University Faisalabad, Faisalabad, Pakistan

S. Malik (✉)
Climate Change Cluster, University of Technology Sydney, Ultimo, NSW, Australia
e-mail: sana.malik@uts.edu.au

M. A. Mehmood et al. (eds.), *Pharmaceutical and Nutraceutical Potential of Cyanobacteria*, https://doi.org/10.1007/978-3-031-45523-0_5

efforts and applications, there is still a need to address challenges associated with the cost-effective, easier, and sustainable cultivation of cyanobacteria to achieve successful commercialization.

Keywords Cyanobacteria · Low-cost cultivation · Cultivation systems · Wastewater · Sustainability

5.1 Introduction

Cyanobacteria are photosynthetic organisms having vast industrial applications and are helpful in CO_2 fixation. Cyanobacteria can be cultivated on a large scale with basic raw materials including carbon dioxide (CO_2), light, and essential growth nutrients, namely, nitrogen, phosphorous, and potassium. Their efficient photosynthetic pathways, shorter cultivation periods, higher level of CO_2 tolerance, lower nutrient requirements, no need for herbicides or pesticides during cultivation, and utilization of nutrients in wastewater (which may lead to eutrophication if not removed) make them an attractive and suitable feedstock for various biotechnological applications (Tan et al. 2020). Cyanobacteria are also a source of several important bioactive compounds that have applications in the pharmaceutical and nutraceutical industries. Decades ago, research efforts have been focused on the conversion of cyanobacterial biomass into biohydrogen, biodiesel, bioethanol, and bio-oil. Recently, a focus shift has been observed in the algal research, where biomass is now being processed for multiple high-value products including pigments, fatty acids, and proteins instead of lipids (for biofuel applications) because lipids-derived biodiesel is not cost-competitive. Regardless of the application of biomass, the selection of cost-effective cultivation is an important parameter that prevents successful commercialization (Saratale et al. 2022).

Cultivating cyanobacteria using wastewater offers a range of significant advantages, making them an attractive and sustainable option for various applications. The extraordinary biosynthetic machinery of cyanobacteria efficiently removes excessive nutrients from the wastewater, leading to effective wastewater recycling (Khan et al. 2023). By growing cyanobacteria in wastewater, we can harness the natural carbon sequestration process, ultimately leading to the reduction of greenhouse gas emissions (Viswanaathan et al. 2022). The biomass produced by cyanobacteria cultivated in wastewater serves as a valuable and cost-effective resource for biofuel production. Unlike other biomass sources, this cyanobacterial biomass is free of lignin, making it easier to convert into biofuels through various eco-friendly processes (Kumar et al. 2020). Harnessing the capabilities of cyanobacteria in wastewater offers a multifaceted approach to addressing environmental challenges, creating renewable resources, and accessing valuable compounds for diverse applications. This holistic approach aligns with the principles of circular economy and sustainable development, making it a promising avenue for future research and implementation (Mohan et al. 2016).

Despite rigorous attempts to cultivate cyanobacteria using pure cultures and strictly controlled processes with sterile medium substrates, successful cultivation remains challenging. One of the primary obstacles is the unavoidable contamination with other types of algae, particularly green algae. Most of the information available on the factors influencing the growth and predominance of cyanobacteria found in the literature comes from freshwater ecosystems such as lakes and reservoirs. Within these environments, cyanobacterial development depends on various physical and chemical factors (Arias et al. 2017). These factors play a paramount role in shaping the photosynthetic efficiency and metabolite composition of cyanobacteria. Among these factors, light intensity, temperature, and nutrient availability emerge as critical determinants of cyanobacterial performance. Adequate levels of natural light are imperative to facilitate optimal photosynthesis, while temperature profoundly influences enzyme activity and cellular processes, directly impacting cyanobacterial growth and productivity (Hu et al. 2021). Furthermore, nutrient availability, especially nitrogen and phosphorus, exerts a significant influence on growth and metabolic pathways, thereby influencing the overall biochemical composition. The interchange of these primary factors, together with pH levels, further modulates algal metabolism. pH is a crucial abiotic factor as it affects biological wastewater treatment. pH levels affect enzyme activity and nutrient uptake, thereby influencing algal growth and productivity (Sutherland et al. 2014).

Certain species of cyanobacteria are documented to exhibit rapid growth and attain higher cell densities when cultivated under heterotrophic or mixotrophic conditions compared to photoautotrophic conditions (Yu et al. 2009). Large-scale industrial microalgae cultivation systems have traditionally utilized open ponds, employing paddle wheels to circulate microalgae cells, nutrients, and water while constantly exposing them to the external environment. However, in recent times, closed systems such as photobioreactors (PBRs) of various designs have gained popularity due to their ability to mitigate contamination risks and offer greater control over crucial growth parameters, ultimately leading to enhanced productivity (Suparmaniam et al. 2019).

5.2 Cultivation Strategies

The selection of a suitable cultivation strategy for cyanobacteria and algae is found to have a great impact on biomass production which contributes toward lowering the overall cost. Three types of cultivation strategies used for cyanobacteria and other unicellular algae include photoautotrophic, mixotrophic, and heterotrophic.

5.2.1 Photoautotrophic Mode of Cultivation

Photoautotrophic is the conventional mode of cyanobacterial cultivation in which sunlight, nutrients, and carbon dioxide are provided for its growth. This mode of cultivation is not economically feasible due to low light penetration which causes lower biomass production (Corrêa et al. 2020). Cyanobacteria also have the potential to grow heterotrophically and mixotrophically in the absence or presence of a light source (Andrade and Costa 2007). The utilization of organic carbon sources boosts biomass yield.

5.2.2 Heterotrophic Mode of Cultivation

Cyanobacteria mainly rely on the autotrophic mode of nutrition where they get energy from light, while some cyanobacteria adopt a heterotrophic mode of nutrition in which organic substrates are utilized as an energy source and can grow in the absence of light. Therefore, the heterotrophic cultivation strategy with the utilization of exogenous carbon can be adopted as the suitable mode for cost-effective biomass production. The economic feasibility is increased due to the use of low-cost sources of organic carbon that include cellulose and starch (Meireles Dos Santos et al. 2017). It is reported that heterotrophic cultivation of *Phormidium* sp. resulted in a higher yield of metabolites with the addition of cassava starch and maltodextrin (Francisco et al. 2014). The heterotrophic cyanobacteria are grown in controlled conditions that resulted in higher biomass production with a cost-efficient harvesting system (Rawat et al. 2013). Moreover, heterotrophically grown species possess higher lipid content as compared to photoautotrophically grown cyanobacteria (Miao and Wu 2006). However, all cyanobacterial species could not be able to grow under these conditions.

5.2.3 Mixotrophic Mode of Cultivation

The mixotrophic cultivation strategy is the most suitable mode of cultivation as it covers the benefits of both heterotrophic and autotrophic cultivation and results in a high yield of biomass and other bioactive compounds (Patel et al. 2020). Glycerol, glucose, or acetate could be used as a source of inorganic carbon for mixotrophic cultivation (Prabha et al. 2022). The economic viability of this cultivation mode depends on the low-cost carbon and energy sources and higher biomass productivity. In addition to higher biomass production, other benefits of this cultivation include a longer exponential growth phase, high lipids, pigments, and carbohydrate production. Several studies reported the positive impact of this cultivation approach on the biomass of various cyanobacterial strains. One of the studies depicts that *Nostoc* sp. MCC41 achieved five times more biomass in the mixotrophic mode as compared to the photobioreactor. Another study reported 2.28 and 4.98 times higher biomass production under mixotrophic conditions as compared to heterotrophic and phototrophic modes of cultivation, respectively (Yu et al. 2009). Cultivation of *Anabaena* PCC 7120 with the supplementation of exogenous glucose also showed higher growth than the photoautotrophic mode of cultivation (Zhan et al. 2017). Nevertheless, the photoautotrophic mode of cultivation is preferred in the case of high-value metabolites such as phycobilin, pigments, and carotenoids on an industrial scale due to low contamination problems.

5.3 Cost-Effective Cultivation Systems

The optimization of suitable culture conditions is the main concern for cost-effective cultivation and the accumulation of valuable metabolites. The increased biomass production is the basic purpose of the optimization of cultivation conditions. Organization and reorganization of the limiting factors and adaptation of different cultivation designs and strategies are crucial for the successful creation of a cost-effective cultivation system for cyanobacteria. The main factors such as the source of nutrients, intended products, CO_2 capture, and processing must be considered for the systematic, robust, and economical biomass production of cyanobacteria. The choice of cultivation system is also a crucial factor that depends on the target metabolite and its commercial value. For example, bioreactors with large surface areas utilize the light effectively so they are suitable for cyanobacterial cultivation if the targeted product is a high-value pigment (Mohan et al. 2020).

The two most popular cultivation systems for cyanobacteria and algae include open raceway ponds and closed photobioreactors. The mode of cultivation system and location are the prerequisites for the cost-effective biomass production of cyanobacteria on a large scale (Veerabadhran et al. 2021). The comparison of these cultivation systems is given in Table 5.1.

Table 5.1 Comparison of open raceway pond and photobioreactor based on different parameters

Parameters	Open raceway pond	Photobioreactor
Utilization of land	Higher	Lower
Area-volume ratio	Low (5–10 m^{-1})	High (20–200 m^{-1})
Capital costs	Low	High
Control of growth parameters	Very difficult	Easy
Biomass production	Low	High
Contamination of other species	Difficult to control	Controllable
Quality of biomass	Poor	Good
Disinfection	Difficult	Easy
Application at industrial scale	Easy	Difficult

5.3.1 Cultivation Using Open Raceway Pond (ORP)

Large-scale cultivation of cyanobacteria usually prefers the open raceway ponds and seawater/wastewater for cost-effective cultivation which is carried out in both natural water sources, such as ponds, lagoons, and lakes, and man-made water tanks (Lundquist et al. 2010). The depth of an open pond typically ranges between 1 and 100 cm and is known as the oldest and most widely used system (Sharma et al. 2011; Singh et al. 2014). It comprises a simple structure that includes a pond constructed from concrete or dug into the ground with a lining of polymer. A few types of open raceway ponds are more suitable for commercial applications which include circular, inclined, and natural systems. The easy and inexpensive construction is the main characteristic of these raceway ponds. Wastewater or runoff from lands is the most common nutrient supply in this case as it makes the upstream process more cost-efficient with the additional benefit of water treatment. This type of cultivation system is regarded as a low-cost strategy for producing more biomass and lipids. Besides these benefits, the requirement of large space for its construction and difficulty in controlling the environmental changes cause the intolerance of this system for cyanobacterial cultivation (Faried et al. 2017). Low biomass production due to the unequal distribution of light that can reach only a limited depth is another downside of this system. The major issue of this system is to control the fluctuating environmental temperature that changes with the season in diurnal manner. At higher temperatures, the cooling of the system relies on the evaporation that causes a massive evaporative loss. Additionally, the contamination risks due to other invasive species make this system favorable for a limited number of cyanobacterial strains (Tan et al. 2018). Despite this, several reports in the literature explain the successful operation of open raceway ponds over other cultivation systems (Haider et al. 2023). Cultivation of *Anabaena* sp. in HRAP achieved biomass of 9 g m^{-2} day^{-1} in winter, while 20 g m^{-2} day^{-1} in the summer season (Moreno et al. 2003), even though more improvements are needed to increase its efficiency and decrease operational costs.

5.3.2 Closed Photobioreactors for High-Quality Biomass Production

Production of high-quality biomass is a challenge on the commercial scale which is required for nutraceutical and pharmaceutical applications. Closed photobioreactors are developed to get around the drawbacks of open raceway ponds and to make better control of the cultivation system. Due to controlled conditions, the closed photobioreactors have high biomass production, larger surface-volume ratio, and lower risks of cross-contamination and are also suitable for those species that are sensitive to environmental parameters and culture conditions. The comparison of the cultivation systems based on biomass production is given in Table 5.2. This system

Table 5.2 Comparison of different algal strains based on biomass production in different cultivation systems

Species	Cultivation medium	Cultivation system	Biomass production	References
P. terebrans BERC10	Urban wastewater	Open raceway pond	1.6 gL^{-1}	Haider et al. (2023)
Spirulina sp. LEB 18	Wastewater	Open raceway pond	1.14 gL^{-1}	Mata et al. (2020)
Spirulina sp. LEB 18	Aquaculture wastewater	Open raceway pond	3.3 gL^{-1}	Cardoso et al. (2020)
Scenedesmus sp.	Domestic wastewater	Open raceway pond	4–17 gm^{-2} day^{-1}	Posadas et al. (2015)
Fistulifera sp.	Standard media	Open raceway pond	0.6 gL^{-1}	Sato et al. (2014)
Spirulina platensis	–	Open raceway pond	0.18–0.32 gm^{-2} day^{-1}	Pushparaj et al. (1997)
Arthrospira platensis	Standard media	Photobioreactor	2.7 gL^{-1} day^{-1}	Carlozzi (2003)
Botryococcus braunii	MBG11	Airlift photobioreactor	2.31 gL^{-1}	Ge et al. (2011)
Spirulina maxima	–	Stirred tank photobioreactor	10.8 gm^{-2} day^{-1}	Belkacemi and Hamoudi (2003)
Synechocystis aquatilis	Modified SOT	Flat plate photobioreactor	30 gm^{-2} day^{-1}	Zhang et al. (2002)
Aphanothece microscopica	Synthetic BGN	Bubble column photobioreactor	0.770 gL^{-1} day^{-1}	Jacob-Lopes et al. (2009)
Botryococcus braunii	BG11	Biofilm photobioreactor	0.71 gm^{-2} day^{-1}	Ozkan et al. (2012)
Spirulina platensis	Standard	Tubular photobioreactor	32.5 gm^{-2} day^{-1}	Masojídek et al. (2003)
Chlorococcum	–	Dome photobioreactor	0.1 gL^{-1} day^{-1}	Sato et al. (2006)

lowers the chances of evaporative and CO_2 loss while encouraging robust growth. Moreover, they can use flu gases produced from other processes to make the process eco-friendlier. Closed photobioreactors are constructed by plexiglass which adds the cost (Singh and Sharma 2012). This system has been successfully utilized for the production of high-quality biomass for food applications (Lundquist et al. 2010). These bioreactors consist of a panel of transparent tubes that are placed vertically or horizontally with the continuous CO_2 supply. Currently, various types of bioreactors are used, but the majority of them have applied forms of airlift, flat-panel, and bubble bioreactors. It is a good idea to design a strain-specific bioreactor according to the target metabolite. The economic viability of the cultivation system can be examined by calculating the net-energy ratio (NER) whose value should be >1 (De Bhowmick et al. 2019). The majority of the findings showed that mainly open raceway ponds meet the threshold level of NER. On the other side, photobioreactors are crucial in case of low contamination. Therefore, a hybrid model of ORP and photobioreactor is required due to the unitability of the ideal cultivation system (Huang et al. 2017).

5.3.2.1 Efficient Cultivation of Cyanobacteria Using Tubular Photobioreactors

Tubular photobioreactors are the most promising and highly efficient due to lower harvest time and higher biomass production. These reactors are one of the most suitable for outdoor cultivation due to the higher surface area for light penetration (Wang et al. 2012). They have parallel tubes that are positioned in a way to maximize exposure to sunlight. The diameter of these tubes lies between 10 mm and 60 mm. Airlift device is used to circulate the media through the reactor by reducing the chances of cell damage and contamination. The chances of photosynthesis inhibition are also reduced by removing excess oxygen airlift devices. The efficiency of the culture depends on the geometry of the photobioreactor to increase the light penetration, gas exchange, and optimization of its flow. With the increase in tube diameter, there is a decrease in volumetric productivity; on the other side, increase in volume has a positive impact on areal productivity. It is reported that *P. tricornutum* achieved areal productivity of 1.5 g^{-1} L^{-1} day^{-1} with a 0.06 m tube diameter (Bahadar and Khan 2013).

Despite the popularity of tubular photobioreactors, there are some downsides to the reactors. The most frequent problem is the photolimitation in which the cells that lie in the middle portion of the tubes receive low light and the growth is compromised. Secondly, the higher temperature in the summer season also causes growth inhibition. O_2 accumulation, photo limitation, CO_2 depletion, and increase in temperature are the major problems of tubular photobioreactors at large scale (Vasumathi et al. 2012).

5.3.2.2 Flat-Panel Photobioreactors

This type of photobioreactor is also considered an efficient system for higher biomass productivity due to its high area-to-volume ratio that is constructed with a semitransparent or transparent material like plexiglass, plastic bags, or glass (Ting et al. 2017). These are categorized as outdoor with natural sunlight and indoor reactors with artificial light. Agitation of the sample is provided by mechanical rotation or air bubbling. The mixing rate positively impacts biomass production as it provides a sufficient supply of CO_2 with the continuous discharge of excess oxygen (Faried et al. 2017). High aeration rates are avoided due to an increase in capital cost on a large scale. It is reported that the growth of *B. braunii* improved with the increase in aeration rate (Ge et al. 2011). The main advantage of flat-panel photobioreactors is the higher biomass production because of the lower chances for the accumulation of dissolved oxygen when compared to tubular photobioreactors.

5.3.2.3 Biofilm-Based Cyanobacterial Cultivation

Higher biomass production and increased photosynthetic activity are the focus of the industrial-scale applications of cyanobacteria. The basic photosynthetic theory explains that the higher limit of photosynthetic activity is 12.6% which corresponds to 120–150 gm^{-2} day^{-1} biomass productivity (Wang et al. 2017). But after extensive research on conventional cultivation methods such as open raceway ponds and photobioreactors, these are still unable to achieve the remarkable breakthrough as the maximum biomass productivity is still about 10–30 gm^{-2} day^{-1} (Brennan and Owende 2010). Moreover, other disadvantages of these cultivation systems are difficult to control that includes the massive requirement of water, lower biomass production, higher energy cost, challenges in scaling up the system, and contamination risks.

The biofilm cultivation system is an alternative to other ones. In this system, algal cells with higher density form a biofilm on the artificial substratum. This cultivation method efficiently lowers the cost and decreases the intensive costs of harvesting. Biofilm-based cultivation system tends the photosynthetic active culture to attach to the specified surface and separate from the aqueous media and overcoming the downside of the conventional cultivation systems that results in lower water usage and high harvesting efficiency with greater biomass productivity (Olivieri et al. 2014). Biofilm-based cyanobacterial cultivation mainly focused on the utilization of wastewater and its treatment. Biofilm-based bioreactors include algal turfs, aerosol bioreactors, bed reactors, and membrane photobioreactors (Mantzorou and Ververidis 2019; Podola et al. 2017). Different studies show the suitability of this bioreactor for algal growth by using different types of wastewater.

In addition, the water content of the harvested biomass is comparable to that of centrifuged biomass. On this system, the biomass productivity reaches 50–80 gm^{-2} day^{-1} which is many times as other systems (Zhang et al. 2016b).

5.4 Evaluation of Outdoor and Indoor Cultivation in Terms of Cost and Biomass Production

The comparison of the different parameters of the open raceway pond and photobioreactor is given (Table 5.3). The cyanobacterial cultivation in raceway ponds as well as in photobioreactors is technically viable on a large scale with the

Table 5.3 Comparison of photobioreactor and open raceway ponds (Chisti 2007)

Parameter	Photobioreactor	Open raceway pond
Annual biomass production (kg)	100,000	100,000
Concentration of biomass in media (kg m^{-3})	4.00	0.14
Volumetric productivity (kg m^{-3} day^{-1})	1.535	0.117
Oil yield (m^3 ha^{-1})	136.9	99.4

same amount of CO_2 fixation. The recovery of biomass from media is important for its further processing which can be separated by centrifugation, filtration, or other means. The biomass recovery from the photobioreactor takes a small cost compared to open raceway ponds because the biomass production in ORP is 30 times lower as compared to photobioreactors. Therefore, only a small amount of broth needs to be processed in the case of a photobioreactor as compared to the large quantity of open raceway pond. It is reported that the lipid content of the biomass is higher that is cultivated in open raceways ponds, but the lipid productivity of the photobioreactor is higher as compared to ORP due to high volumetric productivity in the case of photobioreactors (Chisti 2013).

5.5 Wastewater-Based Cultivation of Cyanobacteria

A large amount of water is a prerequisite for the cultivation of algae and cyanobacteria as these are aquatic organisms and require water for their proliferation. It is estimated that almost 1 metric ton of water is required to produce 1 kg of algal biomass. The concentration of biomass is estimated at 0.5% when cultivated in a photobioreactor, while the water footprint is higher in open raceway ponds due to higher evaporation losses where the ratio of cultivation media to biomass is increased when compared to PBRs. The utilization of large amounts of water for cyanobacterial cultivation is a major concern at an industrial level as the LCA studies have shown overall negative energy values. In this way, both water and nutrients constitute a major portion of the operating costs. Therefore, there is a dire need for a sustainable cultivation system to achieve industrialization of cyanobacteria-based products.

The utilization of wastewater as a substrate for cyanobacterial growth is a cost-effective and sustainable choice. Cyanobacteria and microalgae utilize different nutrients such as phosphorous, nitrogen, and inorganic and organic carbon present in wastewater as their nutrient source. Wastewater contains essential micro- and macronutrients including urea, nitrate, phosphorous, and trace minerals for the growth of cyanobacteria. In addition to low-cost cultivation, cyanobacteria also treat wastewater by absorbing nutrients that cause eutrophication in water bodies. The utilization of wastewater as a growth media provides an eco-friendly, economically viable approach to the production of high-value metabolites.

Mainly wastewater is generated by two sources, industrial and domestic which differ in their composition based on the levels of BOD, COD, and nutrient concentration. Phosphorous, nitrogen, and carbon are the main nutrients present in wastewater, and their stoichiometric ratio influences the growth of cyanobacteria. Industrial wastewater has more potential due to the abundance of organic carbon and nutrients in industrial wastewater as compared to municipal wastewater for the pilot-scale cultivation of cyanobacteria.

Several studies based on the use of algae and cyanobacteria as potential candidates for wastewater treatment show promising results in nutrient recovery from waste streams. Lower energy input is required for the removal of nutrients because photosynthesis aids with nutrient assimilation. However, the response of algal and cyanobacterial strains toward nutrient recovery potential, biomass production, and biochemical composition differs due to the variable nutrient composition of wastewater from different origins (Renuka et al. 2021). Various types and concentrations of contaminants may be present in wastewater depending upon its source, but generally, organic and inorganic nitrogen, phosphorus, inorganic particles, pathogens, and pharmaceutical compounds are present. Various filamentous cyanobacterial species such as *Oedogonium*, *Spirogyra*, *Klebsormidium*, and *Cladophora* have been reported to use wastewater as low-cost media for cultivation (Fig. 5.1).

5.5.1 Improving the Process Economics Through Water and Nutrient Recycling

LCA studies depict that the economics of cyanobacterial cultivation can be improved by the reuse of water and nutrients. This can be achieved by the reuse of water and nutrients in culture media. Because most of the water footprint lies in the cultivation step, so it is important to reuse and recover the residual nutrients at this stage. A large number of nutrient losses occur if the water is not recycled which affects the sustainability and economic viability of the cyanobacterial products (Farooq et al. 2015). Two approaches can be adopted for nutrient reuse: firstly, the spent media can be reused for cyanobacterial recultivation, while the other option involves the utilization of residual biomass after the extraction of target compounds. The residual nutrients in the media can be reused by the supplementation of the exhausted nutrients (González-López et al. 2013). But the main concern to reuse the spent media is the occurrence of several inhibitory organic compounds secreted by the cyanobacterial cells and the presence of other contaminants. Various sterilization techniques can be applied to degrade these contaminants and inhibitory compounds from the media. However, recently reported studies related to the reuse of spent media showed no negative impact in the recultivation of cyanobacteria (Khan et al. 2022). Another finding shows that the recycling of cultivation media in the case of *Chlorella vulgaris* has remained unaffected.

5.6 Factors Affecting the Cyanobacterial Growth

Different biotic and abiotic factors and operational conditions have a great impact on cyanobacteria growth and metabolite production. It is reported that the lipid content and biomass productivity of various microalgal strains of *Chlorophyta* increased under mixotrophic cultivation conditions by using municipal wastewater with the

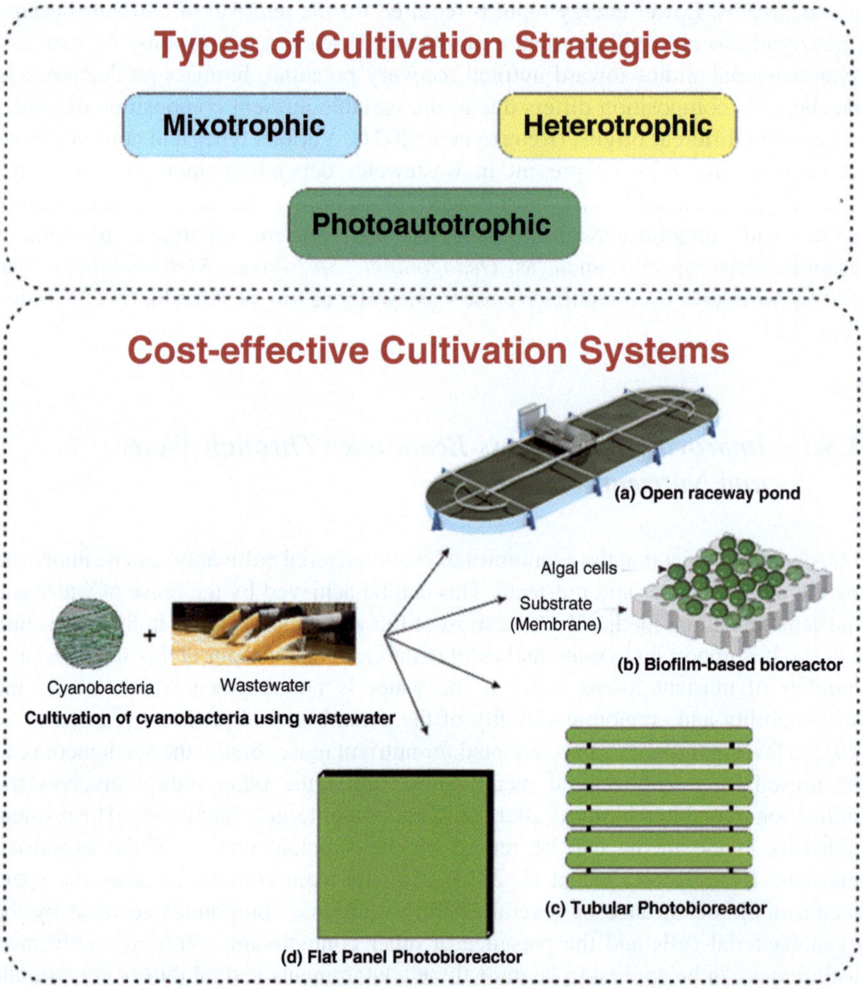

Fig. 5.1 Graphical representation of different cultivation strategies and cost-effective wastewater-integrated cultivation systems

supplementation of glycerol (Park et al. 2012). This depicts that the addition of organic carbon sources increased biomass production. Another study demonstrated that utilization of wastewater had a positive impact on algal growth. Increased biomass and lipid production is reported in the case of *Nannochloropsis oculata* when it was cultivated under increased CO_2 concentration (Dhandayuthapani et al. 2021).

Environmental factors such as light, temperature, and nutrients have a major impact on photosynthetic efficiency and metabolite content. Other factors such as pH and the presence of heavy metals also affect algal metabolism (Mata et al. 2010). The main reasons for the failure of outdoor cultivation include low biomass productivity and negative impact on metabolite concentration due to (1) low light intensity, (2) photoinhibition of superficial layers, (3) temperature variation, (4) evaporation loss that causes salinity variations, (5) formation of reactive oxygen species, and (6) contamination risks. In a broader sense, these environmental factors affect the CO_2 fixation and utilization of carbon in the synthesis of different macromolecules.

5.6.1 Impact of Temperature

The key metabolic activities such as respiration and photosynthesis depend on light and temperature. Temperature is an important environmental factor that has a major impact on the growth, biochemical composition, and nutrient requirement of algal biomass. There are a limited number of algal species reported that can grow at low temperatures (Lundquist et al. 2010). Most of the cyanobacterial strains are extremophiles and grow at high temperatures. *Chlorella minutissima* have optimum temperature to 35 °C (Aleya et al. 2011). Seasonal variations resulted in temperature fluctuations which are responsible for variable lipid productivity and other metabolites. The optimum temperature for increasing lipids differs from species to species. One of the studies reported the impact of temperature on the growth and extra polymeric substance of *Spirulina platensis*. Among different temperature ranges, the cyanobacteria showed the highest growth at 30 °C, while maximum EPS production was analyzed at 33 °C to 35 °C (Trabelsi et al. 2009). It is reported in a recent study that the growth of cyanobacterium *Plectonema terebrans* has increased with the temperature in outdoor conditions as it showed the maximum biomass of 1.9 gL^{-1} at 42 °C during the summer season (Haider et al. 2023).

5.6.2 Impact of pH Variations

pH is a crucial factor as it plays an important role in cyanobacterial growth and metabolism and contamination control and determines the availability and solubility of CO_2. It regulates the distribution of inorganic ions. pH also plays an important role in the activity of various enzymes (Zhang et al. 2016a). Mostly, cyanobacterial strains have optimum pH ranging from neutral to slightly alkaline (Pandey et al. 2010; Ying et al. 2014). *Spirulina* can survive at a highly alkaline pH of 10–12 (Gualtieri 2014). Alkaline pH favors the accumulation of neutral lipids with the additional benefit of reduced contamination risk from bacteria, fungi, protozoa, and other algal species (Shahid et al. 2021). Change in pH from its optimum value may affect growth and causes metabolic inhibition. Higher lipids are observed at neutral pH, while alkaline pH stress favors the accumulation of TAG (Zhang et al. 2016a; Mandotra et al. 2016). Similarly, alkaline pH limits the utilization of CO_2 and carbon is available in the form of HCO_3. However, the ability of carbon fixation of algae is limited so there is a need to find specific strains that are capable to grow on alkaline pH (Yen et al. 2019).

Cyanobacterial strains *Oscillatoria* and *Lyngbya* produced 1.226 gL^{-1} and 1.196 gL^{-1} of biomass and 192 mg g^{-1} and 219 mg g^{-1} of carbohydrate, at pH 9.0, respectively (Kushwaha et al. 2018). Similarly, cultivation of *Nannochloropsis* at pH 9.5 promoted growth and lipid content resulting in 1.04 gL^{-1} of biomass and 151.2 mg g^{-1} of lipids (Peng et al. 2020). Likewise, *S. abundans* was shown to tolerate pH 5.0–8.0, while maximum growth of 7.69 gL^{-1} and maximum lipid yield of 179 mgL^{-1} were observed at pH 8.0 and pH 6.0, respectively (Mandotra et al. 2016).

5.6.3 Impact of Light

Light is an important factor that may limit cyanobacterial growth. Quality and quantity of incident light have a strong impact on the photosynthetic activity, thereby influencing biomass and pigment production (Ho et al. 2015). The problem of light management occurs in every cultivation system. Dark zones arise in culture media due to the self-shading of cells during the growth phase of cyanobacteria which hinders light penetration that resulted in lower biomass production, but this problem can be addressed by lowering the optical path of cultivation media (Gifuni et al. 2019). A decrease of $\leq$0.01 m in the optical path can result in a 5–8% increase in photosynthetic efficiency (Janssen 2016). Photoperiod also plays an important role in the growth of cyanobacteria. The equal distribution of light and dark period is beneficial as compared to the exposure to continuous light that may cause photoinhibition due to the formation of free radicals.

The quality and intensity of the light also have a strong relation to the biochemical composition of cyanobacteria. It is reported that LED light is more suitable for the growth of cyanobacteria as compared to other artificial light sources (Park and Dinh 2019). A detailed study was also conducted to analyze the impact of various light sources on pigment composition, and a 40% increase in productivity of C-PC of *Spirulina platensis* was observed (Ho et al. 2018). LED light may cause stress during photosynthesis. Therefore, cyanobacteria synthesize more pigments to cope with this stress situation. Another study also reported similar results where *Spirulina* showed 2.7 times increased phycocyanin under the influence of LED light (Da Fontoura et al. 2018). On the contrary, the growth under red and yellow light favors the production of saturated fatty acids that is suitable for biodiesel production. The high light intensity is responsible for the increase in lipid content. The study on *Synechococcus* sp. PCC6803 showed a direct relationship between lipid content and light intensity (Cuellar-Bermudez et al. 2015).

The efficient utilization and management of light in the form of photons are also beneficial for the higher growth of cyanobacteria while cultivating using a photobioreactor. The position and location of a bioreactor should also be optimized to receive maximum light. A study showed that the vertically positioned photobioreactors in the east-west direction utilize solar radiations more efficiently in comparison with the horizontally placed photobioreactor in the north-south direction. In addition, various light management approaches such as solar tracking devices are being used to increase the efficiency of photobioreactor by increasing the exposure to solar radiations (Nwoba et al. 2019).

5.7 Conclusion and Prospects

The main challenges for the industrialization of cyanobacterial products include the high cost of cultivation and nutrient supply. Various techniques should be implemented for lowering the cost of upstream processing to make the process economically more viable. The cultivation system integrated with wastewater is a sustainable way to produce cyanobacterial biomass for biotechnological applications. It is also crucial to optimize the growth conditions according to the wastewater composition and environmental conditions before the cultivation at the pilot scale. For high-value metabolites, the cyanobacterial biomass should be from contamination-free sources. The reuse of residual nutrients and water is also a beneficial approach for achieving economic feasibility.

Another challenge is the limited knowledge available related to the economic feasibility of new cyanobacterial strains. Life cycle assessment and techno-economic studies are important to explore the potential strains for future use. Novel approaches should be investigated and introduced for the cost-effective mass production of cyanobacteria that require great effort. These investigations are helpful to address the problems of existing species and unveil the potential of novel strains. Genetic modification is also a conceivable approach to developing the industrially viable strains. These engineering approaches can make a prominent change in the profitability of cyanobacterial biomass-based applications. But the techniques of genetic engineering are expensive; therefore, this requires more time and deeper insights for commercial feasibility.

References

Aleya L, Dauta A, Reynolds CS (2011) Endogenous regulation of the growth-rate responses of a spring-dwelling strain of the freshwater alga, *Chlorella minutissima*, to light and temperature. Eur J Protistol 47:239–244

Andrade MR, Costa JA (2007) Mixotrophic cultivation of microalga *Spirulina platensis* using molasses as organic substrate. Aquaculture 264:130–134

Arias DM, Uggetti E, García-Galán MJ et al (2017) Cultivation and selection of cyanobacteria in a closed photobioreactor used for secondary effluent and digestate treatment. Sci Total Environ 587:157–167

Bahadar A, Khan MB (2013) Progress in energy from microalgae: a review. Renew Sust Energ Rev 27:128–148

Belkacemi K, Hamoudi S (2003) Photobioreactor stability by phase plane technique applied to *Spirulina maxima* growth. Int J Chem React Eng 1

Brennan L, Owende P (2010) Biofuels from microalgae—a review of technologies for production, processing, and extractions of biofuels and co-products. Renew Sust Energ Rev 14:557–577

Cardoso LG, Duarte JH, Andrade BB et al (2020) *Spirulina* sp. LEB 18 cultivation in outdoor pilot scale using aquaculture wastewater: high biomass, carotenoid, lipid and carbohydrate production. Aquaculture 525:735272

Carlozzi P (2003) Dilution of solar radiation through "culture" lamination in photobioreactor rows facing south–north: a way to improve the efficiency of light utilization by cyanobacteria (*Arthrospira platensis*). Biotechnol Bioeng 81:305–315

Chisti Y (2007) Biodiesel from microalgae. Biotechnol Adv 25:294–306

Chisti Y (2013) Constraints to commercialization of algal fuels. J Biotechnol 167:201–214

Corrêa PS, Morais Júnior WG, Martins AA et al (2020) Microalgae biomolecules: extraction, separation and purification methods. Processes 9:10

Cuellar-Bermudez SP, Romero-Ogawa MA, Vannela R et al (2015) Effects of light intensity and carbon dioxide on lipids and fatty acids produced by *Synechocystis* sp. PCC6803 during continuous flow. Algal Res 12:10–16

Da Fontoura PD, Radmann EM, Duarte JH et al (2018) *Spirulina* cultivated under different light emitting diodes: enhanced cell growth and phycocyanin production. Bioresour Technol 256:38–43

De Bhowmick G, Sarmah AK, Sen R (2019) Zero-waste algal biorefinery for bioenergy and biochar: a green leap towards achieving energy and environmental sustainability. Sci Total Environ 650:2467–2482

Dhandayuthapani K, Sarumathi V, Selvakumar P et al (2021) Study on the ethanol production from hydrolysate derived by ultrasonic pretreated defatted biomass of *Chlorella sorokiniana* NITTS3. Chem Data Collections 31:100641

Faried M, Samer M, Abdelsalam E et al (2017) Biodiesel production from microalgae: processes, technologies and recent advancements. Renew Sust Energ Rev 79:893–913

Farooq W, Suh WI, Park MS et al (2015) Water use and its recycling in microalgae cultivation for biofuel application. Bioresour Technol 184:73–81

Francisco ÉC, Franco TT, Wagner R et al (2014) Assessment of different carbohydrates as exogenous carbon source in cultivation of cyanobacteria. Bioprocess Biosyst Eng 37:1497–1505

Ge Y, Liu J, Tian G (2011) Growth characteristics of *Botryococcus braunii* 765 under high CO2 concentration in photobioreactor. Bioresour Technol 102:130–134

Gifuni I, Pollio A, Safi C et al (2019) Current bottlenecks and challenges of the microalgal biorefinery. Trends Biotechnol 37:242–252

González-López C, Cerón-García M, Fernández-Sevilla J et al (2013) Medium recycling for *Nannochloropsis gaditana* cultures for aquaculture. Bioresour Technol 129:430–438

Gualtieri P (2014) Algae: anatomy, biochemistry, and biotechnology. CRC Press

Haider MN, Khan AZ, Usman M et al (2023) Impact of seasons and wastewater cultivation on the biomass and biodiesel production by the *Plectonema terebrans* BERC10 as a candidate for a multiproduct algal biorefinery. Fuel 332:125987

Ho S-H, Nakanishi A, Ye X et al (2015) Dynamic metabolic profiling of the marine microalga *Chlamydomonas* sp. JSC4 and enhancing its oil production by optimizing light intensity. Biotechnol Biofuels 8:1–17

Ho S-H, Liao J-F, Chen C-Y et al (2018) Combining light strategies with recycled medium to enhance the economic feasibility of phycocyanin production with *Spirulina platensis*. Bioresour Technol 247:669–675

Hu L, Shan K, Huang L et al (2021) Environmental factors associated with cyanobacterial assemblages in a mesotrophic subtropical plateau lake: a focus on bloom toxicity. Sci Total Environ 777:146052

Huang Q, Jiang F, Wang L et al (2017) Design of photobioreactors for mass cultivation of photosynthetic organisms. Engineering 3:318–329

Jacob-Lopes E, Scoparo CHG, Lacerda LMCF et al (2009) Effect of light cycles (night/day) on CO2 fixation and biomass production by microalgae in photobioreactors. Chem Eng Process Process Intensif 48:306–310

Janssen M (2016) Microalgal photosynthesis and growth in mass culture. In: Advances in chemical engineering. Elsevier

Khan AZ, Malik S, Mehmood MA et al (2022) Two-stage algal cultivation for the biotransformation of urban wastewater's pollutants into multiple bioproducts in a circular bioeconomy paradigm. Energy Convers Manag 273:116400

Khan F, Malik S, Shahid A et al (2023) Characterization of a newly isolated cyanobacterium *Trichocoleus desertorum* BERC08 as a potential feedstock for the algal biorefinery. Biomass Convers Biorefin 13:5283–5294

Kumar M, Sun Y, Rathour R et al (2020) Algae as potential feedstock for the production of biofuels and value-added products: opportunities and challenges. Sci Total Environ 716:137116

Kushwaha D, Upadhyay S, Mishra PK (2018) Growth of cyanobacteria: optimization for increased carbohydrate content. Appl Biochem Biotechnol 184:1247–1262

Lundquist TJ, Woertz IC, Quinn N et al (2010) A realistic technology and engineering assessment of algae biofuel production, vol 1. Energy Biosciences Institute

Mandotra S, Kumar P, Suseela M et al (2016) Evaluation of fatty acid profile and biodiesel properties of microalga *Scenedesmus abundans* under the influence of phosphorus, pH and light intensities. Bioresour Technol 201:222–229

Mantzorou A, Ververidis F (2019) Microalgal biofilms: a further step over current microalgal cultivation techniques. Sci Total Environ 651:3187–3201

Masojídek J, Papáček Š, Sergejevova M et al (2003) A closed solar photobioreactor for cultivation of microalgae under supra-high irradiance: basic design and performance. J Appl Phycol 15: 239–248

Mata TM, Martins AA, Caetano NS (2010) Microalgae for biodiesel production and other applications: a review. Renew Sust Energ Rev 14:217–232

Mata SN, De Souza ST, Cardoso LG et al (2020) *Spirulina* sp. LEB 18 cultivation in a raceway-type bioreactor using wastewater from desalination process: production of carbohydrate-rich biomass. Bioresour Technol 311:123495

Meireles Dos Santos A, Vieira KR, Basso Sartori R et al (2017) Heterotrophic cultivation of cyanobacteria: study of effect of exogenous sources of organic carbon, absolute amount of nutrients, and stirring speed on biomass and lipid productivity. Front Bioeng Biotechnol 5:12

Miao X, Wu Q (2006) Biodiesel production from heterotrophic microalgal oil. Bioresour Technol 97:841–846

Mohan SV, Nikhil G, Chiranjeevi P et al (2016) Waste biorefinery models towards sustainable circular bioeconomy: critical review and future perspectives. Bioresour Technol 215:2–12

Mohan SV, Hemalatha M, Chakraborty D et al (2020) Algal biorefinery models with self-sustainable closed loop approach: trends and prospective for blue-bioeconomy. Bioresour Technol 295:122128

Moreno J, Vargas MÁ, Rodrıguez H et al (2003) Outdoor cultivation of a nitrogen-fixing marine cyanobacterium, *Anabaena* sp. ATCC 33047. Biomol Eng 20:191–197

Nwoba EG, Parlevliet DA, Laird DW et al (2019) Light management technologies for increasing algal photobioreactor efficiency. Algal Res 39:101433

Olivieri G, Salatino P, Marzocchella A (2014) Advances in photobioreactors for intensive microalgal production: configurations, operating strategies and applications. J Chem Technol Biotechnol 89:178–195

Ozkan A, Kinney K, Katz L et al (2012) Reduction of water and energy requirement of algae cultivation using an algae biofilm photobioreactor. Bioresour Technol 114:542–548

Pandey J, Pathak N, Tiwari A (2010) Standardization of pH and light intensity for the biomass production of *Spirulina platensis*. J Algal Biomass Util 1:93–102

Park J, Dinh TB (2019) Contrasting effects of monochromatic LED lighting on growth, pigments and photosynthesis in the commercially important cyanobacterium *Arthrospira maxima*. Bioresour Technol 291:121846

Park KC, Whitney C, Mcnichol JC et al (2012) Mixotrophic and photoautotrophic cultivation of 14 microalgae isolates from Saskatchewan, Canada: potential applications for wastewater remediation for biofuel production. J Appl Phycol 24:339–348

Patel AK, Choi YY, Sim SJ (2020) Emerging prospects of mixotrophic microalgae: way forward to sustainable bioprocess for environmental remediation and cost-effective biofuels. Bioresour Technol 300:122741

Peng X, Meng F, Wang Y, Yi X, Cui H (2020) Effect of pH, temperature, and CO_2 concentration on growth and lipid accumulation of Nannochloropsis sp. MASCC 11. J Ocean Univ China 19:1183–1192

Podola B, Li T, Melkonian M (2017) Porous substrate bioreactors: a paradigm shift in microalgal biotechnology? Trends Biotechnol 35:121–132

Posadas E, Del Mar MM, Gomez C et al (2015) Influence of pH and CO2 source on the performance of microalgae-based secondary domestic wastewater treatment in outdoors pilot raceways. Chem Eng J 265:239–248

Prabha S, Vijay AK, Paul RR et al (2022) Cyanobacterial biorefinery: towards economic feasibility through the maximum valorization of biomass. Sci Total Environ 814:152795

Pushparaj B, Pelosi E, Tredici MR et al (1997) As integrated culture system for outdoor production of microalgae and cyanobacteria. J Appl Phycol 9:113–119

Rawat I, Kumar RR, Mutanda T et al (2013) Biodiesel from microalgae: a critical evaluation from laboratory to large scale production. Appl Energy 103:444–467

Renuka N, Ratha SK, Kader F et al (2021) Insights into the potential impact of algae-mediated wastewater beneficiation for the circular bioeconomy: a global perspective. J Environ Manag 297:113257

Saratale RG, Ponnusamy V, Sirohi R et al (2022) Microalgae cultivation strategies using cost–effective nutrient sources: recent updates and progress towards biofuel production. Bioresour Technol 361:127691

Sato T, Usui S, Tsuchiya Y et al (2006) Invention of outdoor closed type photobioreactor for microalgae. Energy Convers Manag 47:791–799

Sato R, Maeda Y, Yoshino T et al (2014) Seasonal variation of biomass and oil production of the oleaginous diatom *Fistulifera* sp. in outdoor vertical bubble column and raceway-type bioreactors. J Biosci Bioeng 117:720–724

Shahid A, Malik S, Liu C-G et al (2021) Characterization of a newly isolated cyanobacterium *Plectonema terebrans* for biotransformation of the wastewater-derived nutrients to biofuel and high-value bioproducts. J Water Process Eng 39:101702

Sharma YC, Singh B, Korstad J (2011) A critical review on recent methods used for economically viable and eco-friendly development of microalgae as a potential feedstock for synthesis of biodiesel. Green Chem 13:2993–3006

Singh R, Sharma S (2012) Development of suitable photobioreactor for algae production—a review. Renew Sust Energ Rev 16:2347–2353

Singh B, Guldhe A, Rawat I et al (2014) Towards a sustainable approach for development of biodiesel from plant and microalgae. Renew Sust Energ Rev 29:216–245

Suparmaniam U, Lam MK, Uemura Y et al (2019) Insights into the microalgae cultivation technology and harvesting process for biofuel production: a review. Renew Sust Energ Rev 115:109361

Sutherland DL, Howard-Williams C, Turnbull MH et al (2014) Seasonal variation in light utilisation, biomass production and nutrient removal by wastewater microalgae in a full-scale high-rate algal pond. J Appl Phycol 26:1317–1329

Tan XB, Lam MK, Uemura Y et al (2018) Cultivation of microalgae for biodiesel production: a review on upstream and downstream processing. Chin J Chem Eng 26:17–30

Tan JS, Lee SY, Chew KW et al (2020) A review on microalgae cultivation and harvesting, and their biomass extraction processing using ionic liquids. Bioengineered 11:116–129

Ting H, Haifeng L, Shanshan M et al (2017) Progress in microalgae cultivation photobioreactors and applications in wastewater treatment: a review. Int J Agric Biol Eng 10:1–29

Trabelsi L, Ben Ouada H, Bacha H et al (2009) Combined effect of temperature and light intensity on growth and extracellular polymeric substance production by the cyanobacterium *Arthrospira platensis*. J Appl Phycol 21:405–412

Vasumathi K, Premalatha M, Subramanian P (2012) Parameters influencing the design of photobioreactor for the growth of microalgae. Renew Sust Energ Rev 16:5443–5450

Veerabadhran M, Natesan S, Mubarakali D et al (2021) Using different cultivation strategies and methods for the production of microalgal biomass as a raw material for the generation of bioproducts. Chemosphere 285:131436

Viswanaathan S, Perumal PK, Sundaram S (2022) Integrated approach for carbon sequestration and wastewater treatment using algal–bacterial consortia: opportunities and challenges. Sustainability 14:1075

Wang B, Lan CQ, Horsman M (2012) Closed photobioreactors for production of microalgal biomasses. Biotechnol Adv 30:904–912

Wang J, Liu W, Liu T (2017) Biofilm based attached cultivation technology for microalgal biorefineries—a review. Bioresour Technol 244:1245–1253

Yen H-W, Hu I-C, Chen C-Y et al (2019) Design of photobioreactors for algal cultivation. In: Biofuels from algae. Elsevier

Ying K, Zimmerman W, Gilmour D (2014) Effects of CO and pH on growth of the microalga *Dunaliella salina*. J Microb Biochem Technol 6:167–173

Yu H, Jia S, Dai Y (2009) Growth characteristics of the cyanobacterium *Nostoc flagelliforme* in photoautotrophic, mixotrophic and heterotrophic cultivation. J Appl Phycol 21:127–133

Zhan J, Rong J, Wang Q (2017) Mixotrophic cultivation, a preferable microalgae cultivation mode for biomass/bioenergy production, and bioremediation, advances and prospect. Int J Hydrog Energy 42:8505–8517

Zhang K, Kurano N, Miyachi S (2002) Optimized aeration by carbon dioxide gas for microalgal production and mass transfer characterization in a vertical flat-plate photobioreactor. Bioprocess Biosyst Eng 25:97–101

Zhang H, Zeng R, Chen D et al (2016a) A pivotal role of vacuolar H+-ATPase in regulation of lipid production in *Phaeodactylum tricornutum*. Sci Rep 6:1–17

Zhang T-Y, Hu H-Y, Wu Y-H et al (2016b) Promising solutions to solve the bottlenecks in the large-scale cultivation of microalgae for biomass/bioenergy production. Renew Sust Energ Rev 60:1602–1614

Chapter 6
Storage, Processing, and Stability of Phycobilins

Muhammad Rizwan Tariq, Shinawar Waseem Ali, Zunaira Basharat, Waseem Safdar, Saeed Ahmed, and Asma Saleem Qazi

Abstract Phycobilins are a class of chromophores that are present in certain species of algae including cyanobacteria and red algae. These organisms exhibit a range of characteristic colors, from red to blue-green, which are attributed to the presence of highly efficient light-harvesting pigments. The three major phycobilins are phycocyanobilin (PCB), phycoerythrobilin (PEB), and phycurobilin (PUB). All of these pigments could absorb light within a particular region of the visible spectrum, thereby enabling living organisms to harness a broad range of light energy for the process of photosynthesis. Phycobilins exhibit physiological association with chlorophyll, which serves as the primary photosynthetic pigment in plants, albeit their chemical structure is distinct. Phycobiliproteins are specialized protein complexes that function as auxiliary pigments, collecting light and transferring it to chlorophyll molecules. They are composed of a linear tetrapyrrole chromophore that is explicitly linked to the protein complexes. The possession of phycobilins confers a competitive edge to these organisms in habitats with low light intensity, such as those found in deep water or shaded areas. These organisms are capable of assimilating and utilizing light energy, a feat that poses a challenge for certain photosynthetic organisms. This chapter is focused on extraction, storage, and processing of phycobilins for different biotechnological applications.

M. R. Tariq (✉)
Department of Food Sciences, University of the Punjab, Lahore, Pakistan

Department of Biological Sciences, National University of Medical Sciences, Rawalpindi, Pakistan
e-mail: rizwan.foodsciences@pu.edu.pk

S. W. Ali · Z. Basharat
Department of Food Sciences, University of the Punjab, Lahore, Pakistan

W. Safdar · S. Ahmed · A. S. Qazi
Department of Biological Sciences, National University of Medical Sciences, Rawalpindi, Pakistan
e-mail: waseem.safdar@numspak.edu.pk

M. A. Mehmood et al. (eds.), *Pharmaceutical and Nutraceutical Potential of Cyanobacteria*, https://doi.org/10.1007/978-3-031-45523-0_6

Keywords Phycobilins · Phycocyanobilin · Chlorophyll · Stability · Phycobiliproteins · Photosynthesis

6.1 Introduction

Phycobilins are a class of chromophores that are present in certain species of cyanobacteria and a subset of red algae. These organisms exhibit a range of characteristic colors, from red to blue-green, which are attributed to the presence of highly efficient light-harvesting pigments. The three major phycobilins are phycocyanobilin (PCB), phycoerythrobilin (PEB), and phycurobilin (PUB). Every one of these pigments could absorb light within a particular region of the visible spectrum, thereby enabling living organisms to harness a broad range of light energy for the process of photosynthesis. Phycobilins exhibit physiological association with chlorophyll, which serves as the primary photosynthetic pigment in plants, albeit their chemical structure is distinct (Taniguchi and Lindsey 2023). Phycobiliproteins are specialized protein complexes that function as auxiliary pigments, collecting light and transferring it to chlorophyll molecules. They are composed of a linear tetrapyrrole chromophore that is explicitly linked to the protein complexes. The possession of phycobilins confers a competitive edge to these organisms in habitats with low light intensity, such as those found in deep water or shaded areas (Dagnino-Leone et al. 2022). These organisms are capable of assimilating and utilizing light energy, a feat that poses a challenge for certain photosynthetic organisms.

Phycobilins can also act as photoprotective agents, protecting cells from excessive light and preventing potential damage. Due to their characteristic features and possible applications, phycobilins are used as fluorescent labels in biological research and as natural dyes in food and cosmetics. The basic structure of phycobilins, like that of chlorophyll (Chl), is a tetrapyrrole unit, but rather than being coordinated to a central metal, the four pyrroles form an open chain (Fig. 6.1).

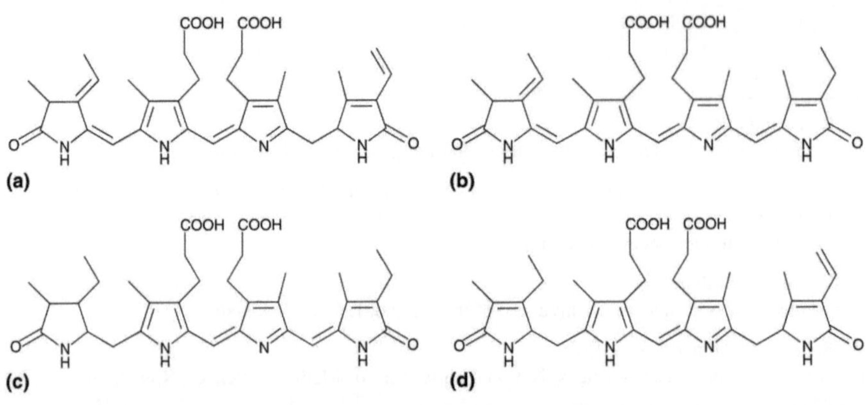

Fig. 6.1 Structure of phycobilin chromophores: (**a**) phycoerythrobilin, (**b**) phycocyanobilin, (**c**) phycoviolobilin, and (**d**) phycourobilin

Similar to carotenoids, the primary purpose of phycobilins is to absorb the energy, where Chl is ineffective, thereby increasing Chl's ability to capture light. Phycobilins absorb green to red wavelengths of light, while carotenoids absorb mostly in the blue-green spectrum (Taniguchi and Lindsey 2023). Species with red-absorbing phycobilins are found in shallower water than species with green chromophores, a direct result of the water column's ability to filter light. Phycobilins are covalently bound to phycobiliproteins through a thioether bond, resulting in water-soluble pigment-protein complexes that can aggregate to form phycobilisomes, which are supercomplexes attached to the thylakoid membrane and directing light energy to the photosynthetic reaction center (Buecker et al. 2022).

These molecules come in a variety of shapes and sizes, as seen by the four phycobilin chromophores depicted in Fig. 6.1. Changes in p-electron conjugation create distinct features of the absorption spectra. Red, blue, purple, and yellow are the colors of phycoerythrobilin, phycocyanobilin, phycoviolobilin, and phycourobilin, respectively. When light-harvesting pigments bind to their cognate proteins, complexes are formed that have unique absorptions (Yuan et al. 2022). The binding of phycoerythrobilin to proteins results in the formation of a specific type of phycoerythrin protein with an absorption maximum between 500 nm and 565 nm. Phycocyanin and allophycocyanin, two phycobiliproteins, are synthesized from phycocyanobilin; they absorb between 595 nm and 640 nm and 650 nm and 655 nm, respectively (Tomazic et al. 2021). The combination of phycoviolobilin and phycocyanobilin yields phycoerythrocyanin, which has an absorption maximum between 570 nm and 575 nm. When phycobilin and phycoerythrobilin are mixed, a new form of phycoerythrin is produced, again with an absorption maximum between 500 nm and 565 nm. All these proteins have high fluorescence quantum yields, making them useful for tagging antibodies and studying their role in energy transfer during photosynthesis (Nobel 2020).

6.2 Storage of Phycobilins

Phycobilins are stored within the cells of algae and cyanobacteria in specialized structures called phycobilisomes. The phycobilin pigments are found in phycobilisome protein complexes. These complexes attach to thylakoid membranes, where they absorb light and transfer energy to chlorophyll molecules (Blot et al. 2009). Core proteins, linkers and chromophores (phycobilins) are phycobilisome components. Chromophores are pigments responsible for absorbing light. Phycobilins such as phycocyanobilin and phycoerythrobilin bind to proteins in the phycobilisome structure. Inside the cell, phycobilisomes are organized to maximize light absorption. They form a concentric ring around the center of the photosynthetic reaction that ensures the efficient transfer of energy to the chlorophyll molecules. It is important to note that the deposition and organization of phycobilins in algae and cyanobacteria may differ slightly. However, the concept of conservation of these pigments in phycobilisomes remains consistent (Wiethaus et al. 2010). Storage of phycobilins is essential for maintaining their quality and functionality. Temperature, humidity, light, and oxygen exposure are environmental factors that can cause phycobilins to break down and lose their therapeutic properties (Beale 2008). Consequently, choosing a suitable storage technique is important to guarantee the long-term stability of phycobilins (Schoefs 2004). This chapter discusses two typical storage methods: lyophilization and vacuum packaging.

6.2.1 Freeze-Drying

The freeze-drying process is a widely used technique for the preservation of phycobilins due to its ability to allow extended storage times without significant degradation or reduced activity. The application of freeze-drying methodology mitigates the oxidation potential, which is a significant factor in the degradation of phycobilins (Peterson et al. 1981). Removing water from the product results in a substantial reduction in the likelihood of bacterial proliferation, which is a critical factor in preserving phycobilins. However, the freeze-drying process is an expensive and laborious undertaking that requires specialized equipment and expertise. Additionally, the suitability of lyophilization may not be universal for all items, and some items may require the inclusion of a cryoprotectant to ensure their preservation during the lyophilization process (Schreiber 1980).

6.2.2 Vacuum Packaging

Packaging uses vacuum sealing to extract air from the package containing food items, creating a hermetic seal that effectively prevents air and moisture from entering into the food package. Vacuum packaging has been found to be a viable technique for preserving phycobilins due to its ability to minimize the presence of oxygen and moisture in the packaging (Buecker et al. 2022; Paull and Chen 2008). This is crucial because these factors are known to contribute to phycobilin degradation. This technique helps preserve the color and nutritional composition of foods.

The use of vacuum packaging is a cost-effective and uncomplicated method of preserving phycobilins that can be used for various applications. Regardless, the use of vacuum packaging is not a substitute for basic storage and handling practices. As a result, commodities should be stored in a cool, dry place that is isolated from light and other conceivable factors that may cause deterioration (Yuan et al. 2022). Implementation of this method has the potential to extend the shelf life of phycobilin-infused products while maintaining their chromatic and nutritional properties. Choosing the right vacuum packaging method is essential because there are a large number of methods, each with varying degrees of suitability for preserving specific products.

6.3 Factors Effecting Phycobilin Stability During Storage

There are several factors which affect the stability of phycobilins during storage, some of which are listed here:

- Phycobilins are affected by various environmental factors such as light, temperature, pH, and oxygen exposure. Degradation of phycobilin can be initiated by various factors such as exposure to light, elevated temperatures, and acidic or alkaline conditions. Prolonged exposure to oxygen can lead to oxidation of phycobilins and production of potentially harmful metabolites (Dumay et al. 2014).
- The stability and shelf life of products containing phycobilins can be significantly affected by their storage conditions (Kovaleski et al. 2022).
- In addition to the environmental factors mentioned above, the stability of phycobilins can also be influenced by the used storage container and packaging. Oxidation of phycobilins due to exposure to air can lead to loss of both their color and nutritional properties. The presence of moisture can lead to degradation or mold formation of objects containing phycobilins. To avoid oxidation and deterioration, it is desirable to use packaging materials that are impermeable to oxygen and moisture (Beyrer et al. 2020).
- It is recommended that the products be stored in a place that is cool and dry and not exposed to light or other potential factors that could cause damage. The use of packaging materials that are impermeable to oxygen and moisture, such as hermetically sealed containers or vacuum sealed bags, could potentially be beneficial.

Observance of optimal techniques to preserve phycobilins can increase shelf life without compromising the color and nutritional properties of these valuable pigments. This is especially important for sectors that rely on phycobilins such as the food, pharmaceutical, and cosmetic industries. Prospective research indicates that persistent research and development efforts will undoubtedly reveal innovative methodologies to enhance the preservation and shelf life of phycobilin. The proposed approach includes exploration of innovative packaging materials, implementation of modified atmosphere packaging, and identification of new storage conditions and strategies to mitigate the impact of environmental factors on phycobilin stability (Falkeborg et al. 2018). Implementation of optimal storage techniques for phycobilins is a viable approach to preserve their essential pigmentation and nutritional properties for extended periods of time. This measure has significant benefits for both the industry sector and the wider community.

6.4 Stability of Phycobilins

The consideration of stability is of paramount importance when selecting a food colorant. The functional stability of R-phycoerythrin is observed to be satisfactory over a broad pH spectrum ranging from 3 to 10. Nevertheless, its color intensity experiences a significant decline upon exposure to thermal treatment exceeding 40 ° C, as reported by Munier et al. (2014). The color of R-PE is known to be sensitive to temperature, as evidenced by the change from red to green observed in roasted dried Nori flakes, as reported by Rakhimberdieva et al. (2007). According to Steglich et al. (2003), the utilization of high-pressure (HP) food processing has the potential to conserve nutrients, colors, and flavors during the processing stage, thereby offering a viable alternative to conventional food treatment methods. The strong tendency of tetrapyrrole chromophores to form complexes with metal ions is a crucial factor that impacts the color stability of PBP in food. Prior studies have indicated that PBPs' color and fluorescence characteristics can be altered by various bioactive metal ions, including but not limited to Mn^{2+}, Ca^{2+}, Zn^{2+}, and Cu^{2+} (Steglich et al. 2003; Zhao et al. 2000). In addition, the binding of micronutrients to R-PE may influence their bioavailability in the gastrointestinal tract, as well as the (patho)physiological processes that involve these metal ions. The aim of this study was to investigate the thermal and high-pressure stability of food-derived R-PE that was isolated from dried Nori flakes (Dumay et al. 2014). A combination of precipitation and chromatographic techniques was used to obtain high purity protein. The investigation revealed that the purified phycobiliprotein exhibited the characteristics of conventional R-PE, having a substantial β-helical composition, as determined by absorption, fluorescence, and CD spectra. Absorption data obtained from high-pressure and high-temperature investigations showed that the stability of R-PE was less adversely affected by pressure treatment compared to heat treatment. Factors affecting stability (Chaiklahan et al. 2012) include pH control, temperature control, light exposure, oxygen exposure, and other environmental factors.

Phycobilins are inherent chromophores present in cyanobacteria, red algae, and other microalgae. Flavonoids are recognized for their antioxidative, anti-inflammatory, and fluorescent characteristics and are extensively employed in various sectors such as food, cosmetics, and pharmaceuticals (Rhie and Beale 1992). The efficacy of phycobilins in the said applications is significantly impacted by their stability, rendering it a crucial consideration.

6.4.1 Effect of Light on the Stability of Phycobilin

Light is a critical factor that influences the stability of phycobilins. Phycobilins exhibit high photosensitivity, particularly toward ultraviolet and blue wavelengths of light. Exposure to light can induce photodegradation, isomerization, and discoloration of phycobilins. Photodegradation is a phenomenon that results in the breakdown of the chemical composition of phycobilins, which leads to the deterioration of their stability and bioactivity. The photodegradation mechanism can lead to the production of free radicals and reactive oxygen species (ROS) due to the absorption of light energy by phycobilin (Kovaleski et al. 2022). The effect of reactive oxygen species (ROS) on phycobilins can lead to a deterioration of their molecular configuration, thereby causing a decrease in their chromaticity, fluorescence, and biological effectiveness. Isomerization is a potential process that can occur when phycobilins are exposed to light.

Phycobilins undergo a process of isomerization, which involves the reorganization of their chemical structure and subsequent modification of their spectral properties (Dumay et al. 2014). The isomerization phenomenon has the potential to lead to the loss of the characteristic color and fluorescence of phycobilins, thus limiting their effectiveness as food dyes and biological markers. Color change is a common result of isomerization and photodegradation of phycobilins (Singhai et al. 1981). Degradation of chromophores in phycobilins and subsequent formation of degradation products can lead to chromatic aberrations. The sensory properties of foods that contain phycobilins can be affected by chromatic aberration, thereby affecting their taste and smell properties (Safdar et al. 2017). In order to prevent photodegradation, isomerization, and coloration, it is imperative to protect phycobilins from light during processing, storage, and utilization. The implementation of opaque containers, light filters, or light deprivation techniques can aid in addressing this issue. Phycobilins have the potential to be protected from light through encapsulation, microencapsulation, or immobilization within a matrix (Fischer and Häder 1992).

The implementation of certain methodologies can facilitate the preservation of phycobilins' stability and bioactivity while circumventing the deleterious effects of light exposure. The stability of PC, akin to numerous bioactive compounds of natural origin, poses a considerable challenge. The major challenges faced by PC in food and medicinal formulations are precipitation and color degradation under conditions of low pH or light exposure. The formation of ($\alpha\beta$)-monomers is facilitated by the association of two subunits, namely, α and β. These monomers rapidly assemble into ($\alpha\beta$)3-trimers and other multimers of higher order, as reported in reference (Cen et al. 2023). Regrettably, macromolecular proteins often exhibit rigidity in response to external perturbations. It has been observed that PC exhibits limited solubility in acidic conditions in the vicinity of its isoelectric point (pI), which ranges from 4.0 to 4.8 (Tamamizu et al. 2023). Under acidic conditions, the protein architecture undergoes aggregation, leading to significant alterations in the configuration and morphology of the enclosed chromophores which results in a shift in the color of PC from blue to green, as reported in references (Tomazic et al. 2021; Nobel 2020).

Under acidic conditions, the process of protein denaturation and color degradation is expedited by the impact of light, as evidenced by previous studies (Nobel 2020; Blot et al. 2009). Numerous food and pharmaceutical systems are frequently subjected to fluctuating light conditions throughout their preparation or storage. Food dyes and dietary supplements are often manufactured at elevated concentrations and may contain biopolymers, metal ions, and acids that can cause denaturation or oxidation of phosphatidylcholine (PC). In order to utilize PC as a viable food colorant or nutritional supplement, its inherent instability in the presence of an acidic environment and exposure to light must be addressed (Xu et al. 2022).

6.4.2 Effect of pH Control on the Stability of Phycobilin

pH is an important factor that strongly affects the stability of phycobilins. The structural integrity and stability of phycobilins can be significantly affected by pH fluctuations, which can lead to structural changes and degradation. The stability of phycobilin is affected by the pH of the medium. The stability of phycobilins is determined by the specific type of phycobilins and their respective optimal pH values. Phycoerythrin shows optimal stability in the pH range of 6.0–7.0, and phycocyanin shows optimal stability in the pH range of 7.0–8.0. At low pH levels, phycobilins are susceptible to acid hydrolysis, which leads to the destruction of their structural integrity, functional properties, and biological activity. Phycobilin tends to accumulate at low pH levels, which leads to reduced solubility, staining, and bioactivity. The potential application of phycobilins as food coloring and biological markers may be affected by aggregate formation. Maintaining a constant pH level during processing, storage, and use is essential to reduce phycobilin degradation and instability that can result from pH fluctuations. Maintaining an optimal pH level is critical for maintaining the stability and biological activity of a particular phycobilin mutant. Buffers help stabilize the pH and maintain the stability of phycobilin (Wang et al. 2022).

6.4.3 Effect of Temperature Control on the Stability of Phycobilin

It is well established that phycobilins are sensitive to temperature fluctuations and exposure to high temperatures can lead to denaturation, thermal degradation, and loss of stability of these pigments. High temperature may destroy the chemical composition of phycobilins and ultimately reduce their biological activity and color properties. The rate of pyrolysis depends on the specific type of phycobilin and is influenced by variables such as temperature, duration, and degradation rate (Yuan et al. 2022). The heat-induced denaturation process is a factor that affects the stability of phycobilins against temperature. The denaturation process causes phycobilins to lose their three-dimensional structure and deactivate their function and destroy their biological activity. Phycobilins are prone to denature when exposed to high temperatures, which leads to loss of bioactivity, color, and solubility (Böcker et al. 2020). Temperature control in different stages of manufacturing, storage, and use is very important to reduce thermal degradation and denaturation of phycobilin compounds. It is necessary to maintain the temperature in the optimal range for the specific phycobilin mutant. To maintain the stability and bioactivity of phycobilins, low temperature methods such as freeze-drying and freeze-thaw can be used (Frank and Cogdell 2012).

6.4.4 Oxygen Exposure Effecting the Stability of Phycobilin

Phycobilins are very sensitive to oxygen exposure, making them vulnerable to oxidation, photooxidation, and instability. Phycobilins undergo oxidative degradation when exposed to oxygen. The chemical structure of phycobilins collapses in reaction with oxygen, and as a result, its biological activity and color are lost. The degree of oxidation depends on the specific type of phycobilin and shows a positive correlation with oxygen concentration and duration. Photooxidation is an additional mechanism by which oxygen affects phycobilin stability. Phycobilins undergo photooxidation when exposed to oxygen and light. Exposure to air and light can degrade the chemical structure of phycobilins, leading to loss of bioactivity and color properties. The rate of light oxidation accelerates with increasing levels of oxygen concentration, light intensity, and duration. To protect phycobilins from oxidation and light oxidation, it is essential to limit exposure to oxygen during preparation, storage, and use of phycobilins (Krynická et al. 2023). Phycobilin stability and preservation of bioactivity can be achieved by introducing an oxygen-free or hypoxic environment. Prevention of photooxidation of phycobilin can be achieved by using light-blocking agents or by keeping the material in a light-free environment (Sampath-Wiley et al. 2008).

6.4.5 Other Environmental Factors Effecting the Stability of Phycobilin

Phycobilins are reactive with light, especially ultraviolet (UV) light (Dailer et al. 2012). Exposure to UV radiation may cause photooxidation of phycobilins, leading to a loss of their bioactivity and stability. Minimizing light exposure during the entire process of production, storage, and utilization is imperative for preserving the stability and bioactivity of phycobilins. Moisture exposure is a potential factor that could influence the stability of phycobilins. Phycobilins exhibit hygroscopic properties, which enable them to absorb moisture from their ambient environment. Exposure to moisture can induce the degradation of phycobilins, leading to a decline in their stability and bioactivity. Minimizing moisture exposure during the preparation, storage, and utilization of phycobilins is imperative for maintaining their stability and bioactivity (Afzal et al. 2017).

Chemical agents are substances that have a distinct chemical composition and can be used for various purposes. The stability of phycobilins can be influenced by acids, bases, and oxidizing agents. Exposure to phycobilins can lead to their degradation and destabilization. It is imperative to avoid exposing phycobilins to external agents during processing, storage, and utilization in order to preserve their stability and bioactivity (Araki et al. 2014). Although the protein chromophore complex in cyanobacteria like *Arthrospira platensis* can be employed as a natural food colorant, its appearance can change depending on the conditions in which the protein-chromophore interactions occur. The purpose of this investigation is to determine how well phycocyanin stabilized by carrageenan holds its color throughout storage at 4 °C and 25 °C and over a range of heating and pH conditions (Karnieli et al. 1999).

6.5 Extraction of Phycobilins

Extraction of phycobilins includes different extraction techniques, and extraction efficiency depends on different variables that can affect the output and properties of the extracted chromophores (Rakhimberdieva et al. 2007). Choosing an appropriate extraction process is the primary step in the phycobilin extraction process. The most commonly used extraction techniques include solvent extraction, acid extraction, enzymatic extraction, and mechanical extraction. The selection of an appropriate approach for the extraction of phycobilins depends on various factors, such as the nature of the organism under consideration, the properties of the phycobilins, and the intended application of the pigments (Schoefs 2004). Notably, each approach has its own set of advantages and disadvantages. The solvent extraction technique involves the separation of phycobilins from cell structures using organic solvents such as methanol, ethanol, or acetone. This relatively simple technique has the potential to extract unwanted substances such as chlorophyll and lipids while providing significant amounts of pigments (Zhao et al. 2000).

Phycobilins are obtained from cellular sources through an acid extraction process that involves the use of acidic solutions such as sulfuric acid or hydrochloric acid. The above procedure shows a high degree of efficiency in producing phycobilin shades of the highest purity. However, it is not without disadvantages, as it can damage cellular structures and requires careful regulation of pH levels. Enzymatic extraction refers to the use of enzymes such as cellulases or pectinases to disintegrate cell walls and release phycobilins (Csatorday 1978). Although this process is relatively gentle, it can also be expensive and time-consuming to produce excellent quality pigments. The mechanical extraction process involves physical disruption of the cells, which can be achieved using techniques such as ultrasound or grinding to facilitate the release of pigments. Although this particular method is characterized by its simplicity and cost-effectiveness, it may not be suitable for delicate organisms and could potentially remove unwanted contaminants (Yuan et al. 2022).

6.5.1 Solvent and Solvent-Assisted Extraction

In their study, Kovaleski et al. (2022) conducted an optimization process for the extraction of phycoerythrin from Michrochaete. This was achieved through the use of five different solvents, namely, citrate (pH 5.0; 0.1 M), acetate (pH 6.0; 0.1 M), carbonate (pH 9.6; 0.1 M), Tris-HCl (pH 7.2; 0.05 M), and sodium phosphate (pH 7.0; 0.1 M). The optimized extraction using acetate buffer (pH 6.0, 0.1 M) resulted in a yield of 65.21 mgg^{-1}. The PE of Microchaete exhibited properties of antioxidant, antibacterial, anticancer, and antifungal activities. The study conducted by Sfriso et al. 2018) examined the fluorescence of PE in the presence of phosphate buffer and EDTA at varying concentrations of 0.1, 1, 10, and 100 mM (Taniguchi and Lindsey 2023). The optimization of the procedure was conducted by Sharmila et al. 2017), wherein multiple buffers were utilized at different pH values and cell disruption methods were employed at varying temperatures. The utilization of sodium phosphate with a pH of 7.2 and the application of freeze-thaw cycles at temperatures ranging from −20 °C to −25 °C have been found to enhance the outcomes. Blot et al. (2009) obtained 90% C-PC through the utilization of sodium phosphate for extraction.

The study conducted by Rodrigues et al. 2018) utilized ultrasonic and PIL 2-HEAA +2-HEAF techniques on *A. platensis* in order to attain a PC concentration of 0.75 gL^{-1} (Kaňa et al. 2009). The employment of PILs in conjunction with mechanical agitation and thermal heating was observed in *Spirulina platensis* to result in a twofold increase in concentration (PC concentration of 1.65 gL^{-1}). The optimal technique for extracting phycobiliproteins from algal matrices remains indeterminate based on the analyzed literature. The empirical evidence indicates that the utilization of microwave and ultrasound techniques for extractions, as well as the adoption of buffers as solvents, may result in cost savings. However, these methods do not facilitate the development of extraction procedures that exhibit high selectivity (Beale 2008). The *Spirulina* genus exhibits the most notable extraction yields, indicating its superior ease of processing. The findings suggest that the employment of ultrasonication and ionic liquids is the optimal approach for the extraction of phycocyanin. Nevertheless, the limited number of studies available precludes the establishment of heuristic principles for determining the most effective mechanical methods or solvents. According to Wu et al. (2006), ionic liquids are typically characterized by a higher degree of selectivity as solvents. However, the studies reviewed in this context did not include an assessment of their selectivity.

6.5.2 Conventional Techniques

The required purities for PE vary depending on the intended application in industries such as energy, food, cosmetic, or pharmaceutical. These variations have a significant impact on the production cost and the final product price; as noted by Buecker et al. (2022), irrespective of its efficacy, the process of extraction frequently lacks selectivity. A lack of selectivity is indicative of diminished extract purity. The cost of purified phycobiliprotein solutions is relatively high, despite their established use in markets such as natural food colorants. However, there is potential for these solutions to be utilized in emerging markets with significant economic and industrial importance, including energy, medical, pharmaceutical, and cosmetic applications. According to Yang et al. (2022), the purity index of a protein can be determined by calculating the A565 nm/A280 nm ratio. A ratio of 0.7 indicates food-grade purity, while a ratio of 3.9 represents reactive-grade purity. A ratio greater than 4.0 is indicative of analytical grade purity.

Multicellular, macroscopic algae that exhibit green, red, or brown pigmentation are commonly referred to as seaweeds (Tamamizu et al. 2023). According to Francavilla et al. (2015), the high photosynthetic efficiency, biomass conversion rate, ease of handling, and rapid growth rate of certain marine organisms make them a promising biotechnological raw material for marine biorefineries. According to Beale (2008), the presence of polysaccharides such as agar and cellulose in the cell walls of macroalgae serves to prevent cell rupture during the extraction of bioactive components. The cryogenic technique of utilizing liquid nitrogen has been observed to enhance the processes of maceration and grinding. Certain techniques for cell disruption may require greater investment of time, resources, and financial capital. The utilization of a cryogenic mill operational unit is a viable alternative to the utilization of lab-scale liquid nitrogen, as the latter is not suitable for larger sizes. The application of ultrasonication involves the utilization of sound waves with frequencies exceeding 20 kHz to compress and decompress biomass. This method necessitates reduced time and lower temperature, as reported by Beale and Weinstein (1991).

In the study of Kuzminov et al. (2012), maceration was employed as the primary step in a complex biorefinery cascade to extract phycobiliproteins from *Gracilaria gracilis*, resulting in a production of 7 mg PE. g^{-1} d.w. and 2 PC. Thomas and Marsman (1959) conducted an assessment of various extraction methods, namely, maceration, ultrasonic bath, ultrasonic probe, high pressure, and freeze-thawing, for the purpose of extracting R-PE from identical algae samples. The results obtained from the process of maceration using mortar and pestle indicated that 3.58 ± 0.03 mg PE. g^{-1} and 0.62 ± 0.02 mg PC. g^{-1} were obtained. The research on red macroalgae involved the utilization of various extraction methods such as maceration with freezing-thawing, homogenization and ultrasonication, and maceration and ultrasonication. The results indicated that the latter method produced 77% and 93% R-PE and R-PC, respectively, as reported by Mittal et al. (2019). The study conducted by Guillard et al. (2015) investigated the utilization of ultrasound-assisted extraction and enzymatic hydrolysis techniques for the extraction of *Grateloupia turuturu*. Notwithstanding the intricacy of enzymatic activity, the enzymes' specificity in cleaving biomass bonds (3.6 mg g^{-1} at 22 °C) typically yields exceptional efficacy. Sharmila et al. (2017) utilized various techniques such as maceration, freeze-thawing, lysozyme treatment, and sonication to extract phycobiliproteins from *Kappaphycus alvarezii*. The researchers assessed the efficacy of three distinct freeze-thaw temperatures and determined that the optimal extraction occurred at a temperature range of −20 °C to −25 °C (Naeher et al. 2013).

Heocha (1958) extracted phycobiliproteins from *Spirulina* sp., a cyanobacterium which is utilized in food. Although most of the research has focused on recovering phycobiliproteins from macroalgae, the researcher used ultrasonic to extract them and obtained a substantially higher yield of phycocyanin (11.3 mg mL^{-1}). The yield of the technique under research was compared to standard water extraction at 4 °C (9.8 mg mL^{-1}) and 25 °C (5.7 mg mL^{-1}). PE analysis shows that the species under study produces 0.8 mg mL^{-1} phycobiliprotein. Tamamizu et al. (2023) examined *Pseudanabaena* strains' PC and PE synthesis. *P. amphigranulata* produced 86 ± 15 mgL^{-1}, while *P. catenata* produced 25.5 ± 5.1 mgL^{-1}. A mortar and pestle maceration followed three biomass cryopreservation cycles in liquid nitrogen.

Before cell rupture, species for phycobiliprotein retrieval should consider cell pretreatment. Taniguchi and Lindsey (2023) compared fresh and freeze-dried biomass using freeze-thawing and ultrasonography. Lipids from freshly collected *Arthrospira platensis* GL and *Porphyridium cruentum* algae were extracted via freeze-thawing. *A. platensis* GL yielded 81% phosphatidylcholine (PC), while *Porphyridium cruentum* yielded 71% PE regardless of extraction procedure. Adjali et al. (2022) used fresh *Porphyridium cruentum* and five freeze-thaw cycles to retrieve 86.6% of R-PE.

6.5.3 Factors Affecting Extraction

Once an extraction method has been chosen, several variables can affect the quantity and quality of extracted phycobilins. Several variables affect the extraction process, including the type and amount of solvents used, the pH level of the extraction buffer, the duration and temperature of the extraction procedure, and the presence of any interfering compounds. The choice of solvent and its concentration can have a significant effect on the quantity and quality of extracted pigments. Some solvents, such as methanol or ethanol, have the ability to induce denaturation of pigments and reduce their fluorescence (Antecka et al. 2022). Dimethyl sulfoxide (DMSO) has the potential to hinder downstream applications such as HPLC, among other solvents. The acidity level of the extraction buffer is a critical determinant.

The structural characteristics of phycobilins are subject to significant alterations at low and high pH levels due to their high sensitivity to changes in pH. Precise pH control is imperative in maintaining the pH of an extraction buffer at an optimal level that is conducive to the stability and solubility of the pigments (Adjali et al. 2022). Although higher temperatures can result in quicker extraction times, it is important to note that this may also lead to denaturation of the pigments. Prolonged extraction durations may lead to increased output; however, they may also introduce undesired impurities (Pez Jaeschke et al. 2021).

6.6 Processing of Phycobilins

Typically, a downstream process necessitates a series of two to three steps for the extraction of chemicals from biomass that has been generated either extracellularly or intracellularly. The downstream procedures are contingent upon the biomolecule (s) being targeted for recovery, which encompasses physical, chemical, and optical characteristics, as well as the morphology of the raw material (Ssebiyonga et al. 2013). Martins and Ventura (2020) have emphasized the disruptive effects of mechanical and chemical treatments on cells, resulting in the release of biomolecules (Zinicovscaia et al. 2021). Various methods including maceration, freeze-thaw, ultrasounds, and microwaves have been evaluated for their efficacy in extracting phycobiliproteins. Apart from mechanical methods, the retrieval of phycobiliproteins involves the utilization of solvents in solid-liquid pigment extraction and enzymatic hydrolysis. The investigation of mechanical-chemical extraction methods is infrequent, as noted by Shanab et al. (2012). The initial stages of retrieving an intracellularly produced valuable substance from any organism involve the processes of cell disruption and subsequent release of cell components (D'Alessandro and Antoniosi Filho 2016). Various methods such as maceration, freeze-thaw, ultrasound, microwave irradiation, and enzymatic hydrolysis have been reported to facilitate the release of R-PE from algae. Typically, these techniques encompass the utilization of solvents such as water or other substances for solid-liquid extraction (Shahid et al. 2021a).

The drying process involves two distinct stages, namely, primary drying and secondary drying. During the primary drying process, the surface moisture present on the droplets is removed, resulting in the formation of a dried husk or crust. In the secondary drying process, the formation of a crust serves to prevent the rapid evaporation of residual moisture present in the drop (Nobel 2020). Secondary drying, which involves the complete removal of residual moisture from the interior of the droplet, can be achieved by increasing the exposure time of the droplets in the drying chamber or by increasing the temperature of the drying gas. After the drying process, the droplets are subjected to particle separation techniques such as cyclone or bag filtration to isolate individual particles, which are then collected (Li et al. 2019). The separator removes any large or clumped particles and reintroduces them to the drying chamber for further processing. After the separation process is completed, the phycobilin powder is extracted from the separator and can be further processed or can be disseminated in a condensed state (Purvis et al. 2019).

6.7 Effects of Processing on Phycobilins

Phycobilins are a class of organic pigments found in various microalgae, red algae, and cyanobacteria. Due to their antioxidant, anti-inflammatory, and fluorescent properties, they are used in a variety of commodities, including food, personal care, and pharmaceuticals (Purvis et al. 2019). The effectiveness of phycobilins in various applications can be affected by processing, which can affect their stability, solubility, and biological activity. This review will discuss the impact of processing on phycobilins. Phycobilins can be obtained from natural sources by various methods such as cell disruption or solvent extraction. The extraction process can have an impact on both the quantity and quality of extracted phycobilins (Zimba 2012). Extraction conditions that are harsh in nature, such as exposure to high temperatures or the use of strong solvents, have the potential to degrade phycobilins and subsequently reduce their yield. Optimization of extraction conditions is necessary to increase the production and quality of phycobilins (Zimba 2012).

The presence of proteins, lipids, and pigments in phycobilin extract can affect their stability and biological activity. Purification is often required to obtain a highly pure phycobilin extract and remove contaminants. The purification process can also affect the stability and biological activity of phycobilins. Phycobilins can be denatured or aggregated by certain purification techniques such as column chromatography, which can lead to reduced solubility and biological activity. Drying is a common method to preserve and stabilize phycobilin extracts. The drying process can affect the stability and bioactivity of phycobilins (Tamary et al. 2012). Phycobilins can be denatured or degraded by the high-temperature drying process, leading to loss of biological activity. The use of mild drying methods such as freeze-drying and spray drying is necessary to maintain the stability and bioactivity of phycobilins. Proper storage conditions are imperative for the preservation of stability and bioactivity of phycobilin extracts (Paull and Chen 2008). Phycobilins may experience degradation when subjected to unfavorable storage conditions, leading to a reduction in their bioactivity. Phycobilins are susceptible to oxidation, denaturation, and degradation upon exposure to light, oxygen, and elevated temperatures. In order to preserve the stability and bioactivity of phycobilin extracts, it is recommended to store them in environments that are cool, dry, and devoid of light (Shen et al. 2008).

6.8 Purification and Characterization

Following the extraction process, it is necessary to subject the phycobilins to a purification procedure aimed at eliminating any contaminants or impurities present in the sample. There exist various techniques for purification, such as gel filtration chromatography, ion exchange chromatography, and HPLC (Zhang et al. 2023). The process of spray drying is a commonly employed method for the manufacture of phycobilin powders derived from naturally occurring sources such as cyanobacteria, red algae, and other microalgae (Kovaleski et al. 2022). Phycobilins are hydrophilic pigments that have extensive applications in the domains of food science, cosmetic chemistry, and pharmaceuticals. The utilization of spray drying presents a proficient and economical approach for the production of phycobilin powders that exhibit superior stability and solubility properties (Antecka et al. 2022). Spray drying is a process which is commonly used in the food and pharmaceutical industries to convert liquid substances into dry powders (Khan et al. 2022). After the initial preparation, the concentrated phycobilin solution is introduced into the spray drying system. Using a nozzle or any other atomizing device, the solution is fragmented into tiny droplets. Subsequently, the droplets are introduced into a drying chamber where they are exposed to hot air or alternative gases to facilitate the drying process (Montoya et al. 2021).

6.8.1 Chromatographic Techniques

The required purities for PE vary depending on the intended application in industries such as energy, food, cosmetic, or pharmaceutical. These variations have a significant impact on the production cost and the final product price, as noted by Antecka et al. (2022). The cost of purified phycobiliprotein solutions is significant, both in established markets such as natural food colorants and in emerging markets with substantial economic and industrial implications, including energy, medical, pharmaceutical, and cosmetic applications. The predominant method of purification is chromatography, which encompasses ion-exchange, expanded-bed absorption, or reverse-phase techniques. Additionally, a purity index can be determined by calculating the A565 nm/A2. Frequently, the process of purification entails the utilization of a variety of techniques in order to attain elevated levels of purity. An illustrative instance involves the application of precipitation, succeeded by chromatography. The authors Agrawal et al. (2022) were able to attain a notable purity index (3.3) of R-PE derived from *Gracilaria gracilis* through the implementation of DEAE-Sepharose fast flow chromatography for purification purposes. The application of ammonium sulfate prior to chromatography is a widely adopted practice owing to its ability to eliminate amino acids, thereby augmenting the purity of PE as reported by Patel et al. (2022).

In a study conducted by Dagnino-Leone et al. (2022), they employed a two-step precipitation method utilizing ammonium sulfate, with initial precipitation at 20% and subsequent precipitation at 40%, followed by extraction on DEAE-Cellulose in *Porphyridium marinum*. This approach resulted in the attainment of a significantly elevated PE purity level of 5.0. In their study, they utilized solely ammonium sulfate precipitation (55%) to isolate a highly pure (PE purity of 5.2) phycobiliprotein from the red alga *Portieria hornemannii*. The efficacy of ultrafiltration as a pretreatment step prior to anion exchange chromatography (using SOURCE 15Q) was assessed in the context of microalgae *Porphyridium cruentum* (Pez Jaeschke et al. 2021). The outcome of this evaluation was the attainment of analytical grade B-PE at the commercial scale, with a purity index of 5.1. The study conducted by Shahid et al. (2021b) aimed to investigate the disparity in the purification of PE through the utilization of solely ammonium sulfate precipitation and its combination with anion-exchange chromatography (DEAE-Cellulose). The purity index was observed to escalate from 1.2 to 2.9, indicating a significant improvement in the purity of the protein. Additionally, a protein with a 0.7 nm ratio is classified as having food-grade purity, while a ratio of 3.9 is categorized as reactive grade, and a ratio exceeding 4.0 is considered analytical grade, as per the findings of Montoya et al. (2021).

In a previous study, Nobel (2020) utilized anion-exchange chromatography to purify the pigment extract (PE) from *Gracilaria corticate*, a red seaweed commonly found in Indian waters. The researchers investigated the potential of PE as a natural coloring agent in carbonated beverages and determined that it exhibited stability in cool, sweetened, and carbonated drinks. Mandal et al. (2020) discovered that the implementation of gel filtration, specifically Sephacryl S-300, prior to anion-exchange chromatography proved to be effective in extracting PE from *Lyngbya arboricola* and *Synechococcus* sp. The purity index for *L. arboricola* was A560/A280 = 5.2, while for *Synechococcus* sp., it was A542/A280 = 3.4.

The utilization of expanded bed adsorption chromatography is a viable method for the retrieval of proteins, obviating the necessity for preliminary clarification. The technique was employed by Issa et al. (2020) on *P. cruentum*, resulting in a 66% recovery of PE. In a study, conducted by Lee et al. (2019), there is a comparison between expanded bed and anion-exchange chromatography techniques in *Porphyra yezoensis*, which is considered as the most significant aquaculture species in China. The results indicated that the expanded bed adsorption method yielded a higher quantity, while the anion-exchange chromatography method resulted in a higher purity ratio. According to Zhao et al. (2019), the application of this method proved to be effective in the process of refining PE from the residual matter of *Pyropia haitanensis*, resulting in a concentration of 247.13 mgL^{-1} and a purity index of 4.1. Hydroxyapatite was utilized by Falkeborg et al. (2018), as a chromatographic resin for the purification of PE. This resin is known for its cost-effectiveness and was able to achieve an optimal purity index of 6.7. Dumay et al. (2014) conducted a study on *Porphyra yezoensis* Ueda, wherein they employed chromatography with hydroxyapatite as the adsorbent material. The researchers also carried out continuous precipitation with ammonium sulfate and achieved a purity ratio of 5.5 of PE and 5.1 of PC.

The cyanobacterium *Nostoc* sp. has demonstrated to be a proficient provider of phycocyanin. In their study, Cheng and Jiang (1992) utilized ammonium sulfate precipitation and gel filtration chromatography (Sephacryl S-100 HR) for the purification process, resulting in a PE purity of 7.2. An additional purification technique, hydrophobic interaction chromatography, was incorporated in a subsequent investigation resulting in a purity level of 11.5. Steglich et al. (2003) extracted phycobiliproteins using saturated ammonium and subsequently isolated via gel filtration (Sephadex G-200). The isolated PE and PC were further purified using reversed-phase-high-performance liquid chromatography (RP-HPLC). The results indicated a higher extraction and purity ratio for PE and an improvement in the protein's antioxidant activity. In a study conducted by Singhai et al. (1981), an electrophoretic elution technique was utilized as an alternative to chromatography to achieve an optimal purity of 5.9 from *Halymenia floresia*. Ultrafiltration is another purification method that is employed. Rhie and Beale (1992) employed the technique of ultrafiltration to isolate PE from high molecular weight polysaccharides in *Porphyridium cruentum.*

The process involved the use of two membranes, one with a molecular weight cutoff of 300,000 Da and the other with a cutoff of 10,000 Da. As a result, the purity index of the isolated PE was determined to be 2.3. Brown et al. (1990) conducted a study in which they investigated the yield of R-PE and R-PC through precipitation with ammonium sulfate. The results indicated that this method yielded good results for both R-PE and R-PC, with a yield of 100% and 81.1%, respectively. However, the method was found to be nonselective for any of the PBPs. In contrast, the use of poly(acrylic acid) sodium salts as precipitation agents, coupled with an ultrafiltration step, yielded an R-PE precipitation after extraction from *Gracilaria gracilis*, with a yield of 79.5%.

6.8.2 Aqueous Biphasic Systems

The ABS process entails a liquid-liquid extraction method, whereby a biphasic system is established through the combination of two hydrophilic and immiscible polymers or one polymer and one salt. The initial research conducted in this particular area was carried out by Benavides and Rito-Palomares in 2004. This study investigated the impact of varying molecular weights of polyethylene glycol (PEG): 1000, 1450, 3350, and 8000 g.gmol^{-1} on its purity (Csatorday 1978). The results indicated that PEG 1450 exhibited the highest purity for PE, with a TLL of 24.9% (w/w) at a pH of 8.0. Subsequently, it was demonstrated that PEG 1000 (TLL 50% w/w and pH 7.0) yielded the highest level of purity for PE, while PEG 1450 was found to be optimal for PC. Subsequently, Antelo and colleagues (2007) conducted further experimentation on traditional antilock braking systems (ABS). The literature of Yang et al. (2010) discovered that the utilization of PEG 1000 resulted in a higher yield for PE. Additionally, this substance caused two alterations in the system, namely, an increase in volume ratio (from 1.0 to 4.5) and a decrease in TLL (from 50% to 45%). These modifications ultimately led to the attainment of a purity level of 3.2.

The ABS methodology has the potential to be utilized in conjunction with other techniques, including isoelectric precipitation. In their study, Hernandez-Mireles and Rito-Palomares 2006) employed three distinct techniques to extract polyphenols (PE), namely, cell disruption through sonification, isoelectric precipitation facilitated by the addition of hydrochloric acid, and PEG/phosphate ABS extraction. Blot et al. (2009) achieved a high level of purity (4.2) using a four-step process that involved cell disruption via bead milling, isoelectric precipitation, ABS, and ultrafiltration. In 2020, Nobel employed acrylamide-based surfactants (ABS) consisting of copolymers and dextran to enhance the purity and stability of C-phycocyanin. The utilization of sodium phosphate for extraction resulted in an extract of 0.52 purity, which was subsequently subjected to purification, leading to a fourfold increase in purity. In their study, Böcker et al. (2020) conducted experiments on various surfactants with the aim of isolating and preserving the structural integrity of R-PE. Through their investigation, they determined that benzyldodecyldimethyl-ammonium bromide was the most suitable surfactant for this purpose. Subsequently, the aforementioned author conducted a study on the utilization of microfluidic devices for enhancing the purification process of phycoerythrin, which yielded favorable outcomes (Issa et al. 2020).

6.9 Applications of Phycobilins

Phycobilins exhibit a diverse array of applications across multiple domains, encompassing biotechnology, food science, cosmetic formulation, and medicinal research. The distinctive characteristics of these compounds, including their notable water solubility, robust fluorescence, and potent antioxidant properties, render them highly desirable for a diverse range of applications.

6.9.1 Biotechnological Applications

Phycobilins possess a wide range of biotechnological applications, rendering them a valuable tool in various research domains. Biotechnology refers to the application of living organisms, cells, and biological processes in the development of novel products or technologies. Phycobilins find various applications in biotechnology, such as bioimaging, flow cytometry, and fluorescence microscopy. The application of phycobilins as fluorescent probes is a crucial utilization of these molecules within the field of biotechnology. Fluorescent probes are utilized to visualize cells, biomolecules, and biological processes. Phycobilins are commonly employed as natural fluorescent probes due to their strong fluorescence properties, which render them ideal for imaging purposes (Chaiklahan et al. 2012). Phycobilins possess the ability to selectively target and visualize specific biomolecules within cellular structures through their conjugation with proteins, peptides, and nucleic acids. Phycobilins are employed in flow cytometry, a powerful technique for analyzing the properties of cells and other particles that are suspended in a liquid medium. Flow cytometry is a widely used technique in scientific and therapeutic contexts for the analysis of cell size, shape, and function, as well as for the identification and classification of specific cell types (Zimba 2012). Phycobilins are employed as fluorescent markers in flow cytometry to facilitate accurate identification and sorting of cells. The application of biotechnology in utilizing phycobilins for bioimaging is a significant area of study. Bioimaging refers to the application of imaging methodologies for the purpose of visualizing biological structures and processes. Phycobilins are utilized as fluorescent probes in bioimaging applications to visualize cells and biological processes, such as protein synthesis and DNA replication. In addition, they are utilized in fluorescence resonance energy transfer (FRET) assays, which is a technique employed for investigating molecular interactions and cellular signalling cascades (Zimba 2012).

6.9.2 Food

Phycobilins possess unique properties that render them suitable for deployment in the food industry, where they serve diverse functions such as natural food additives and colorants. Food additives are utilized to augment the texture, flavor, and longevity of food items, whereas food colorants are employed to enhance the visual appeal of food and render it more attractive to consumers. Phycobilins are employed as natural food colorants owing to their robust and consistent hues, which vary from blue to red contingent upon the particular type of phycobilin. Natural colors are being utilized more frequently in the food industry as a substitute for synthetic colors, which have been associated with adverse health outcomes by some individuals. Phycobilins are generally considered to be safe for consumption due to their natural origin from sources such as algae and cyanobacteria (Shanab et al. 2012). Phycobilins are commonly employed as food additives owing to their antibacterial and antioxidant properties. The utilization of phycobilins as a natural preservative for food products is attributed to their inherent properties that effectively extend the product's shelf life and inhibit spoilage (Blot et al. 2009). Phycobilins have the potential to serve as a dietary supplement for fortifying meals with essential vitamins and minerals, thereby enhancing the nutritional value of the food product. Phycobilins have been identified as a noteworthy source of protein in the food industry. Certain types of phycobilins are considered a favorable protein source for food products due to their elevated protein content. These proteins can be extracted and incorporated as ingredients in various products such as protein bars, drinks, and supplements (Sampath-Wiley et al. 2008).

6.9.3 Cosmetics

The distinctive properties of phycobilins make them well suited for deployment as natural colorants, UV protectants, and antioxidants, rendering them highly versatile for a range of applications within the cosmetics industry. Cosmetics refers to products utilized for enhancing the appearance of skin, hair, and nails. One of the significant applications of phycobilins is their utilization as natural colorants in cosmetic products (Shen et al. 2008). Phycobilins are utilized as a natural pigment source in cosmetics such as lipstick, eye shadow, and blush. Phycobilins possess potent and stable colors that can range from blue to red, depending on their type. As a result, they serve as a promising natural substitute for synthetic colorants that may cause skin irritation and other adverse effects (Rakhimberdieva et al. 2007).

Phycobilins are also employed in the cosmetics industry as natural UV filters. Exposure to ultraviolet radiation emitted by the sun can have detrimental effects on the integumentary system, leading to premature ageing and heightened susceptibility to cutaneous malignancies. Phycobilins have been identified as a promising natural constituent for the formulation of sunscreens and other ultraviolet (UV) protection products, owing to their demonstrated ability to effectively absorb UV radiation and provide a protective barrier against skin damage (Karnieli et al. 1999). The utilization of phycobilins in cosmetics is noteworthy due to their antioxidant properties. Antioxidants are known to protect the skin from free radicals, which have the potential to cause damage to the skin and accelerate the ageing process. Phycobilins possess a high concentration of antioxidants, rendering them a valuable natural constituent for various skincare products such as antiaging creams and lotions. Phycobilins, being a natural protein source, are employed in hair care commodities such as shampoos and conditioners. Due to the considerable protein content of certain phycobilins, they are an ideal component for hair care therapies aimed at fortifying and repairing impaired hair (Yuan et al. 2022).

6.9.4 Medicine

Phycobilins have potential applications in the medical domain as organic agents with anti-inflammatory, anticancer, and antiviral properties. Ongoing research in this field suggests that phycobilins hold significant potential as a prospective reservoir of novel pharmaceuticals and therapies. One of the significant medical applications of phycobilins is their capacity to mitigate inflammation. Inflammation is a common occurrence in various disorders, including but not limited to diabetes, cardiovascular disease, and arthritis. The anti-inflammatory properties of phycobilins have been demonstrated, suggesting their potential as a natural and potentially safer alternative to synthetic anti-inflammatory medications (Yang et al. 2022). Ongoing research is being conducted on the potential anticancer properties of phycobilins. Recent research has suggested that certain types of phycobilins may possess the potential to serve as a novel avenue for the development of cancer therapies, as they have been found to exhibit inhibitory effects on the proliferation and mitosis of malignant cells (Kovaleski et al. 2022). Phycocyanin, a type of phycobilin present in cyanobacteria, has exhibited promising results in inducing apoptosis in cancer cells and impeding carcinogenesis. The antiviral properties of phycobilins are currently under investigation. The emergence of novel viral infections such as COVID-19 has led to an increase in the need for novel antiviral therapeutics. Phycobilins have been shown to possess antiviral properties against several viruses, such as the herpes simplex virus, hepatitis C virus, and influenza virus. The nutritional advantages and potential medicinal applications of phycobilins are currently under investigation. Certain types of phycobilins are considered to be valuable sources of natural supplements and functional foods owing to their elevated levels of vitamins, minerals, and other essential nutrients (Kovaleski et al. 2022; Yuan et al. 2022).

6.10 Conclusion and Prospects

This book chapter comprehensively addresses the topics of phycobilin production, storage, processing, and stability, as well as their multifaceted applications across various industries. This section will discuss potential avenues for future research, provide a summary of the key concepts presented in this chapter, and examine the implications of phycobilins for both the business sector and society as a whole. Phycobilins, which are natural pigments, are found in cyanobacteria and red algae. The health-enhancing properties of the substance under consideration include antioxidant, anti-inflammatory, and immunomodulatory activities. *Spirulina* is widely recognized as the primary origin of phycobilins. Various methodologies have been employed to cultivate *Spirulina*, such as open ponds, closed photobioreactors, and hybrid systems. Ensuring adequate storage is imperative for maintaining the stability and bioactivity of phycobilins. Various preservation methods have been employed to maintain the stability of phycobilins, such as freezing, lyophilization, and vacuum packaging. Research has been conducted on the impact of processing on the stability of phycobilins. Various processing techniques, including extraction and spray drying, have been employed to generate extracts that are abundant in phycobilins from *Spirulina*. The stability of phycobilins is influenced by a variety of environmental factors, such as pH, temperature, light intensity, and oxygen exposure. Phycobilins exhibit potential applications across diverse sectors such as the food, pharmaceutical, cosmetic, biotech, and biosensor industries. Future research directions entail conducting further investigations to comprehensively explore the potential applications of phycobilins across diverse industries. Further investigation is required to explore the bioactive properties of phycobilins and assess their potential therapeutic applications in various medical conditions. Further research is necessary to explore methods for environmentally conscious and sustainable production of phycobilins and to enhance the production and processing methodologies for generating phycobilin-enriched extracts that exhibit elevated bioactivity and stability. The effects on commerce and the broader community are the following: phycobilins have the potential to significantly impact various industries such as the food business, pharmaceuticals, cosmetics, and biotechnology. The utilization of phycobilins across diverse industries presents an opportunity to provide natural, nonhazardous, and environmentally advantageous alternatives to synthetic constituents, thereby promoting eco-friendly and sustainable practices. Furthermore, the health-enhancing properties of phycobilins may offer substantial advantages to consumers, thereby contributing to the development of a more health-conscious society.

References

Adjali A, Clarot I, Chen Z et al (2022) Physicochemical degradation of phycocyanin and means to improve its stability: a short review. J Pharm Anal 12:406–414. https://doi.org/10.1016/j.jpha.2021.12.005

Agrawal S, Siddiqui SA, Chaudhary DR, Rathore MS (2022) Bio-prospection and compositional multivariate analysis revealed the industrial potential of selected seaweeds along Gujarat seacoast. Bioresour Technol Rep 19:101130. https://doi.org/10.1016/J.BITEB.2022.101130

Afzal I, Shahid A, Ibrahim M, et al (2017) Microalgae: a promising feedstock for energy and high-value products. Algae Based Polym Blends, Compos Chem Biotechnol Mater Sci 55–75. https://doi.org/10.1016/B978-0-12-812360-7.00003-3

Antecka A, Klepacz-Smółka A, Szeląg R et al (2022) Comparison of three methods for thermostable C-phycocyanin separation and purification. Chem Eng Process Process Intensif 171:108563. https://doi.org/10.1016/j.cep.2021.108563

Araki M, Akimoto S, Mimuro M, Tsuchiya T (2014) Artificially acquired chlorophyll b is highly acceptable to the thylakoid-lacking cyanobacterium, Gloeobacter violaceus PCC 7421. Plant Physiol Biochem 81:155–162. https://doi.org/10.1016/J.PLAPHY.2014.01.006

Beale SI (2008) Photosynthetic pigments: perplexing persistent prevalence of "superfluous" pigment production. Curr Biol 18:R342. https://doi.org/10.1016/J.CUB.2008.02.064

Beale SI, Weinstein JD (1991) Chapter 5 Biochemistry and regulation of photosynthetic pigment formation in plants and algae. New Compr Biochem 19:155–235. https://doi.org/10.1016/S0167-7306(08)60112-X

Beyrer M, Pina-Perez MC, Martinet D, Andlauer W (2020) Cold plasma processing of powdered spirulina algae for spore inactivation and preservation of bioactive compounds. Food Control 118:107378. https://doi.org/10.1016/J.FOODCONT.2020.107378

Blot N, Wu XJ, Thomas JC et al (2009) Phycourobilinin trichromatic phycocyanin from oceanic cyanobacteria is formed post-translationally by a phycoerythrobilin lyase-isomerase. J Biol Chem 284:9290–9298. https://doi.org/10.1074/JBC.M809784200

Böcker L, Hostettler T, Diener M et al (2020) Time-temperature-resolved functional and structural changes of phycocyanin extracted from Arthrospira platensis/spirulina. Food Chem 316: 126374. https://doi.org/10.1016/j.foodchem.2020.126374

Brown SB, Houghton JD, Vernon DI (1990) New trends in photobiology biosynthesis of phycobilins. Formation of the chromophore of phytochrome, phycocyanin and phycoerythrin. J Photochem Photobiol B Biol 5:3–23. https://doi.org/10.1016/1011-1344(90)85002-E

Buecker S, Grossmann L, Loeffler M et al (2022) Influence of storage temperature on the stability of heat treated phycocyanin-λ-carrageenan complexes in liquid formulations. Green Chem 24: 4174–4185. https://doi.org/10.1039/d2gc00809b

Cen C, Zhang K, Zhang T et al (2023) Decoding algae under abnormal temperature variability with odorant as a responsive indicator. Sci Total Environ 881:163390. https://doi.org/10.1016/J.SCITOTENV.2023.163390

Chaiklahan R, Chirasuwan N, Bunnag B (2012) Stability of phycocyanin extracted from spirulina sp.: influence of temperature, pH and preservatives. Process Biochem 47:659–664. https://doi.org/10.1016/j.procbio.2012.01.010

Cheng L, Jiang L (1992) Circular dichroism and stereochemistry of the phycobilins and their derivatives. J Photochem Photobiol B Biol 15:343–353. https://doi.org/10.1016/1011-1344(92)85140-P

Csatorday K (1978) Fluorescence from sensitizing phycobilin chromophores in the blue-green alga Anacystis nidulans. BBA Bioenergetics 504:341–343. https://doi.org/10.1016/0005-2728(78)90181-0

D'Alessandro EB, Antoniosi Filho NR (2016) Concepts and studies on lipid and pigments of microalgae: a review. Renew Sust Energ Rev 58:832–841. https://doi.org/10.1016/J.RSER.2015.12.162

Dagnino-Leone J, Figueroa CP, Castañeda ML et al (2022) Phycobiliproteins: structural aspects, functional characteristics, and biotechnological perspectives. Comput Struct Biotechnol J 20: 1506–1527. https://doi.org/10.1016/J.CSBJ.2022.02.016

Dailer ML, Smith JE, Smith CM (2012) Responses of bloom forming and non-bloom forming macroalgae to nutrient enrichment in Hawai'i, USA. Harmful Algae 17:111–125. https://doi.org/10.1016/J.HAL.2012.03.008

Dumay J, Morançais M, Munier M et al (2014) Phycoerythrins: valuable proteinic pigments in red seaweeds. Elsevier
Falkeborg MF, Roda-Serrat MC, Burnæs KL, Nielsen ALD (2018) Stabilising phycocyanin by anionic micelles. Food Chem 239:771–780. https://doi.org/10.1016/j.foodchem.2017.07.007
Fischer M, Häder DP (1992) UV effects on photosynthesis and phycobiliprotein composition in the flagellate Cyanophora paradoxa. FEMS Microbiol Lett 101:121–131. https://doi.org/10.1016/0378-1097(92)90833-A
Francavilla M, Manara P, Kamaterou P, et al (2015) Cascade approach of red macroalgae Gracilaria gracilis sustainable valorization by extraction of phycobiliproteins and pyrolysis of residue. Bioresour Technol 184:305–313. https://doi.org/10.1016/J.BIORTECH.2014.10.147
Frank HA, Cogdell RJ (2012) Light capture in photosynthesis. Elsevier
Guillard C, Dumay J, Donnay-Moreno C, et al (2015) Ultrasound-assisted extraction of R-phycoerythrin from Grateloupia turuturu with and without enzyme addition. Algal Res 12: 522–528. https://doi.org/10.1016/J.ALGAL.2015.11.002
Heocha CÓ (1958) Comparative biochemical studies of the phycobilins. Arch Biochem Biophys 73:207–219. https://doi.org/10.1016/0003-9861(58)90256-X
Hernandez-Mireles T and Rito-Palomares M (2006) Improved recovery of B-phycoerythrin produced by the red microalga Porphyridium cruentum. J Chem Technol Biotechnol 81:989–996. https://doi.org/10.1002/jctb
Issa A, Ali E, Abdel-Basset R et al (2020) The impact of nitrogen concentrations on production and quality of food and feed supplements from three cyanobacteria and potential application in biotechnology. Biocatal Agric Biotechnol 24:101533. https://doi.org/10.1016/J.BCAB.2020.101533
Kaňa R, Prášil O, Mullineaux CW (2009) Immobility of phycobilins in the thylakoid lumen of a cryptophyte suggests that protein diffusion in the lumen is very restricted. FEBS Lett 583:670–674. https://doi.org/10.1016/J.FEBSLET.2009.01.016
Karnieli A, Kidron GJ, Glaesser C, Ben-Dor E (1999) Spectral characteristics of cyanobacteria soil crust in semiarid environments. Remote Sens Environ 69:67–75. https://doi.org/10.1016/S0034-4257(98)00110-2
Khan AZ, Malik S, Mehmood MA et al (2022) Two-stage algal cultivation for the biotransformation of urban wastewater's pollutants into multiple bioproducts in a circular bioeconomy paradigm. Energy Convers Manag 273:116400. https://doi.org/10.1016/J.ENCONMAN.2022.116400
Kovaleski G, Kholany M, Dias LMS et al (2022) Extraction and purification of phycobiliproteins from algae and their applications. Front Chem 10:1–19. https://doi.org/10.3389/fchem.2022.1065355
Krynická V, Skotnicová P, Jackson PJ et al (2023) FtsH4 protease controls biogenesis of the PSII complex by dual regulation of high light-inducible proteins. Plant Commun 4:100502. https://doi.org/10.1016/J.XPLC.2022.100502
Kuzminov FI, Karapetyan NV, Rakhimberdieva MG et al (2012) Investigation of OCP-triggered dissipation of excitation energy in PSI/PSII-less Synechocystis sp. PCC 6803 mutant using non-linear laser fluorimetry. Biochim Biophys Acta Bioenerg 1817:1012–1021. https://doi.org/10.1016/J.BBABIO.2012.03.022
Lee WK, Lim YY, Ho CL (2019) pH affects growth, physiology and agar properties of agarophyte Gracilaria changii (Rhodophyta) under low light intensity from Morib, Malaysia. Reg Stud Mar Sci 30. https://doi.org/10.1016/J.RSMA.2019.100738
Li S, Ji L, Shi Q et al (2019) Advances in the production of bioactive substances from marine unicellular microalgae Porphyridium spp. Bioresour Technol 292:122048. https://doi.org/10.1016/J.BIORTECH.2019.122048
Mandal MK, Chanu NK, Chaurasia N (2020) Cyanobacterial pigments and their fluorescence characteristics: applications in research and industry. Adv Cyanobacterial Biol 55–72. https://doi.org/10.1016/B978-0-12-819311-2.00005-X

Martins M, Ventura SPM (2020) Emerging seaweed extraction techniques using ionic liquids. Elsevier Inc.

Mittal R, Sharma R, Raghavarao KSMS (2019) Aqueous two-phase extraction of R-Phycoerythrin from marine macro-algae, Gelidium pusillum. Bioresour Technol 280:277–286. https://doi.org/10.1016/j.biortech.2019.02.044

Montoya EJO, Dorion S, Atehortua-Garcés L, Rivoal J (2021) Phycobilin heterologous production from the Rhodophyta Porphyridium cruentum. J Biotechnol 341:30–42. https://doi.org/10.1016/j.jbiotec.2021.09.001

Munier M, Jubeau S, Wijaya A, et al (2014) Physicochemical factors affecting the stability of two pigments: Rphycoerythrin of Grateloupia turuturu and B-phycoerythrin of Porphyridium cruentum. Food Chem 150:400–407. https://doi.org/10.1016/J.FOODCHEM.2013.10.113

Naeher S, Schaeffer P, Adam P, Schubert CJ (2013) Maleimides in recent sediments - using chlorophyll degradation products for palaeoenvironmental reconstructions. Geochim Cosmochim Acta 119:248–263. https://doi.org/10.1016/J.GCA.2013.06.004

Nobel PS (2020) Photochemistry of photosynthesis. In: Physicochemical and environmental plant physiology. Springer

Patel AK, Albarico FPJB, Perumal PK et al (2022) Algae as an emerging source of bioactive pigments. Bioresour Technol 351:126910. https://doi.org/10.1016/J.BIORTECH.2022.126910

Paull RE, Chen NJ (2008) Postharvest handling and storage of the edible red seaweed Gracilaria. Postharvest Biol Technol 48:302–308. https://doi.org/10.1016/J.POSTHARVBIO.2007.12.001

Peterson RB, Dolan E, Calvert HE, Ke B (1981) Energy transfer from phycobiliproteins to photosystem I in vegetative cells and heterocysts of anabaena Variabilis. BBA-Bioenergetics 634:237–248. https://doi.org/10.1016/0005-2728(81)90142-0

Pez Jaeschke D, Rocha Teixeira I, Damasceno Ferreira Marczak L, Domeneghini Mercali G (2021) Phycocyanin from spirulina: a review of extraction methods and stability. Food Res Int 143: 110314. https://doi.org/10.1016/j.foodres.2021.110314

Purvis K, Brittain K, Joseph A et al (2019) Third-order nonlinear optical properties of phycobiliproteins from cyanobacteria and red algae. Chem Phys Lett 731:136599. https://doi.org/10.1016/J.CPLETT.2019.136599

Rakhimberdieva MG, Vavilin DV, Vermaas WFJ et al (2007) Phycobilin/chlorophyll excitation equilibration upon carotenoid-induced non-photochemical fluorescence quenching in phycobilisomes of the cyanobacterium Synechocystis sp. PCC 6803. Biochim Biophys Acta Bioenerg 1767:757–765. https://doi.org/10.1016/J.BBABIO.2006.12.007

Rhie GE, Beale SI (1992) Biosynthesis of phycobilins. Ferredoxin-supported NADPH-independent heme oxygenase and phycobilin-forming activities from Cyanidium caldarium. J Biol Chem 267:16088–16093. https://doi.org/10.1016/s0021-9258(18)41970-9

Rodrigues RDP, de Castro FC, Santiago-Aguiar RS de, Rocha MVP (2018) Ultrasound-assisted extraction of phycobiliproteins from spirulina (Arthrospira) platensis using protic ionic liquids as solvent. Algal Res 31:454–462. https://doi.org/10.1016/j.algal.2018.02.021

Safdar W, Zan X, Shamoon M et al (2017) Effects of twenty standard amino acids on biochemical constituents, docosahexaenoic acid production and metabolic activity changes of Crypthecodinium cohnii. Bioresour Technol 238:738–743. https://doi.org/10.1016/j.biortech.2017.04.024

Sampath-Wiley P, Neefus CD, Jahnke LS (2008) Seasonal effects of sun exposure and emersion on intertidal seaweed physiology: fluctuations in antioxidant contents, photosynthetic pigments and photosynthetic efficiency in the red alga Porphyra umbilicalis Kützing (Rhodophyta, Bangiales). J Exp Mar Bio Ecol 361:83–91. https://doi.org/10.1016/J.JEMBE.2008.05.001

Schoefs B (2004) Determination of pigments in vegetables. J Chromatogr A 1054:217–226. https://doi.org/10.1016/J.CHROMA.2004.05.105

Schreiber U (1980) Reversible uncoupling of energy transfer between phycobilins and chlorophyll in Anacystis nidulans. Light stimulation of cold-induced phycobilisome detachment. Biochim Biophys Acta Bioenerg 591:361–371. https://doi.org/10.1016/0005-2728(80)90167-X

Sfriso AA, Gallo M, Baldi F (2018) Phycoerythrin productivity and diversity from five red macroalgae. J Appl Phycol 30:2523–2531. https://doi.org/10.1007/s10811-018-1440-3
Shahid A, Malik S, Liu CG et al (2021a) Characterization of a newly isolated cyanobacterium Plectonema terebrans for biotransformation of the wastewater-derived nutrients to biofuel and high-value bioproducts. J Water Process Eng 39:101702. https://doi.org/10.1016/J.JWPE.2020.101702
Shahid A, Usman M, Atta Z et al (2021b) Impact of wastewater cultivation on pollutant removal, biomass production, metabolite biosynthesis, and carbon dioxide fixation of newly isolated cyanobacteria in a multiproduct biorefinery paradigm. Bioresour Technol 333:125194. https://doi.org/10.1016/J.BIORTECH.2021.125194
Shanab SMM, Mostafa SSM, Shalaby EA, Mahmoud GI (2012) Aqueous extracts of microalgae exhibit antioxidant and anticancer activities. Asian Pac J Trop Biomed 2:608–615. https://doi.org/10.1016/S2221-1691(12)60106-3
Sharmila Banu VM, Santhosh S, Hemalatha V et al (2017) Optimization study on extraction & purification of phycoerythrin from red algae Kappaphycus Alvarezii. Asian J Pharm Clin Res 10:297–302. https://doi.org/10.22159/ajpcr.2017.v10i2.15598
Shen G, Schluchter WM, Bryant DA (2008) Biogenesis of phycobiliproteins: I. cpcS-I and cpcU mutants of the cyanobacterium Synechococcus sp. PCC 7002 define a heterodimeric phyococyanobilin lyase specific for β-phycocyanin and allophycocyanin subunits. J Biol Chem 283:7503–7512. https://doi.org/10.1074/JBC.M708164200
Singhai GS, Mohanty P, Govindjee (1981) Effect of preheating intact algal cells on pigments revealed by absorption and fluorescence spectra. Z Pflanzenphysiol 103:217–228. https://doi.org/10.1016/S0044-328X(81)80154-7
Ssebiyonga N, Erga SR, Hamre B et al (2013) Light conditions and photosynthetic efficiency of phytoplankton in Murchison Bay, Lake Victoria, Uganda. Limnologica 43:185–193. https://doi.org/10.1016/J.LIMNO.2012.09.005
Steglich C, Mullineaux CW, Teuchner K et al (2003) Photophysical properties of Prochlorococcus marinus SS120 divinyl chlorophylls and phycoerythrin in vitro and in vivo. FEBS Lett 553:79–84. https://doi.org/10.1016/S0014-5793(03)00971-2
Tamamizu K, Sakamoto T, Kurashige Y et al (2023) Scytonemin redox status in a filamentous cyanobacterium visualized by an excitation-laser-line-scanning spontaneous Raman scattering spectral microscope. Spectrochim Acta A Mol Biomol Spectrosc 290:122258. https://doi.org/10.1016/J.SAA.2022.122258
Tamary E, Kiss V, Nevo R et al (2012) Structural and functional alterations of cyanobacterial phycobilisomes induced by high-light stress. Biochim Biophys Acta Bioenerg 1817:319–327. https://doi.org/10.1016/J.BBABIO.2011.11.008
Taniguchi M, Lindsey JS (2023) Absorption and fluorescence spectra of open-chain tetrapyrrole pigments–bilirubins, biliverdins, phycobilins, and synthetic analogues. J Photochem Photobiol C: Photochem Rev 55:100585. https://doi.org/10.1016/J.JPHOTOCHEMREV.2023.100585
Thomas JB, Marsman JW (1959) On the pigment system of the red alga Porphyra Lacineata. BBA Biochim Biophys Acta 35:316–323. https://doi.org/10.1016/0006-3002(59)90380-4
Tomazic N, Overkamp KE, Wegner H et al (2021) Exchange of a single amino acid residue in the cryptophyte phycobiliprotein lyase GtCPES expands its substrate specificity. Biochim Biophys Acta Bioenerg 1862:148493. https://doi.org/10.1016/J.BBABIO.2021.148493
Wang M, Zhan Y, Chen C et al (2022) Amplified cyanobacterial bloom is derived by polyphosphate accumulation triggered by ultraviolet light. Water Res 222:118837. https://doi.org/10.1016/J.WATRES.2022.118837
Wiethaus J, Busch AWU, Dammeyer T, Frankenberg-Dinkel N (2010) Phycobiliproteins in Prochlorococcus marinus: biosynthesis of pigments and their assembly into proteins. Eur J Cell Biol 89:1005–1010. https://doi.org/10.1016/J.EJCB.2010.06.017

Wu JT, Chiang YR, Huang WY, Jane WN (2006) Cytotoxic effects of free fatty acids on phytoplankton algae and cyanobacteria. Aquat Toxicol 80:338–345. https://doi.org/10.1016/J.AQUATOX.2006.09.011

Xu Z, Zhu B, Xue X et al (2022) Study on immediate and long-term growth inhibition of Microcystis aeruginosa by non-thermal plasma. Chem Eng J 429:132397. https://doi.org/10.1016/J.CEJ.2021.132397

Yang D, Qing Y, Min C (2010) Incorporation of the chlorophyll d-binding light-harvesting protein from Acaryochloris marina and its localization within the photosynthetic apparatus of Synechocystis sp. PCC6803. Biochim Biophys Acta Bioenerg 1797:204–211. https://doi.org/10.1016/J.BBABIO.2009.10.006

Yang R, Ma T, Shi L et al (2022) The formation of phycocyanin-EGCG complex for improving the color protection stability exposing to light. Food Chem 370:130985. https://doi.org/10.1016/j.foodchem.2021.130985

Yuan B, Li Z, Shan H et al (2022) A review of recent strategies to improve the physical stability of phycocyanin. Curr Res Food Sci 5:2329–2337. https://doi.org/10.1016/j.crfs.2022.11.019

Zhang Y, Zhang L, Hu J et al (2023) The structural characterization and color stabilization of the pigment protein-phycoerythrin glycosylated with oligochitosan. Food Hydrocoll 136:108241. https://doi.org/10.1016/J.FOODHYD.2022.108241

Zhao KH, Deng MG, Zheng M et al (2000) Novel activity of a phycobiliprotein lyase: both the attachment of phycocyanobilin and the isomerization to phycoviolobilin are catalyzed by the proteins PecE and PecF encoded by the phycoerythrocyanin operon. FEBS Lett 469:9–13. https://doi.org/10.1016/S0014-5793(00)01245-X

Zhao H, Liu G, Han G (2019) High-performance laminated luminescent solar concentrators based on colloidal carbon quantum dots. Nanoscale Adv 1:4888–4894. https://doi.org/10.1039/c9na00527g

Zimba PV (2012) An improved phycobilin extraction method. Harmful Algae 17:35–39. https://doi.org/10.1016/j.hal.2012.02.009

Zinicovscaia I, Cepoi L, Rudi L et al (2021) Accumulation of dysprosium, samarium, terbium, lanthanum, neodymium and ytterbium by Arthrospira platensis and their effects on biomass biochemical composition. J Rare Earths 39:1133–1143. https://doi.org/10.1016/J.JRE.2020.07.019

Chapter 7
Nonconventional and Novel Strategies to Produce *Spirulina* Biomass

Waseem Safdar, Asma Saleem Qazi, Saeed Ahmed, Mohammad Rizwan Tariq, and Haroon Ahmed

Abstract Spirulina, a type of blue-green algae, has been gaining increasing attention in recent years for its potential as a sustainable and nutritious food source. This fascinating microorganism is a photosynthetic cyanobacterium that grows in both freshwater and saltwater environments and has a distinctive spiral shape, which gives it its name. *Spirulina* has a long history of use as a food source in various cultures, particularly for its high protein content and as a rich source of essential vitamins, minerals, and antioxidants. In recent years, nonconventional and novel strategies for producing *Spirulina* biomass have been explored to further improve its sustainability and scalability. One such approach is mixotrophic cultivation, which involves providing both light and an organic carbon source for the algae to grow. Another innovative technique is the use of membrane photobioreactors (MPBRs). These bioreactors combine photobioreactors and membrane filtration technology to produce high-density cultures of *Spirulina*. Phototaxis-based cultivation is another novel approach that involves manipulating the movement of *Spirulina* cells using light gradients. This method concentrates the cells in a specific area, making it easier to harvest the biomass. Cocultivation is yet another strategy that has shown promise for enhancing *Spirulina* cultivation. By growing *Spirulina* together with other microorganisms, such as bacteria or other microalgae, the nutrient uptake and utilization can be improved, leading to better growth rates and biomass quality. Through continued research and responsible cultivation practices, *Spirulina* can play a significant role in meeting the challenges of a growing global population while promoting a healthier and more sustainable future. As we unlock the full potential of *Spirulina* and explore innovative cultivation methods, we move closer to realizing a

W. Safdar (✉) · A. S. Qazi · S. Ahmed
Department of Biological Sciences, National University of Medical Sciences, Abid Majeed Road, The Mall, Rawalpindi, Pakistan
e-mail: waseem.safdar@numspak.edu.pk

M. R. Tariq
Department of Food Science and Technology, Faculty of Agricultural Sciences, University of the Punjab, Quid-i-Azam Campus, Lahore, Pakistan

H. Ahmed
Department of Biosciences, COMSATS University Islamabad (CUI), Islamabad, Pakistan

M. A. Mehmood et al. (eds.), *Pharmaceutical and Nutraceutical Potential of Cyanobacteria*, https://doi.org/10.1007/978-3-031-45523-0_7

world where nutritious and environmentally friendly food sources are accessible to all.

Keywords Spirulina · Mixotrophic cultivation · Membrane photobioreactors · Phototaxis-based cultivation · Cocultivation

7.1 Introduction

Spirulina is a type of blue-green algae that has been gaining increasing attention in recent years for its potential as a sustainable and nutritious food source. It is a photosynthetic microorganism that grows in freshwater and saltwater environments and has a distinctive spiral shape, which gives it its name. *Spirulina* has been used for centuries as a food source in various cultures, particularly in parts of Africa and Central and South America. It is known for its high protein content and is a rich source of essential vitamins, minerals, and antioxidants (Soni et al. 2017). *Spirulina* is considered to be one of the most nutrient-dense foods on the planet, with a wide range of potential health benefits. In addition to its nutritional value, *Spirulina* is also highly sustainable and has a low environmental impact compared to other protein sources (Sow and Ranjan 2021). It requires less water and land than traditional livestock farming and produces significantly fewer greenhouse gas emissions. Nonconventional and novel strategies for producing *Spirulina* biomass have been developed, which have the potential to further improve its sustainability and scalability as a food source.

Furthermore, *Spirulina* has also been studied for its potential in various other applications beyond food. For instance, it has been used in the production of cosmetics, as a natural coloring agent, and even as a biofuel. Its ability to remove heavy metals and other pollutants from water has also led to its use in wastewater treatment (Soni et al. 2017). While *Spirulina* has many potential benefits, there are also some concerns about its safety and efficacy. It is important to ensure that *Spirulina* is grown and processed safely and hygienically to prevent contamination with harmful substances (Sow and Ranjan 2021). Additionally, more research is needed to fully understand the potential risks and benefits of consuming *Spirulina*, particularly in large quantities or for extended periods (Alfredo de Jesús and Brenda Paloma 2022). Overall, *Spirulina* is a fascinating microorganism that has captured the attention of scientists and food industry professionals alike. Its high nutrient content, sustainability, and versatility make it a promising candidate for meeting the growing demand for nutritious and environmentally friendly food sources (Masojı'dek and Torzillo 2008).

Spirulina is classified as a cyanobacterium, which is a type of photosynthetic bacteria that is capable of producing oxygen through photosynthesis. It is commonly found in tropical and subtropical regions and can grow in a wide range of water conditions, including freshwater, saltwater, and alkaline water (Heinsoo 2014). *Spirulina* has been used for centuries as a food source, particularly in regions where access to other sources of protein is limited. It has a high protein content, contains all essential amino acids, and is also a rich source of vitamins and minerals such as iron, calcium, and vitamin B12. Additionally, it contains antioxidants such as phycocyanin, which has been shown to have anti-inflammatory properties (Rakhimberdieva et al. 2007).

Research has also shown that *Spirulina* may have several potential health benefits, such as improving immune function, reducing inflammation, and even lowering blood pressure and cholesterol levels (Koller et al. 2012). Some studies have even suggested that *Spirulina* may have anticancer properties, although more research is needed in this area. *Spirulina* is typically sold as a dietary supplement in the form of tablets or powder and can also be found in various food products such as energy bars and smoothies (Wuang et al. 2016). However, as with any dietary supplement, it is important to purchase *Spirulina* from a reputable source and to follow dosage instructions carefully. In terms of sustainability, *Spirulina* has several advantages over traditional livestock farming (Simpósio et al. 2017). It requires significantly less land and water and produces fewer greenhouse gas emissions. Additionally, *Spirulina* can be grown year-round and has a high yield per unit of land, making it a potentially valuable crop for meeting the increasing demand for sustainable food sources to harness its nutritional and environmental benefits (Barry et al. 2015):

- **Food and nutrition:** *Spirulina* is commonly used as a food supplement and has gained popularity as a superfood due to its high nutrient content. It is a rich source of protein, vitamins, and minerals and has been shown to have antioxidant and anti-inflammatory properties (Safdar et al. 2017a). *Spirulina* can be consumed as a dietary supplement in the form of tablets or powder or incorporated into food products such as smoothies, energy bars, and snacks. It can also be used as a food coloring agent due to its blue-green pigment.
- **Agriculture and aquaculture:** *Spirulina* has potential applications in agriculture and aquaculture. It can be used as a fertilizer due to its high nitrogen content and has been shown to improve soil fertility and plant growth. Additionally, *Spirulina* can be used as a feed source for aquaculture, providing a sustainable and nutritious alternative to traditional fish feeds.
- **Cosmetics and personal care:** *Spirulina* is also used in the cosmetics industry due to its antioxidant and anti-inflammatory properties. It can be used in skincare products such as creams and masks and has been shown to improve skin hydration and elasticity. *Spirulina* extract has also been used in hair care products as a natural conditioning agent.
- **Wastewater treatment:** *Spirulina* is effective in removing heavy metals and other pollutants from water. It can be used in wastewater treatment systems as a natural and sustainable alternative to traditional treatment methods.
- **Pharmaceutical applications:** *Spirulina* has potential applications in the pharmaceutical industry, particularly in the development of new drugs. It has been shown to have anti-inflammatory, immunomodulatory, and anticancer properties and has been studied as a potential treatment for conditions such as allergies, cancer, and HIV/AIDS.

- **Animal feed:** *Spirulina* can be used as a feed source for livestock and pets. It has been shown to improve the growth and health of animals and can be used as a sustainable alternative to traditional animal feeds such as soybean and fish meal.
- **Energy production:** *Spirulina* can be used as a source of biomass for the production of biofuels such as methane and hydrogen. It has been shown to have a higher yield per unit of land compared to traditional biofuel crops such as corn and sugarcane.
- **Environmental remediation:** *Spirulina* has been used in environmental remediation projects to clean up contaminated sites. It is effective in removing heavy metals and other pollutants from soil and can be used in combination with other plants and microorganisms to restore degraded ecosystems.
- **Industrial applications:** *Spirulina* can be used in various industrial processes such as wastewater treatment, paper production, and textile manufacturing. Its high protein content and antioxidant properties make it a potential ingredient in food additives and nutraceuticals.
- **Space exploration**: *Spirulina* has been studied as a potential food source for space missions due to its high nutrient content, sustainability, and ability to grow in extreme conditions.

7.2 Mixotrophic Cultivation

Mixotrophic cultivation is a type of cultivation in which microorganisms such as algae, fungi, and bacteria are grown using a combination of autotrophic and heterotrophic modes of nutrition. In the case of *Spirulina*, mixotrophic cultivation involves providing both light and an organic carbon source for the algae to grow. The advantage of mixotrophic cultivation over purely autotrophic or heterotrophic cultivation is that it can potentially increase the productivity and yield of the algae (Safdar et al. 2017b). This is because the organic carbon source can provide the algae with additional nutrients and energy, allowing it to grow more quickly and efficiently than with light alone. Additionally, mixotrophic cultivation can help to reduce the cost of production, as the organic carbon source can be sourced from waste streams or low-cost substrates (Molina Grima et al. 2003). There are several methods for mixotrophic cultivation of *Spirulina*, including adding an organic carbon source such as glucose or sucrose to the growth medium (Fig. 7.1). This can be done by adding the organic carbon source directly to the growth medium or by feeding it to the algae intermittently (Soni et al. 2017).

Spirulina can be grown in wastewater streams or the presence of organic waste such as molasses or dairy whey. *Spirulina* can be grown in combination with other microorganisms such as bacteria or fungi, which can provide additional nutrients and organic carbon (Simpósio et al. 2017; Zhu et al. 2020; Duran Quintero et al. 2021; Harini and Rajkumar 2022). Research has shown that mixotrophic cultivation of *Spirulina* can significantly increase the productivity and yield of the algae compared to purely autotrophic or heterotrophic cultivation. Additionally, mixotrophic

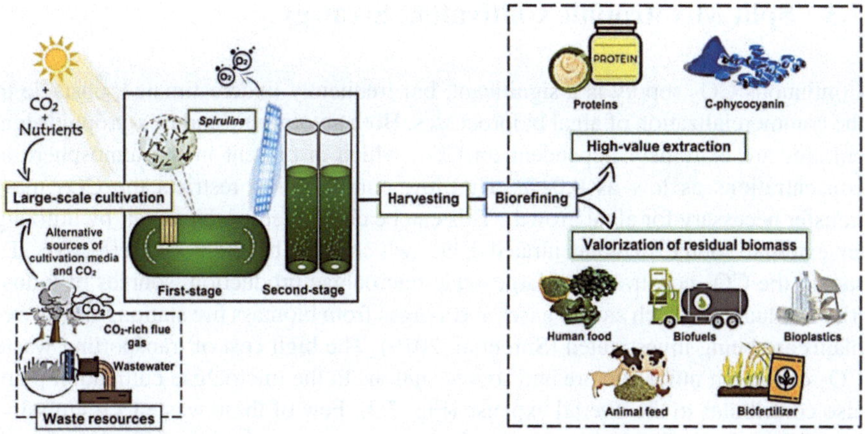

Fig. 7.1 Mixotrophic cultivation of *Spirulina* for high-value products (adapted with permission from Thevarajah et al. (2022) via permission license number 5597571082901)

cultivation has the potential to reduce the cost of production and improve the sustainability of *Spirulina* cultivation by utilizing waste streams and low-cost substrates (Thomas and Govindjee 1960):

- **Increased biomass and lipid productivity:** Mixotrophic cultivation can increase the growth rate and biomass productivity of *Spirulina* by providing the algae with additional nutrients and energy (Babu et al. 2022). This can also lead to increased lipid productivity, which is important for the production of biofuels and other value-added products.
- **Improved nutritional quality:** Mixotrophic cultivation can improve the nutritional quality of *Spirulina* by providing additional vitamins, minerals, and amino acids. This can make *Spirulina* a more valuable source of nutrition for human and animal consumption.
- **Enhanced stress tolerance:** Mixotrophic cultivation can help to improve the stress tolerance of *Spirulina*, allowing it to grow in a wider range of environmental conditions. This can be especially useful for commercial cultivation, where algae must be able to tolerate fluctuations in temperature, light, and nutrient availability.
- **Reduced carbon dioxide emissions:** Mixotrophic cultivation can help to reduce carbon dioxide emissions by utilizing waste streams or low-cost substrates as an organic carbon source. This can make *Spirulina* cultivation more sustainable and eco-friendly.

7.3 Split Mixotrophic Cultivation Strategy

Continuous CO_2 supply is a significant, but frequently underestimated, obstacle to the commercialization of algal bioprocesses. Both autotrophic and mixotrophic algal cultures are extremely dependent on CO_2, which is present in the atmosphere at concentrations as low as 0.04%. This low concentration restricts the CO_2 mass transfer necessary for algal growth. This can be circumvented, however, by utilizing an external source of concentrated CO_2, which may be more cost-effective. To satisfy the CO_2 constraints of large-scale microalgae production, sources of industrial residual CO_2 such as flue gases and syngas from biomass fire stations and power plants are being investigated (Sim et al. 2019). The high cost of transporting waste CO_2 collection infrastructure and source stations to the microalgae cultivation plant also contributes to the overall expense (Fig. 7.2). Few of these waste CO_2 emission sources are located near large-scale algal cultivations, making it difficult to locate algal cultivation facilities near these sources. As a result, support strategies and additional research to lessen reliance on external sources of CO_2 were always

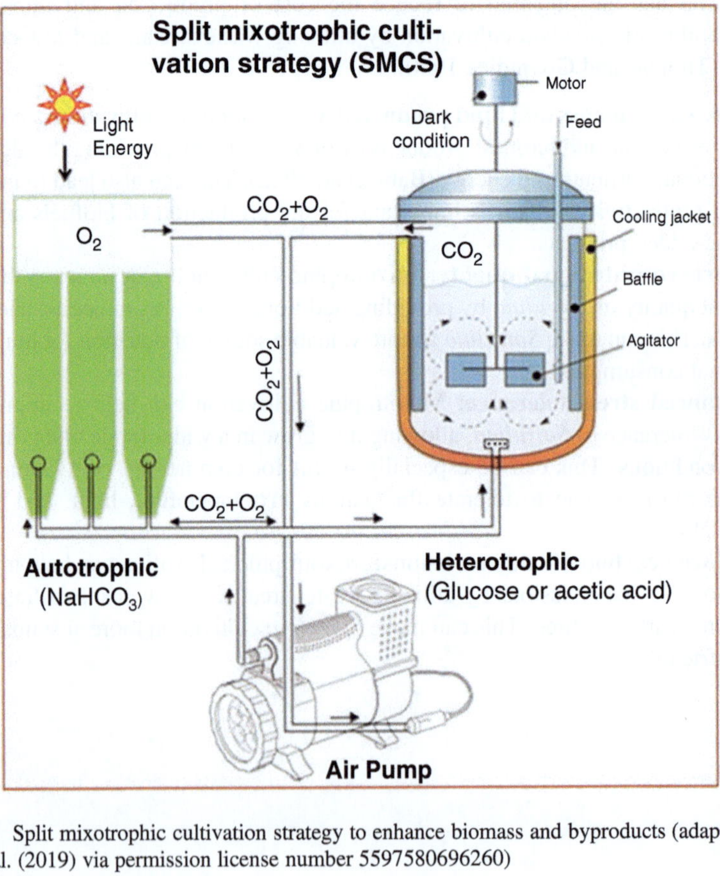

Fig. 7.2 Split mixotrophic cultivation strategy to enhance biomass and byproducts (adapted from Sim et al. (2019) via permission license number 5597580696260)

envisioned. By supplying endogenous CO_2 produced by heterotrophic cellular respiration, a compartmentalized mixotrophic coculture strategy may mitigate this issue (Soni et al. 2017).

Extensive mixotrophy involves both autotrophic and heterotrophic systems and consequently can have a negative impact on both processes. The mixotrophic cultivation method has also brought to light important issues such as chlorophyll loss, shading effects, and inhibition of organic carbon absorption, thereby successfully addressing these obstacles and providing additional technical and economic advantages (Safdar et al. 2017c). Not only for the optimization of algal bioprocesses but also for the environmental benefits provided by their capacity to sequester carbon, mixotrophic cultivation has emerged as an important research topic (Castillo et al. 2021). It has numerous advantages over autotrophic and heterotrophic cultivation alone, such as increased crop yield and biomass (from organic and inorganic carbon), crop quality, and robustness, among others. Mixotrophy is a dual system that converts light energy via autotrophic processes (Pereira et al. 2019). Through photosynthesis, it is converted into chemical energy, heterotrophic processes cause the catabolism of organic compounds, and respiration provides the energy required for cell division.

This indicates that photosynthesis and oxidative organic carbon metabolism coexist in a mixed trophic mode of growth. Therefore, mixotrophic growth was calculated as the sum of photoautotrophic (henceforth autotrophic) growth and heterotrophic growth. In mixotrophy, organic carbons such as glucose and acetate provide cellular biosynthesis with additional energy in the form of acetyl-CoA, NADPH, and other intermediates (Safdar et al. 2017a). Under conditions of light, however, carbon dioxide produced by cellular respiration can be recycled for photosynthesis. Consequently, it is evident that cell growth increases in terms of mass and number compared to solitary systems such as phototrophic and/or heterotrophic growth modes, possibly due to the synergistic effect of the two metabolisms. In addition to being less dependent on external CO_2, mixotrophic growth consumes considerably less mineral and organic matter per unit of biomass than dark heterotrophic growth (Marting Vidaurre et al. 2023). Therefore, combined feeding was determined to be the most cost-effective microalgae cultivation method. Not only is there an increase in biomass and crop yield, but there is also a maintenance of CO_2 absorption capacity and an increase in the carbon footprint of commercial algal processes, which heterotrophic modes cannot provide (Mousavian et al. 2023).

Direct production of gases (CO_2 and O_2) by autotrophic and heterotrophic organisms provides advantage to mixotrophs while isolating their liquid environment. Other liquid-phase factors also contribute to the cause. With these objectives in mind, models that mitigated these obstacles and enabled more effective combined feeding than conventional feeding modes were investigated by (Crivellaro Gonçalves et al. 2022). The purpose of this study was to validate the synergistic effects of gas exchange when separating autotrophic and heterotrophic liquid streams and integrating them under SMCS. In addition, the enhancement of growth conditions to increase mixotrophic effects, such as photosynthesis and respiration efficiency and DIC and DO concentrations, is determined (Zhu et al. 2020). In addition to the growth benefits of SMCS, the technical advantages of this model over conventional mixed nutrition agriculture must also be confirmed. Importantly, the SMCS model employs two reactors (for the partitioning process), and each reactor represents only one trophic state (half) of the mixotrophic composite complex. Therefore, the average biomass yield of the two reactors should not be comparable to the conventional mixed feed with two nutrient modes (Sim et al. 2019). This paper presents a proof of concept for more effective mixotrophy and examines the synergy of growth strategies that contribute to this growth. This will increase our understanding of mixotrophic phenomena so that we can develop large-scale mixotrophic phytoplankton bioprocesses that are cost-effective (FAO Fisheries 2008; Soni et al. 2017).

7.4 Eco-Design of *Spirulina* Solar Cultivation

In addition, *Spirulina* contains lipids and polysaccharides, which are of interest in disciplines such as green chemistry (including biomaterials and biofertilizers) and energy. Multiple studies have demonstrated that *Spirulina* biomass can be utilized as a feedstock for the production of biogas, biodiesel, and hydrogen (Safdar et al. 2017d). Moreover, numerous authors cite encouraging findings regarding the biosorption capacity of *Spirulina* for the remediation of synthetic dye-contaminated water on a large scale. Similarly, *Spirulina* has been shown to be capable of treating highly polluted effluent produced by anaerobic digestion (Duran Quintero et al. 2021). Microalgal biomass (here, prokaryotic cyanobacteria and eukaryotic microalgae) has been shown to reduce the latter due to its greater photosynthetic efficiency (e.g., 10–20% vs. 0.5%) when compared to that of conventional plants. to offer It has garnered significant attention over the past decade (Yu et al. 2023). In addition, these microorganisms can absorb dissolved CO_2 to satisfy their metabolic requirements for inorganic carbon, making them an additional tool in the fight against climate change. Nonetheless, in the global context of environmental degradation (greenhouse effect, ecosystem pollution, depletion of limited resources, etc.) and the pursuit of sustainable industrial development, operations associated with the growth process of microalgae should demonstrate strong sustainable environmental performance (Usai et al. 2023) compared to what can be accomplished by creating conventional or comparable products (Shekhar Sarker et al. 2023).

Life cycle assessment techniques are used to identify "hot spots" in spirulina-related processes, from biomass growth to the production of alternative products such as oral supplements, phycocyanin, and sustenance (Schreiber 1979). Few investigations have utilized it as biogas. However, these previous studies were restricted to constant operating conditions and did not investigate the variability of operating parameters (Overkamp et al. 2014). In actuality, when a process contains multiple parameters related to a variable value, the evaluation of only a small number of scenarios seems to underestimate the heterogeneity of the system and leads to incomplete results and analyses, resulting in the creation of limited and prospective options for decision-makers (Dineshkumar et al. 2016). To reduce our environmental impact the primary purpose of this research is to combine *Spirulina* biomass growth models with LCA methodologies in order to identify important operational aspects that improve environmental balance. To facilitate appropriate decision-making in particular cultural contexts, these models should have parameters that represent particular operating conditions (Duran Quintero et al. 2021). Our background system consists of *Spirulina* cultivation in greenhouse-controlled interior and outdoor ponds (ORP and CRP). Our functional unit (FU) is the manufacture of 1 kg of 20% dry weight *Spirulina* biomass encapsulated in a 20% dry weight slurry (Pereira et al. 2019). For the sake of simplification, we will now refer to it as 1 kg of biomass (dw).

Figure 7.3 depicts the LCA "cradle to gate" system (Fig. 7.3). The foreground system is depicted in the figure's central frame. The solar farming mode is evaluated using a surface area of 388.5 square meters. All simulations of microalgae production took into account the annual climate conditions in Nantes, France (Soni et al. 2017). The supplementary data (I) provide an exhaustive description of the background processes. Above and below the central frame are processes that provide input flows (such as water, electricity, and nutrients) and output flows (such as effluent treatment) to the foreground system (Duran Quintero et al. 2021).

Simulations of annual microalgal biomass production are used to compile background system inventories. Annual production is estimated using three stages for provided parameters (production period, geographic location, pool composition):

(i) Calculation of the maximal biomass surface productivity per hour based on the dynamics of solar radiation
(ii) Determine the incubation temperature based on the greenhouse's dynamic heat exchange
(iii) Calculation of "corrected" hourly surface productivity using the influence of temperature on microalgal growth

7.5 Heterotrophic Cultivation

Heterotrophic cultivation is a type of cultivation in which microorganisms such as algae, fungi, and bacteria are grown using an organic carbon source as their sole energy and carbon source. In the case of *Spirulina*, heterotrophic cultivation involves growing the algae in the absence of light and providing it with an organic carbon source, such as glucose or sucrose, to fuel its growth (Pereira et al. 2019). Compared to autotrophic cultivation, which relies on light as the primary energy source, heterotrophic cultivation has several advantages for *Spirulina* production. Heterotrophic cultivation can result in higher biomass productivity compared to autotrophic cultivation. This is because algae grown heterotrophically can allocate more resources toward growth and reproduction instead of synthesizing photosynthetic pigments and defense compounds. Heterotrophic cultivation can also result in

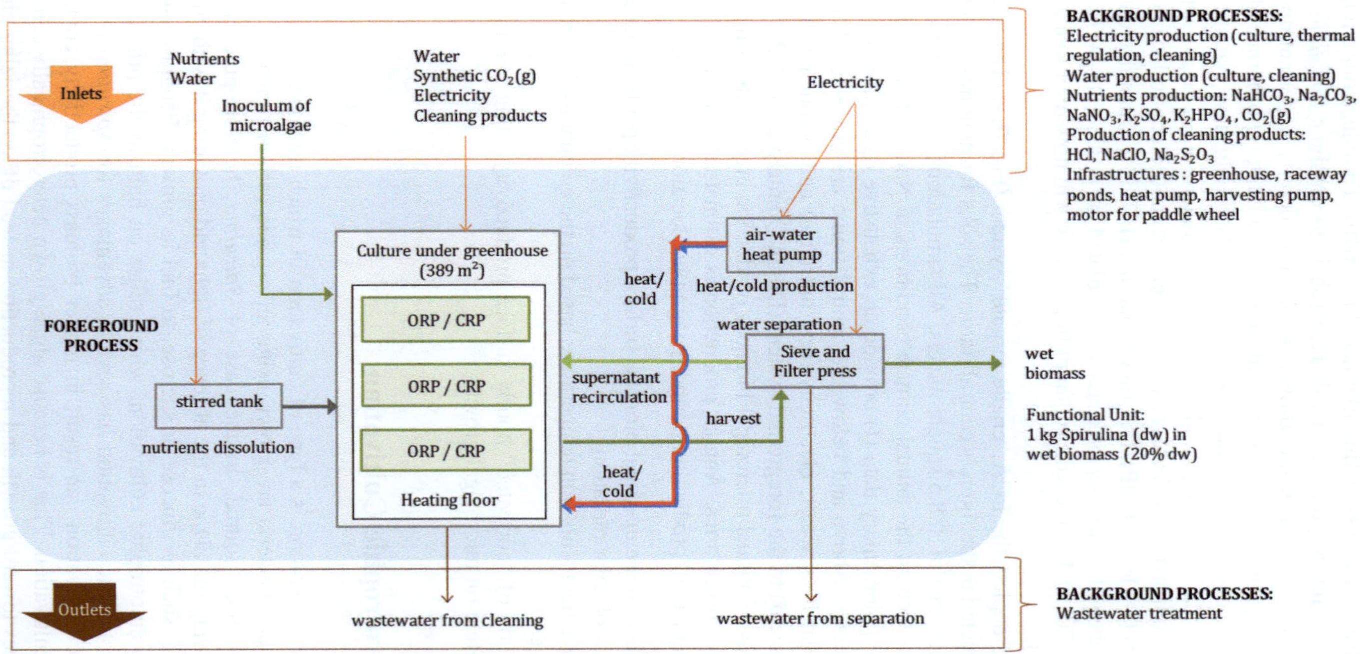

Fig. 7.3 Representation of the *Spirulina* production process (adapted from Duran Quintero et al. (2021) via permission license number 5597581099009)

higher lipid productivity compared to autotrophic cultivation (Alfredo de Jesús and Brenda Paloma 2022). This is because algae grown heterotrophically can channel more of their carbon and energy toward lipid synthesis instead of carbohydrate synthesis. Heterotrophic cultivation can reduce the risk of contamination from other photosynthetic microorganisms, which can compete for nutrients and light with *Spirulina* (Jung et al. 2015).

However, heterotrophic cultivation also has some disadvantages (Simpósio et al. 2017). The most significant of these is the higher cost of production, as it requires a constant supply of organic carbon sources and energy for the cultivation process. Additionally, the heterotrophic growth of *Spirulina* can result in lower nutritional quality compared to autotrophic cultivation due to the absence of light (Grahl 2019). Heterotrophic cultivation of *Spirulina* is typically done in closed systems, such as bioreactors, to control the growth environment and optimize production parameters. It is often used in combination with other cultivation strategies, such as mixotrophic or photoheterotrophic cultivation, to optimize biomass and lipid productivity while maintaining nutritional quality (Thevarajah et al. 2022).

Heterotrophic cultivation of *Spirulina* can be done using a variety of organic carbon sources, such as glucose, sucrose, molasses, and acetic acid. Glucose and sucrose are among the most commonly used carbon sources, as they are readily available and can be easily metabolized by the algae (de Oliveira Moraes et al. 2013). The cultivation process typically involves inoculating a sterile culture of *Spirulina* into a nutrient-rich medium containing the organic carbon source. The culture is then incubated in a controlled environment with adequate aeration and mixing to promote growth (Lafarga et al. 2021). Heterotrophic cultivation can be done on a large scale in industrial bioreactors, which can be designed to optimize production parameters such as temperature, pH, and dissolved oxygen levels (Musio et al. 2022). The process can also be done on a smaller scale using low-cost materials such as plastic bags or bottles, which can be used to grow the algae in a more cost-effective manner. Overall, heterotrophic cultivation is a promising strategy for improving the productivity and lipid content of *Spirulina* biomass (Wang et al. 2023). However, it is important to carefully consider the cost and environmental impact of using organic carbon sources for cultivation as well as the potential impact on nutritional quality and other factors. As research in this area continues, it is likely that heterotrophic cultivation will become an increasingly important strategy for commercial *Spirulina* production (Bortolini et al. 2022).

7.5.1 Advantages and Disadvantages of Heterotrophic Cultivation

There are several types of heterotrophic cultivation that can be used to grow *Spirulina* algae, each with its own advantages and disadvantages (Lucas et al. 2023; Hamidi et al. 2023; Rosenau et al. 2023):

- **Batch cultivation:** In batch cultivation, a fixed amount of nutrient-rich medium containing the organic carbon source is used to cultivate *Spirulina* in a closed system such as a bioreactor. The culture is allowed to grow until it reaches a certain density, at which point the biomass is harvested and the process is repeated. Batch cultivation is relatively simple and easy to manage but can result in lower biomass and lipid productivity compared to other methods.
- **Fed-batch cultivation:** In fed-batch cultivation, a nutrient-rich medium containing the organic carbon source is added to the culture incrementally over time to maintain optimal growth conditions. This allows for better control over nutrient levels and can result in higher biomass and lipid productivity compared to batch cultivation.
- **Continuous cultivation:** In continuous cultivation, a nutrient-rich medium containing the organic carbon source is added continuously to the culture in a bioreactor, while the biomass is harvested continuously to maintain optimal growth conditions. This allows for more efficient use of resources and can result in higher biomass and lipid productivity compared to batch or fed-batch cultivation.
- **Two-stage cultivation:** In two-stage cultivation, *Spirulina* is first grown under autotrophic conditions to produce a large biomass, which is then transferred to a heterotrophic medium containing the organic carbon source for further growth and lipid accumulation. This approach can result in higher biomass and lipid productivity compared to heterotrophic cultivation alone while still maintaining nutritional quality. The choice of heterotrophic cultivation method will depend on a variety of factors, including the desired biomass and lipid productivity, cost and availability of resources, and environmental impact. Each method has its own advantages and disadvantages, and careful consideration must be given to the specific requirements of each cultivation system to ensure optimal results.

Heterotrophic cultivation systems for *Spirulina* can range from simple and low-cost setups to more complex and advanced industrial bioreactors. Here are some examples of heterotrophic cultivation systems that can be used for *Spirulina*:

- **Higher production costs:** Heterotrophic cultivation can be more expensive than autotrophic cultivation, due to the need for a carbon source and the associated costs of energy, equipment, and nutrients.
- **Environmental impact:** Heterotrophic cultivation may have a negative environmental impact compared to autotrophic cultivation, due to the carbon source being derived from organic waste or fossil fuels, which can result in greenhouse gas emissions and other pollutants.
- **Complexity:** Heterotrophic cultivation can be more complex and require more specialized knowledge compared to autotrophic cultivation, due to the need to maintain optimal growth conditions and avoid contamination.
- **Nutritional quality:** While heterotrophic cultivation can result in higher protein and pigment content in the biomass, it can also result in lower nutritional quality compared to autotrophic cultivation, due to changes in the fatty acid composition and other nutritional parameters.

- **Susceptibility to contamination:** Heterotrophic cultivation can be more susceptible to contamination by bacteria, fungi, and other microorganisms, due to the use of organic carbon sources that can also support the growth of other microorganisms. This can lead to reduced biomass productivity and quality.
- **Ethical concerns:** Heterotrophic cultivation using organic waste or byproducts may raise ethical concerns regarding the use of food or agricultural resources for nonfood purposes and the potential competition with other uses of these resources.
- **Regulatory hurdles:** Heterotrophic cultivation using waste or byproducts may face regulatory hurdles related to waste management and food safety regulations, which can add complexity and costs to the cultivation process.

Heterotrophic cultivation has several benefits and drawbacks for *Spirulina* cultivation, and the choice between heterotrophic and autotrophic cultivation will depend on various factors such as the specific application, production costs, availability of resources, and environmental impact. The use of a mixed cultivation strategy that combines both autotrophic and heterotrophic growth may offer the benefits of both methods while minimizing their drawbacks, but further research is needed to optimize such strategies for *Spirulina* cultivation (FAO Fisheries 2008; Sow and Ranjan 2021).

7.5.2 Plastic Bags or Bottles

One of the simplest and most low-cost methods for heterotrophic cultivation of *Spirulina* is using plastic bags or bottles (Fig. 7.4). In this setup, the culture is inoculated into a nutrient-rich medium containing the organic carbon source, and the

Fig. 7.4 Cultivation of *Spirulina* in (**a**) bottles and (**b**) plastic bags (adapted from Delavar and Wang 2022 via permission license number 5597591230800)

bags or bottles are sealed and placed in a controlled environment with adequate aeration and mixing (Susanna et al. 2019). This method is relatively easy to set up and manage but may have limited scalability and may not be suitable for large-scale commercial production (Barry et al. 2015; Zhu et al. 2020; Thevarajah et al. 2022).

7.5.3 *Photobioreactors (PBRs)*

Photobioreactors (PBRs) are closed systems that allow for more precise control over growth conditions and can result in higher biomass and lipid productivity compared to open systems such as plastic bags or bottles. PBRs can be designed in various configurations such as tubular, flat-panel, or airlift and can be made from materials such as glass, plastic, or stainless steel. The design of the PBR will depend on factors such as the desired production capacity, cost, and availability of materials (Chen et al. 2011). Photobioreactors are utilized specifically for fermentation under natural or synthetic light (Fig. 7.5). Artificial illumination is costly due to its high energy consumption. Only outdoor photobioreactors capable of solar energy harvesting are chosen. For instance, photobioreactors are utilized to produce or convert certain essential substances, such as phytoplankton growth.

Photobioreactors may be constructed of glass or, more frequently, transparent plastic. Structured solar receivers are comprised of tubes or planar surfaces. Maintaining adequate sunlight penetration is necessary (Alfredo de Jesús and Brenda Paloma 2022). Photobioreactors typically operate continuously between 25 and 40 degrees Celsius. To circulate the culture in the solar receiver, a centrifugal compressor or airlift can be used. It is crucial for the cells to continue to circulate without aggregating. CO2 is introduced to the bioreactor in order to provide oxygen.

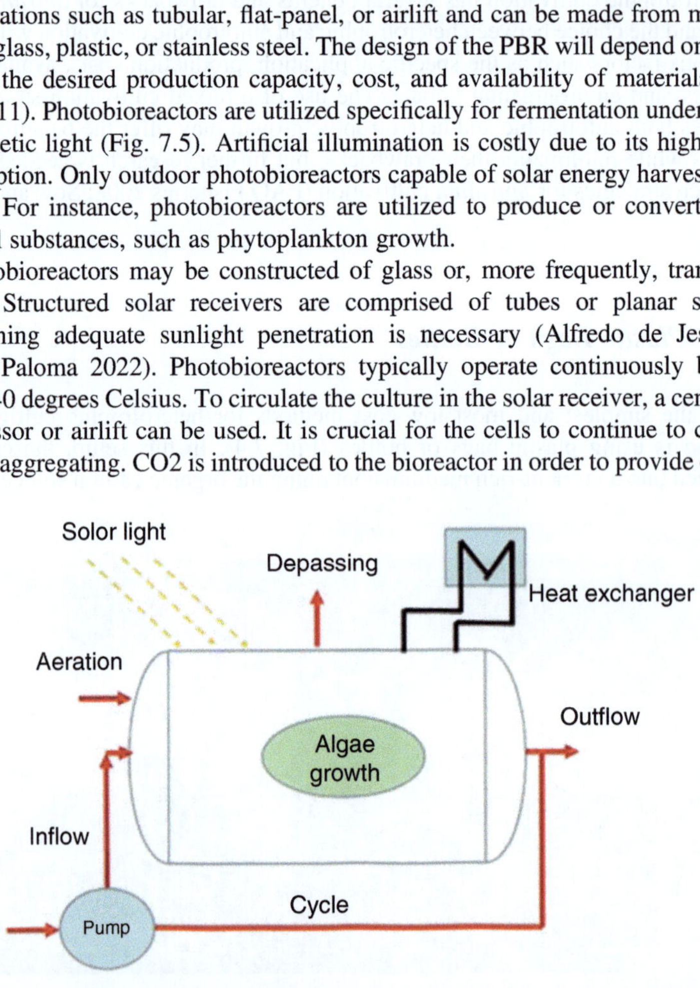

Fig. 7.5 Schematic of a photobioreactor for algae growth (adapted from Delavar and Wang (2022) via permission license number 5597591230800)

Daytime phytoplankton growth consumes carbon dioxide and releases oxygen (Benner et al. 2022). A heat exchanger may be utilized to chill the cylinder and prevent its temperature from increasing. Increasing the oxygen concentration in the bioreactor by accumulating oxygen can inhibit the proliferation of phytoplankton. Consequently, excess oxygen is extracted from the port. Microbes proliferate and produce products overnight. Photobioreactors are an appropriate technique for CO2 conversion and biofuel production (Deprá et al. 2019).

7.5.4 Closed Cultivation Systems

Open road ponds are simpler and less expensive than closed systems, also known as photobioreactors (PBRs). There are numerous designs available, including internal airlift PBR, flat-panel PBR, tubular PBR, and sealed floor PBR. Each plan has benefits and drawbacks. PBRs are typically made of glass or plastic, and the culture is agitated by pumping or airlifting. Because most microalgae are fragile and have a low shear stress tolerance, air agitation is preferred to pumped circulation. Not only does PBR require agitation to homogenize the culture medium but also to extract dissolved oxygen (degassing) (Haines et al. 2022). During photosynthesis, microalgae produce oxygen, which can accumulate in solution, particularly in PBR, where oxygen cannot freely escape into the atmosphere. Dissolved oxygen levels above the saturation limit of air can inhibit the growth of numerous microalgal species. If the objective is to produce a high-value composite, they may be preferable despite their substantially higher construction and operating costs. Less susceptible to contamination, closed systems can support monoculture production (Saeid and Chojnacka 2016). In addition to its exceptional gas release rate and high gas utilization efficiency, it also has a low CO2 loss and almost no water evaporation. Importantly, the high surface-to-volume ratio results in substantially higher biomass concentrations (2–5 g/L) than in open ponds, making harvesting and dewatering easier.

7.5.5 Stirred-Tank Bioreactors

Stirred-tank bioreactors are commonly used in industrial-scale heterotrophic cultivation of *Spirulina* (Fig. 7.6). In this setup, the culture is grown in a closed vessel with mechanical agitation to ensure adequate mixing and aeration (Benner et al. 2022). The bioreactor can be designed to optimize production parameters such as temperature, pH, and dissolved oxygen levels and can be scaled up or down depending on the desired production capacity.

Fig. 7.6 A closed-plate gas-lift reactor launched from one side (left) is combined with a distinct photobioreactor, such as an open thin-film cascade photobioreactor fired from above (right) if the mixing is sufficient (turbulence) (Benner et al. 2022). [Permission is not required as per Open Access policy]

7.5.6 Closed and Semiclosed Outdoor Photobioreactors

The closed photobioreactor is an adaptable system that can be optimized in accordance with the biological and physiological characteristics of the microalgal species of interest. Figure 7.7 depicts an overview of several closed or semiclosed outdoor microalgae bulk cultivation systems (Benner et al. 2022). The simplest cultivation systems are depicted in Figure 7.7, which consists of a transparent bag containing a microalgae suspension and a stream of CO_2-enriched air. In comparison to open systems, photobioreactors offer numerous benefits, such as reproducible culture conditions in terms of environmental impact, a decreased risk of contamination, minimal CO_2 losses, and a smaller footprint (Susanna et al. 2019). Closed systems are more difficult to clean, conduit materials can reduce sunlight penetration, and surplus oxygen produced by cultivation can inhibit plant growth, thereby diminishing the effectiveness of the system (Fig. 7.7). It should be chilled to 20 °C and degassed. In addition, construction costs are approximately $100 per square meter higher than those of open ponds, which are approximately $100 per square meter (Sow and Ranjan 2021).

7.5.7 Hybrid Systems

Hybrid cultivation systems (Fig. 7.8) that combine autotrophic and heterotrophic cultivation can also be used to improve the productivity and lipid content of

Fig. 7.7 Scalable examples of closed or semiclosed photobioreactors for microalgae cultivation in large-scale production facilities (Masojı'dek and Torzillo 2008). (**a**) Hanging plastic bags containing air pockets ($^{+}CO_2$) (Canakkale Onekismat University, Turkey, Faculty of Marine Sciences). (**b**) Horizontal tubular photobioreactor with bidirectional flow (CNR Ecosystem Institute, Florence, Italy). (**c**) A vertical tubular photobioreactor inserted in a greenhouse and created by IGV GmbH (Germany's Salata GmbH). (**d**) "Green wall panel" vertical plate photobioreactor (CNR Ecosystem Laboratory, Florence, Italy). (**e**) A 100 L annular column photobioreactor composed of two glass cylinders nestled within one another to form a culture chamber. The LED light source is installed in a cylinder that can be installed both indoors and outdoors to incorporate natural and artificial light (Laboratory of Microbiology, Trebon, Czech Republic). (**f**) Ecoduna GmbH's innovative flat-plate photobioreactor "Hanging Gardens" (Brunn a/L, Austria). It consists of 12 parallel panels (0.03 × 2 × 6 m) mounted on a movable structure that accommodates to the sun's movement. Internal baffles enable the culture medium to circulate, while air and CO_2 are injected from below to produce a gas-lifting effect. [Printed with permission license number 5597600578713]

Spirulina biomass. In these systems, *Spirulina* is first grown under autotrophic conditions to produce a large biomass, which is then transferred to a heterotrophic medium containing the organic carbon source for further growth and lipid

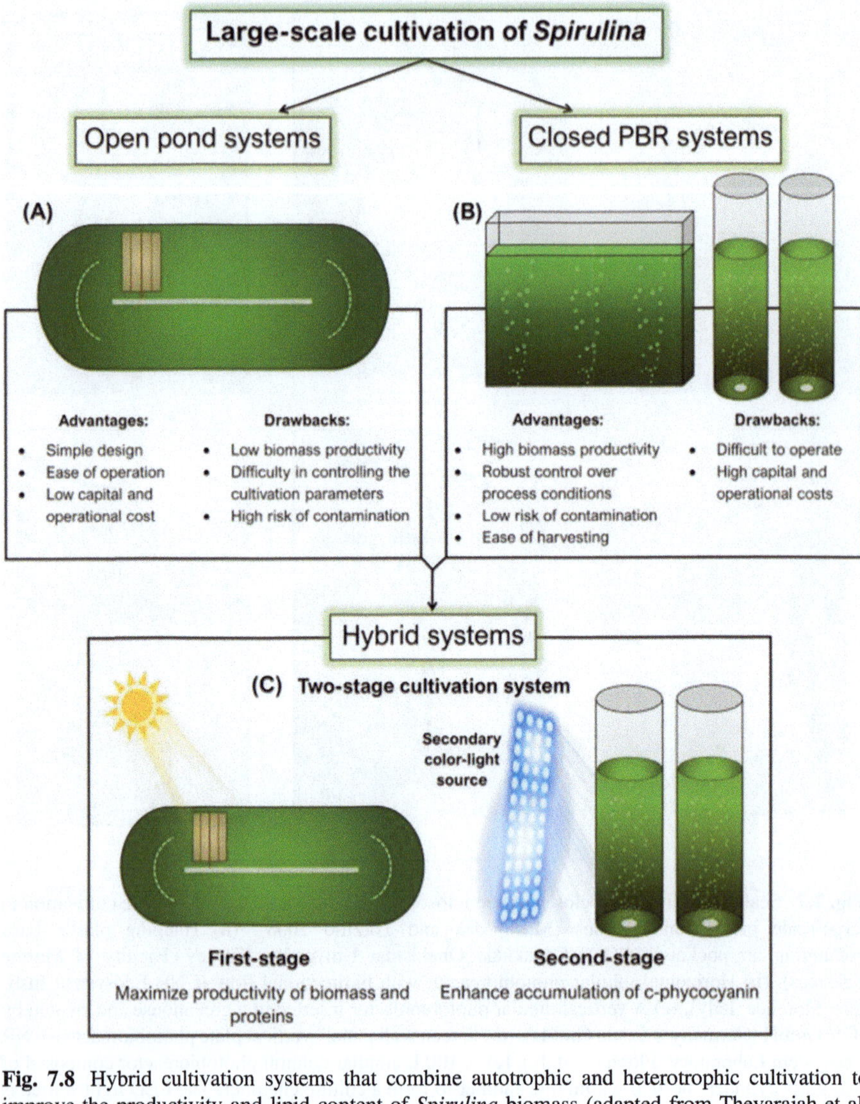

Fig. 7.8 Hybrid cultivation systems that combine autotrophic and heterotrophic cultivation to improve the productivity and lipid content of *Spirulina* biomass (adapted from Thevarajah et al. (2022) via permission license number 5597580015306)

accumulation. This approach can result in higher biomass and lipid productivity compared to heterotrophic cultivation alone while still maintaining nutritional quality (Deprá et al. 2019). Overall, the choice of a heterotrophic cultivation system will depend on various factors such as the desired production capacity, cost, availability of materials, and scalability. It is important to carefully consider these factors when selecting a cultivation system to ensure optimal results.

To combat disadvantages such as construction costs, operational costs, pollution, and CO2 loss, hybrid systems incorporating closed and open PBR are being developed. There are polybags, integrated transport systems, and external tubular loops in hybrid bioreactors. The thin-layer cascade reactor is a hybrid system with a modest culture depth and slope that facilitates recirculation of the culture. The thin-film cascade system is a novel design that must be optimized for scalability (Wicker et al. 2023a). For the synthesis of high-value metabolites from microalgae, two-step cultivation is an economically viable strategy. For instance, a hybrid railway pond incorporating aerial PBR with an open system for the cultivation of *Haematococcus pluvialis* was designed to maximize biomass productivity in the first stage and astaxanthin accumulation in the second stage. In order to increase protein and C-PC yields, employing this strategy for large-scale *Spirulina* cultivation may be beneficial. In addition, Garca-López et al. (Castillo et al. 2021) used low-cost Zarrouk blue and mid-blue LEDs in a second phase to elicit C-PC in *A. maxima* in a 2000 L pond following biomass enrichment with membrane technology. It was evaluated for two-stage outdoor cultivation. This experiment successfully induced C-PC with maximal biomass productivity spanning between 13.63 and 18.97 g/m^2/dL (Alishah Aratboni et al. 2019). Therefore, it is evident that exposure to a blue-emitting light source substantially increases C-PC accumulation. Adopting a two-step culture strategy for the large-scale cultivation of *Spirulina* for protein and C-PC production may be fruitful if additional research on C-PC induction using color diodes and other light sources is conducted.

7.6 Membrane Photobioreactors (MPBRs)

Membrane photobioreactors (MPBRs) are a type of bioreactor that combine the advantages of photobioreactors and membrane filtration technology to produce high-density cultures of microalgae such as *Spirulina* (Zhang et al. 2019). In MPBRs, the algae are grown in a transparent chamber (photobioreactor) equipped with membranes that separate the biomass from the culture medium and provide a surface area for the attachment and growth of the algae (Fig. 7.9). The main advantages of MPBRs for *Spirulina* cultivation include the following (Kaňa et al. 2009; Zhang et al. 2019; Tomazic et al. 2021; Zhou et al. 2023):

- **Higher biomass productivity:** MPBRs can achieve higher biomass productivity compared to conventional photobioreactors, due to the enhanced light exposure and efficient gas exchange.
- **Efficient nutrient utilization:** MPBRs allow for efficient nutrient utilization by the algae, due to the continuous filtration and recycling of the culture medium.
- **Reduced contamination risks:** MPBRs can reduce the risk of contamination by external microorganisms, due to the enclosed and controlled environment.

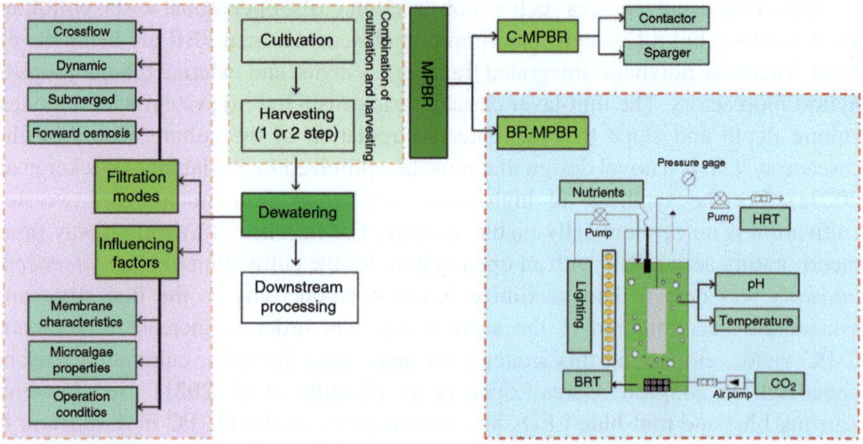

Fig. 7.9 Schematic illustration of algal cultivation through membrane photobioreactor (adapted from Zhang et al. (2019) via permission license number 5597600765124)

- **Enhanced product quality:** MPBRs can result in higher-quality biomass, due to the efficient removal of metabolites and other impurities by the membrane filtration.
- **Reduced water consumption:** MPBRs can reduce the water consumption in *Spirulina* cultivation, due to the continuous recycling of the culture medium.

However, there are also some limitations and challenges associated with MPBRs, such as the following:

- **High investment and maintenance costs:** MPBRs can be expensive to set up and maintain, due to the complex equipment and the need for high-quality membranes.
- **Limited scalability:** MPBRs may not be suitable for large-scale *Spirulina* cultivation, due to the limitations in the size of the membrane modules and the associated costs.
- **Membrane fouling:** MPBRs can experience membrane fouling, which can reduce filtration efficiency and biomass productivity. This requires regular cleaning and replacement of the membranes, which can add to the operational costs.
- **Light attenuation:** MPBRs may experience light attenuation due to the accumulation of biomass on the membrane surface, which can reduce light penetration and limit the growth of the algae.

Recent studies have explored various strategies to enhance the performance and sustainability of MPBRs for *Spirulina* cultivation (Tomazic et al. 2021; Glover et al. 2023):

- **Membrane material optimization:** Researchers have tested various types of membranes with different pore sizes, hydrophilicity, and surface properties to optimize the filtration efficiency and reduce membrane fouling. For example, hydrophilic membranes with smaller pore sizes have been shown to achieve higher biomass productivity and longer membrane lifespan compared to hydrophobic membranes.

- **Integrated wastewater treatment:** MPBRs can be integrated with wastewater treatment systems to utilize organic nutrients and reduce the environmental impact of wastewater discharge. Some studies have shown that *Spirulina* can effectively remove nitrogen, phosphorus, and other pollutants from wastewater while producing high-quality biomass.
- **CO_2 supply optimization:** MPBRs can be supplied with CO_2 from various sources, such as flue gases from power plants or industrial processes. Researchers have explored the use of membrane contactors to improve the efficiency of CO_2 transfer and reduce the energy consumption of the CO_2 supply.
- **Light intensity and spectrum optimization:** MPBRs can be optimized for the light intensity and spectrum to enhance the photosynthetic efficiency and biomass productivity of *Spirulina*. Some studies have shown that blue light can enhance the growth and lipid content of *Spirulina*, while red light can improve pigment content and antioxidant activity.
- **Process modeling and control:** MPBRs can be optimized using process modeling and control techniques, such as computational fluid dynamics and artificial intelligence. These methods can provide insights into the hydrodynamics, mass transfer, and biomass growth in MPBRs and help to optimize the operational parameters and improve the process efficiency (Zhang et al. 2019).

Light, nutrients, and a carbon source are the three fundamental requirements for microalgal growth. To maximize biomass production, substantial quantities of each of these elements must be present to meet the culture's growth requirements. The most common method of CO_2 delivery in conventional microalgae cultivation systems is direct air sparging. Due to the low transfer efficiency, only a small quantity of dissolved CO_2 is produced, while the remaining CO_2 is discharged into the atmosphere (Glover et al. 2023). By flooding the bioreactor with CO_2-rich air or flue gas, the yield of microalgae can be increased, but the carbonation efficiency is still very low. Using a membrane as a contactor or sparger to aid in the release of dissolved CO_2 is an efficient technique. The PBR membrane contactor employs a hydrophobic membrane as a substrate to enhance the microalgae cultures' CO_2 contact surface. Since the membrane contactor is operated by a concentration gradient between the gas and liquid phases, the membrane applied permits CO_2 to be delivered to the culture medium at low gas pressures (Alfredo de Jesús and Brenda Paloma 2022).

According to the summary of preliminary research, membrane contactor systems are substantially superior to transport systems for CO_2 separation. In general, PBR membrane contactors favor highly porous and hydrophobic membranes. This is due to the fact that more CO_2 gas can diffuse into the liquid phase, while the membrane perforations prevent water from traveling through (Sow and Ranjan 2021). Due to their large specific surface area, hollow fiber membranes are commonly used as gas exchangers to enhance the growth rate of microalgae in contact membrane PBR systems. Recent research has centered on maximizing CO_2 absorption efficiency and overcoming drawbacks like membrane moisture and contamination. We compared the effectiveness of hydrophilic and hydrophobic hollow fiber membrane photobioreactors in batch and semicontinuous modes against the growth of *Synechococcus elongatus*. The outcomes demonstrated that the hydrophobic membranes are performed admirably (Chowdury et al. 2020; Sow and Ranjan 2021; Alfredo de Jesús and Brenda Paloma 2022).

To overcome some of the limitations of traditional MPBRs, researchers have been exploring innovative designs and strategies. Some of the recent advancements in MPBR technology for *Spirulina* cultivation include the following:

- Photovoltaic-powered MPBRs: Photovoltaic (PV) panels can be used to generate electricity to power the MPBR system, reducing the reliance on external power sources (Zhang et al. 2019). This improves sustainability and reduces the operating costs of the system.
- Integrated CO_2 capture and delivery systems: CO_2 capture and delivery systems can be integrated with the MPBR, allowing for a steady supply of CO_2 for photosynthesis. This can be achieved using carbon capture and utilization technologies or by integrating the MPBR with industrial processes that generate CO_2 emissions.

Researchers have been developing new membrane materials and designs that offer improved filtration efficiency, reduced fouling, and enhanced biomass productivity. For example, membrane materials with improved hydrophilicity or nanoporous membranes have been shown to enhance CO_2 diffusion and nutrient uptake. Hybrid cultivation systems that combine the advantages of heterotrophic and phototrophic cultivation have been developed. In these systems, *Spirulina* is initially grown in heterotrophic conditions to achieve high biomass densities, before being transferred to a phototrophic environment for further growth and pigment synthesis. This approach allows for higher biomass productivity and pigment content while reducing the risk of contamination (Jung et al. 2015).

Some of the benefits of membrane photobioreactors (MPBRs) for *Spirulina* cultivation include the following:

- Higher biomass productivity: MPBRs offer a controlled environment that can enhance the growth of *Spirulina* and increase biomass productivity compared to open pond systems. The enclosed system of MPBRs reduces the risk of contamination from external sources such as insects, dust, and other microorganisms.
- Better quality control: MPBRs offer better quality control and monitoring of cultivation parameters such as pH, temperature, and nutrient levels, allowing for more precise and consistent production of *Spirulina*. MPBRs can be designed to use renewable energy sources such as solar power, reducing the carbon footprint of cultivation and making it more sustainable (Kaňa et al. 2009).

However, there are also some drawbacks associated with MPBRs for *Spirulina* cultivation, including the following (Kaňa et al. 2009; Zhang et al. 2019; Zhou et al. 2023):

- High capital costs: The initial investment required to set up an MPBR system can be significantly higher compared to open pond systems, which can make it less accessible to small-scale growers.
- Energy consumption: Although MPBRs can be designed to use renewable energy sources, they still require energy to operate the pumps, lights, and other equipment, which can be a significant operating cost.
- Membrane fouling: Membrane fouling, where the membranes become clogged with organic matter and other impurities, can reduce the efficiency of the MPBR and require frequent cleaning.
- Limited scale-up: The scale-up of MPBR systems can be challenging due to the technical complexity and high costs involved, making it difficult to implement large-scale production.

Overall, while MPBRs offer several advantages over open pond systems for *Spirulina* cultivation, their adoption is still limited by the high capital costs and technical complexity involved. However, ongoing research and development are focused on addressing these challenges and making MPBRs a more viable option for sustainable and high-density cultivation of Spirulina.

7.7 Phototaxis-Based Cultivation

Phototaxis-based cultivation is a novel approach for growing *Spirulina* that involves using light gradients to direct the movement of the cells toward an optimal growth zone. In this approach, *Spirulina* cells are exposed to a light source that is directed toward a certain area of the cultivation vessel, creating a light gradient that encourages the cells to move toward the desired location (Sim et al. 2019; Zhang et al. 2023). One potential avenue for advancing phototaxis-based cultivation is the development of more efficient and cost-effective light sources and sensors that can precisely control light intensity and direction. Additionally, the use of advanced imaging and data analytic tools can help growers monitor and optimize the growth and movement of *Spirulina* cells in real time, further increasing productivity and quality. Another area of research is the exploration of different light wavelengths and their effects on *Spirulina* growth and physiology. Recent studies have shown that blue light can enhance *Spirulina* biomass productivity, while red light can increase the production of specific pigments and antioxidants. By fine-tuning the light spectrum, growers may be able to further improve the nutritional and functional properties of *Spirulina* biomass (Zhang et al. 2023).

Phototaxis-based cultivation systems for *Spirulina* typically involve the use of enclosed photobioreactors or flat-panel reactors with controlled lighting conditions. These systems are designed to create a light gradient that directs the movement of

Spirulina cells toward an optimal growth zone, where they can be more efficiently harvested and processed. One example of a phototaxis-based cultivation system is the "concentration-controlled phototaxis reactor" which uses a light gradient to direct the movement of *Spirulina* cells toward a high-concentration zone, where they can be more effectively harvested (Sim et al. 2019). This system consists of a vertically oriented photobioreactor with a light source at the bottom, which creates a gradient of light intensity from the bottom to the top. *Spirulina* cells are introduced at the top of the reactor and are drawn toward the light source by phototaxis, ultimately accumulating at the bottom of the reactor where they can be harvested. Another example of a phototaxis-based cultivation system is the "light-guide panel reactor" which uses a flat-panel made of transparent material to guide the movement of *Spirulina* cells toward a concentrated growth zone (Zhang et al. 2023). This system consists of a flat-panel reactor with a light source at the edge of the panel, which creates a gradient of light intensity across the panel. *Spirulina* cells are introduced at one end of the panel and are drawn toward the light source by phototaxis, ultimately accumulating in a concentrated zone at the opposite end of the panel.

Both of these systems have been shown to improve biomass productivity and reduce energy consumption compared to traditional open pond systems while also reducing the risk of contamination and providing more precise control over cultivation parameters. However, they can also be more complex and expensive to implement and may require specialized expertise to operate and maintain. Despite the potential benefits of phototaxis-based cultivation systems, there are also some challenges and limitations to consider. One key issue is the need to maintain consistent light gradients over time, which can be difficult to achieve with some types of lighting sources and reactor designs. This can lead to reduced efficiency and biomass productivity as well as increased operational costs (Davis et al. 2016; Zhu et al. 2020; Gao et al. 2022).

Another challenge is the potential for photoinhibition, which occurs when *Spirulina* cells are exposed to high-intensity light for extended periods of time. This can result in reduced photosynthetic efficiency and biomass productivity as well as cellular damage and reduced nutritional quality. To mitigate this risk, growers may need to carefully balance light intensity, duration, and wavelength as well as incorporate shading or light-diffusing materials into the reactor design. Finally, it is worth noting that phototaxis-based cultivation systems may not be suitable for all production scenarios, particularly those that require large-scale or outdoor cultivation. In these cases, traditional open pond systems may be more cost-effective and practical, despite their lower efficiency and greater susceptibility to contamination and environmental fluctuations (Susanna et al. 2019).

Phototaxis-based cultivation systems for *Spirulina* typically involve the use of photobioreactors that are designed to create and maintain controlled light gradients within the culture medium. These gradients can be created using a variety of approaches, including the use of light-emitting diodes (LEDs), optical fibers, or other light-diffusing materials. One common approach involves the use of vertical column photobioreactors, which consist of a tall cylindrical vessel filled with a *Spirulina* culture medium (AlFadhly et al. 2022). Light sources are typically mounted at the top and bottom of the reactor, creating a light gradient that causes the *Spirulina* cells to migrate up or down the column in response to changes in light intensity. This movement helps to ensure that all cells receive optimal exposure to light, maximizing photosynthetic efficiency and biomass productivity (Soni et al. 2017). Another approach involves the use of horizontal photobioreactors, which utilize a flat or shallow culture vessel to create a horizontal light gradient. In these systems, light sources are typically mounted at one or both ends of the vessel, creating a gradient that causes the *Spirulina* cells to move along the surface of the culture medium (Susanna et al. 2019; Aziz et al. 2020; Portillo et al. 2021; Kumar et al. 2022).

Benefits and drawbacks of phototaxis-based cultivation for *Spirulina* include the following (Zhang et al. 2023):

- **Higher productivity:** Phototaxis-based cultivation systems can lead to higher productivity compared to other cultivation methods due to their ability to optimize light gradients and other environmental factors.
- **Improved nutrient uptake:** By promoting cell movement and gas exchange, phototaxis-based systems can help to ensure optimal nutrient uptake, leading to improved growth rates and biomass quality.
- **Reduced energy consumption:** By utilizing controlled light gradients, phototaxis-based systems can help to reduce the amount of energy required to produce *Spirulina* biomass, potentially leading to lower operating costs and improved sustainability.
- **Enhanced product quality:** Phototaxis-based systems can help to improve the nutritional quality and purity of *Spirulina* biomass, leading to increased marketability and value.
- **Complexity and cost:** Phototaxis-based cultivation systems can be more complex and expensive to implement than other cultivation methods, requiring specialized equipment and expertise.
- **Maintenance requirements:** Phototaxis-based systems may require more frequent cleaning and maintenance compared to other systems to prevent biofouling and other issues.
- **Scale-up challenges:** Scaling up phototaxis-based systems to commercial production levels can be challenging due to the complexity of the technology and the need for precise control over environmental variables.
- **Limited flexibility:** Phototaxis-based systems may be less flexible than other cultivation methods, as they require specific light and nutrient gradients to be maintained to achieve optimal results.

7.8 Cocultivation for Spirulina

Cocultivation refers to the cultivation of multiple species of microorganisms together in a single system. In the context of *Spirulina* cultivation, cocultivation typically involves the addition of other microorganisms, such as bacteria or other microalgae, to the *Spirulina* culture (Koller et al. 2012; Chu 2017). There are several potential benefits to the cocultivation of Spirulina, including the following:

- **Improved nutrient uptake:** Cocultivation with other microorganisms can help to increase nutrient uptake and utilization by *Spirulina*, leading to improved growth rates and biomass quality.
- **Reduced contamination**: By promoting the growth of beneficial microorganisms, cocultivation can help to reduce the risk of contamination by harmful pathogens or other unwanted organisms.
- **Enhanced product quality:** Cocultivation can help to improve the nutritional and functional properties of *Spirulina* biomass, leading to increased marketability and value.
- **Compatibility issues:** Not all microorganisms are compatible with *Spirulina*, and some may even be harmful to its growth and quality.
- **Control over species ratios:** Cocultivation can be challenging to control and optimize, as different microorganisms may have different growth rates and nutrient requirements.
- **Risk of contamination:** Cocultivation can increase the risk of contamination by harmful pathogens or other unwanted organisms.

Another potential benefit of the cocultivation of *Spirulina* is the potential for bioremediation. By coculturing *Spirulina* with other microorganisms, it may be possible to remove or mitigate the effects of pollutants or other environmental contaminants. For example, some bacteria have been shown to effectively degrade hydrocarbons and other pollutants, which could be used in conjunction with *Spirulina* cultivation to remediate contaminated water or soil. There are also several specific cocultivation strategies that have been developed for *Spirulina*, including cocultivation with other microalgae: Several studies have explored the use of cocultivation of *Spirulina* with other microalgae species, such as *Chlorella* or *Scenedesmus* (Sim et al. 2019; Forghani et al. 2022). This approach can help to improve overall productivity and nutrient utilization, as different species may have complementary nutrient requirements (Koller et al. 2012; Chu 2017; Saral et al. 2022).

There are several cocultivation systems that have been developed for *Spirulina*, including sequential batch cultivation: In this system, *Spirulina* has cultured alone for a period of time, and then other microorganisms (such as bacteria or other algae) are introduced for a subsequent cultivation period. This approach allows for targeted control over the composition of the culture and can be used to optimize nutrient utilization and productivity (Zhu et al. 2020; Abdelfattah et al. 2023):

- **Simultaneous cultivation:** In this system, *Spirulina* is grown together with other microorganisms from the beginning of the cultivation period. This approach can help to improve overall productivity and biomass quality by promoting nutrient cycling and reducing nutrient waste (Aziz et al. 2020).
- **Consortia cultivation:** In this system, a diverse community of microorganisms are cultivated together with *Spirulina* in a complex, dynamic ecosystem. This approach aims to replicate natural microbial communities and can help to improve nutrient cycling and overall ecosystem function (Wicker et al. 2023b).
- **Integrated cultivation:** In this system, *Spirulina* is cultivated in conjunction with other agricultural or aquaculture systems. For example, *Spirulina* cultivation can be integrated with fish farming, where the waste products from the fish can be used as a nutrient source for the *Spirulina*. Each of these cocultivation systems has its own unique advantages and challenges, and the optimal approach will depend on the specific application and production goals. However, cocultivation represents a promising strategy for improving *Spirulina* cultivation and expanding its potential applications (Okeke et al. 2022; Padi et al. 2023).

Consortia cultivation is a relatively new approach to *Spirulina* cultivation that involves the cocultivation of multiple microorganisms in a single system. This approach is based on the idea that different microorganisms can work together synergistically to improve the overall efficiency and productivity of the cultivation process (Markou 2020). Consortia cultivation can involve the cocultivation of multiple strains of *Spirulina* as well as other microorganisms such as bacteria, fungi, and algae. The goal is to create a balanced ecosystem that promotes the growth of each individual organism while minimizing competition and other negative interactions. Some studies have shown that consortia cultivation can lead to higher biomass yields compared to monoculture cultivation. For example, a study by Wang et al. (2023) reported that cocultivation of *Spirulina* with the microalga *Chlorella vulgaris* led to a 30% increase in biomass yield compared to monoculture cultivation. Another study by Kim et al. (Soni et al. 2017) reported that the cocultivation of *Spirulina* with the bacterium *Rhizobium radiobacter* led to a 50% increase in biomass yield compared to monoculture cultivation.

The choice of microorganisms for consortia cultivation depends on several factors, including their compatibility, nutritional requirements, growth rates, and metabolic capabilities. In some cases, the microorganisms may have complementary nutritional requirements, such that one microorganism produces a nutrient that another requires for growth. In other cases, the microorganisms may have synergistic metabolic pathways, such that one microorganism produces a metabolite that another can use as a substrate for growth (Chen et al. 2011; Susanna et al. 2019). One of the main advantages of consortia cultivation is the potential for improved nutrient utilization. Different microorganisms have different abilities to utilize different nutrients, and by cocultivating multiple microorganisms, it may be possible to maximize the utilization of available nutrients in the cultivation system. This can help reduce the need for external nutrient inputs and increase the overall efficiency of the cultivation process.

By cocultivating multiple microorganisms, it may be possible to create a more resilient ecosystem that is better able to cope with fluctuations in environmental conditions, such as temperature, light intensity, pH, and nutrient availability. This can help reduce the risk of crop failure and increase the overall stability and productivity of the cultivation system (Pereira et al. 2019). Despite these potential advantages, consortia cultivation also presents several challenges. One of the main challenges is the need to optimize the composition and balance of the microbial community. Different microorganisms have different growth rates and metabolic capabilities, and if the composition of the microbial community is not well balanced, it can lead to competition and other negative interactions between microorganisms. This can reduce the overall efficiency and productivity of the cultivation system. Another challenge of consortia cultivation is the risk of contamination and other disruptions. Cocultivating multiple microorganisms increases the complexity of the cultivation system and makes it more vulnerable to contamination by unwanted microorganisms (AlFadhly et al. 2022).

Integrated cultivation is a strategy that combines two or more different cultivation methods to create a more efficient and sustainable system. In the context of *Spirulina* biomass production, integrated cultivation can involve combining different cultivation methods, such as phototrophic and heterotrophic cultivation, to achieve optimal biomass yields (Benner et al. 2022). One example of integrated cultivation for *Spirulina* is the use of wastewater as a nutrient source for phototrophic cultivation. The wastewater can provide a rich source of nutrients, while phototrophic cultivation can help to remove pollutants from the water (Chen et al. 2011). Additionally, the carbon dioxide produced during heterotrophic cultivation can be used as a carbon source for phototrophic cultivation, further improving the efficiency of the system. Other examples of integrated cultivation for *Spirulina* include the use of photobioreactors in conjunction with open ponds or the use of cocultivation in combination with heterotrophic or phototrophic cultivation. The benefits of integrated cultivation for *Spirulina* include improved resource utilization, improved system stability, enhanced biomass yields, and better environmental impact.

Integrated cultivation is a promising approach for *Spirulina* biomass production that can help to improve the sustainability and efficiency of cultivation methods. However, careful management and optimization of the system are required to ensure optimal yields and minimize potential drawbacks (Susanna et al. 2019; Aziz et al. 2020; Saxena et al. 2022; Ugya and Meguellati 2022). Integrated cultivation is a strategy that combines two or more different cultivation methods to create a more efficient and sustainable system. In the context of *Spirulina* biomass production, integrated cultivation can involve combining different cultivation methods, such as phototrophic and heterotrophic cultivation, to achieve optimal biomass yields. One example of integrated cultivation for *Spirulina* is the use of wastewater as a nutrient source for phototrophic cultivation. The wastewater can provide a rich source of nutrients, while phototrophic cultivation can help to remove pollutants from the water. Additionally, the carbon dioxide produced during heterotrophic cultivation can be used as a carbon source for phototrophic cultivation, further improving the efficiency of the system (Heinsoo 2014). Other examples of integrated cultivation for

Spirulina include the use of photobioreactors in conjunction with open ponds or the use of cocultivation in combination with heterotrophic or phototrophic cultivation. The benefits of integrated cultivation for *Spirulina* include the following:

- Improved resource utilization: By combining different cultivation methods, integrated cultivation can help to maximize the use of resources, reducing waste and improving the efficiency of the system.
- Improved system stability: By combining different cultivation methods, integrated cultivation can create a more stable system that is better able to withstand environmental fluctuations.

7.9 Conclusion and Prospects

Nonconventional and novel strategies for *Spirulina* biomass production include various cultivation systems such as membrane photobioreactors, phototaxis-based systems, cocultivation, sequential and simultaneous batch cultivation, consortia cultivation, integrated cultivation, and the use of 3D printing and vertical photobioreactors. Membrane photobioreactors use membranes to retain the cells inside the bioreactor and increase the cell density and productivity. Phototaxis-based systems utilize the natural movement of the cells toward light to increase productivity. Cocultivation involves the simultaneous cultivation of *Spirulina* with other microorganisms, while sequential and simultaneous batch cultivation use different batch protocols to optimize biomass production. Consortia cultivation utilizes interactions between multiple microorganisms to enhance productivity. Integrated cultivation combines different cultivation systems for optimal results.

Mixotrophic cultivation involves growing *Spirulina* in the presence of organic carbon sources such as glucose or acetate. This can enhance the growth rate and biomass productivity of *Spirulina*, making it a more efficient cultivation method. In heterotrophic cultivation method, *Spirulina* is grown in the absence of light and carbon dioxide, using organic carbon sources such as glucose or glycerol. This can lead to higher biomass yields and faster growth rates compared to conventional cultivation methods. Membrane photobioreactor cultivation system uses a semipermeable membrane to separate the culture medium from the biomass, allowing for continuous harvesting of the *Spirulina* biomass. This method has been shown to increase biomass productivity and reduce contamination compared to open pond systems.

Phototaxis-based cultivation method involves manipulating the movement of *Spirulina* cells using light gradients to concentrate the cells in a particular area, making it easier to harvest the biomass. This technique has the potential to increase biomass yields and reduce energy consumption compared to conventional cultivation methods. Cocultivation involves growing *Spirulina* with other microorganisms, such as bacteria or yeast, to enhance growth and increase biomass yields. This method has been shown to improve the nutritional value of *Spirulina* and reduce contamination. Overall, these nonconventional and novel strategies offer promising alternatives to conventional *Spirulina* cultivation methods, with the potential to increase biomass yields, improve nutritional quality, and reduce contamination.

References

Abdelfattah A, Ali SS, Ramadan H et al (2023) Microalgae-based wastewater treatment: mechanisms, challenges, recent advances, and future prospects. Environ Sci Ecotechnol 13. https://doi.org/10.1016/J.ESE.2022.100205

AlFadhly NKZ, Alhelfi N, Altemimi AB et al (2022) Tendencies affecting the growth and cultivation of genus spirulina: an investigative review on current trends. Plan Theory 11:1–21. https://doi.org/10.3390/plants11223063

Alfredo de Jesús MR, Brenda Paloma GL (2022) Growth of Spirulina maxima in an optimized concentric internal tube photobioreactor. Iran J Biotechnol 20:44–53. https://doi.org/10.30498/ijb.2022.273070.2985

Alishah Aratboni H, Rafiei N, Garcia-Granados R et al (2019) Biomass and lipid induction strategies in microalgae for biofuel production and other applications. Microb Cell Factories 18:1–17. https://doi.org/10.1186/s12934-019-1228-4

Aziz MMA, Kassim KA, Shokravi Z et al (2020) Two-stage cultivation strategy for simultaneous increases in growth rate and lipid content of microalgae: a review. Renew Sust Energ Rev 119: 109621. https://doi.org/10.1016/j.rser.2019.109621

Babu SS, Gondi R, Vincent GS et al (2022) Microalgae biomass and lipids as feedstock for biofuels: sustainable biotechnology strategies. Sustainability 14:1–31. https://doi.org/10.3390/su142215070

Barry AN, Starkenburg SR, Sayre RT (2015) Strategies for optimizing algal biology for enhanced biomass production. Front Energy Res 3:1–5. https://doi.org/10.3389/fenrg.2015.00001

Benner P, Meier L, Pfeffer A et al (2022) Lab-scale photobioreactor systems: principles, applications, and scalability. Bioprocess Biosyst Eng 45:791–813. https://doi.org/10.1007/s00449-022-02711-1

Bortolini DG, Maciel GM, de Fernandes IAA et al (2022) Functional properties of bioactive compounds from Spirulina spp.: current status and future trends. Food Chem Mol Sci 5. https://doi.org/10.1016/J.FOCHMS.2022.100134

Castillo T, Ramos D, García-Beltrán T et al (2021) Mixotrophic cultivation of microalgae: an alternative to produce high-value metabolites. Biochem Eng J 176:108183. https://doi.org/10.1016/j.bej.2021.108183

Chen C, Yeh K, Aisyah R et al (2011) Cultivation, photobioreactor design and harvesting of microalgae for biodiesel production: a critical review. Bioresour Technol 102:71–81. https://doi.org/10.1016/j.biortech.2010.06.159

Chowdury KH, Nahar N, Deb UK (2020) The growth factors involved in microalgae cultivation for biofuel production: a review. Comput Water Energy Environ Eng 09:185–215. https://doi.org/10.4236/cweee.2020.94012

Chu WL (2017) Strategies to enhance production of microalgal biomass and lipids for biofuel feedstock. Eur J Phycol 52:419–437. https://doi.org/10.1080/09670262.2017.1379100

Crivellaro Gonçalves G, Torres Nazari M, Gerhardt Magro FG et al (2022) Effects of culture medium recycling in the chemical composition of Spirulina platensis biomass cultivated in semi-continuous mode. Ciência e Nat 44:e52. https://doi.org/10.5902/2179460x69474

Davis R, Markham J, Kinchin C et al (2016) Process design and economics for the production of algal biomass: algal biomass production in open pond systems and processing through dewatering for downstream conversion, vol 128. National Renewable Energy Laboratory de Oliveira Moraes I, de Oliveira Moraes Arruda R, Maresca NR et al (2013) Spirulina platensis: process optimization to obtain biomass. Food Sci Technol 33:179–183. https://doi.org/10.1590/S0101-20612013000500026

Delavar MA, Wang J (2022) Advanced methods and mathematical modeling of biofilms: applications in health care, medicine, food, aquaculture, environment, and industry. Elsevier, pp 1–251. https://doi.org/10.1016/C2020-0-03059-8

Deprá MC, Mérida LGR, De Menezes CR et al (2019) Chemical engineering research and design a new hybrid photobioreactor design for microalgae culture. Chem Eng Res Des 144:1–10. https://doi.org/10.1016/j.cherd.2019.01.023

Dineshkumar R, Narendran R, Sampathkumar P (2016) Cultivation of Spirulina platensis in different selective media. Indian J Geo-Marine Sci 45:1749–1754

Duran Quintero C, Ventura A, Lépine O, Pruvost J (2021) Eco-design of spirulina solar cultivation: key aspects to reduce environmental impacts using life cycle assessment. J Clean Prod 299: 126741. https://doi.org/10.1016/j.jclepro.2021.126741

FAO Fisheries and Aquaculture Circular No. 1034 (2008) A review on culture, production and use of spirulina as food for humans and feeds for domestic animals and fish

Forghani B, Mayers JJ, Albers E, Undeland I (2022) Cultivation of microalgae - Chlorella sorokiniana and Auxenochlorella protothecoides - in shrimp boiling water residues. Algal Res 65:102753. https://doi.org/10.1016/J.ALGAL.2022.102753

Gao S, Edmundson S, Huesemann M (2022) Oxygen stress mitigation for microalgal biomass productivity improvement in outdoor raceway ponds. Algal Res 68:102901. https://doi.org/10.1016/J.ALGAL.2022.102901

Glover CJ, McDonnell A, Rollins KS et al (2023) Assessing the environmental impact of resource recovery from dairy manure. J Environ Manag 330:117150. https://doi.org/10.1016/J.JENVMAN.2022.117150

Grahl S (2019) Food product development with spirulina (Arthrospira platensis) – sensory profiling, product perception and consumer acceptance, 150

Haines M, Vadlamani A, Daniel Loty Richardson W, Strous M (2022) Pilot-scale outdoor trial of a cyanobacterial consortium at pH 11 in a photobioreactor at high latitude. Bioresour Technol 354:127173. https://doi.org/10.1016/J.BIORTECH.2022.127173

Hamidi M, Mohammadi A, Mashhadi H, Mahmoudnia F (2023) Evaluation of effective environmental parameters on lipid, protein and beta-carotene production in Spirulina platensis microalga. Results Eng 18:101102. https://doi.org/10.1016/J.RINENG.2023.101102

Harini AB, Rajkumar R (2022) Development of sustainable bioproducts from microalgae biomass: current status and future perspectives. Bioresources 17:7285–7312. https://doi.org/10.15376/biores.17.4.Harini

Heinsoo D (2014) Cultivation of Spirulina on conventional and urine based medium in a household scale system, 51

Jung JY, Yang JW, Kim K et al (2015) Cost-efficient cultivation of Spirulina platensis by chemical absorption of CO2 into medium containing NaOH. Korean J Chem Eng 32:2285–2289. https://doi.org/10.1007/s11814-015-0073-3

Kaňa R, Prášil O, Mullineaux CW (2009) Immobility of phycobilins in the thylakoid lumen of a cryptophyte suggests that protein diffusion in the lumen is very restricted. FEBS Lett 583:670–674. https://doi.org/10.1016/j.febslet.2009.01.016

Koller M, Salerno A, Tuffner P et al (2012) Characteristics and potential of micro algal cultivation strategies: a review. J Clean Prod 37:377–388. https://doi.org/10.1016/j.jclepro.2012.07.044

Kumar A, Ramamoorthy D, Verma DK et al (2022) Antioxidant and phytonutrient activities of Spirulina platensis. Energy Nexus 6:100070. https://doi.org/10.1016/J.NEXUS.2022.100070

Lafarga T, Sánchez-Zurano A, Villaró S et al (2021) Industrial production of spirulina as a protein source for bioactive peptide generation. Trends Food Sci Technol 116:176–185. https://doi.org/10.1016/j.tifs.2021.07.018

Lucas BF, Alberto Vieira Costa J, Brunner TA (2023) Attitudes of consumers toward spirulina and açaí and their use as a food ingredient. LWT 178. https://doi.org/10.1016/J.LWT.2023.114600

Markou G (2020) Overview of microalgal cultivation, biomass processing and application. Elsevier

Marting Vidaurre NA, Jurišić V, Bieling C et al (2023) Social assessment of miscanthus cultivation in Croatia: assessing farmers' preferences and willingness to cultivate the crop. GCB Bioenergy 2023:916–931. https://doi.org/10.1111/gcbb.13071

Masojı'dek J, Torzillo G (2008) Mass cultivation of freshwater microalgae. Ecol Eng 1:2226–2235

Molina Grima E, Belarbi EH, Acién Fernández FG et al (2003) Recovery of microalgal biomass and metabolites: process options and economics. Biotechnol Adv 20:491–515. https://doi.org/10.1016/S0734-9750(02)00050-2

Mousavian Z, Safavi M, Salehirad A et al (2023) Improving biomass and carbohydrate production of microalgae in the rotating cultivation system on natural carriers. AMB Express 13:39. https://doi.org/10.1186/s13568-023-01548-5

Musio B, Mourad Hussein Ahmed EMF, Antonicelli M et al (2022) A spectroscopic study to assess heavy metals absorption by a combined hemp/spirulina system from contaminated soil. Environ Adv 7:100144. https://doi.org/10.1016/J.ENVADV.2021.100144

Okeke ES, Ejeromedoghene O, Okoye CO et al (2022) Microalgae biorefinery: an integrated route for the sustainable production of high-value-added products. Energy Convers Manag X 16: 100323. https://doi.org/10.1016/J.ECMX.2022.100323

Overkamp KE, Gasper R, Kock K et al (2014) Insights into the biosynthesis and assembly of cryptophycean phycobiliproteins. J Biol Chem 289:26691–26707. https://doi.org/10.1074/jbc.M114.591131

Padi RK, Braud L, Kevin M, Murphy F (2023) Prospects for commercial microalgal biorefineries: integrated pilot demonstrations and process simulations based techno-economic assessment of single and multi-product value chains. Algal Res 103190. https://doi.org/10.1016/J.ALGAL.2023.103190

Pereira MIB, Chagas BME, Sassi R et al (2019) Mixotrophic cultivation of Spirulina platensis in dairy wastewater: effects on the production of biomass, biochemical composition and antioxidant capacity. PLoS One 14:1–17. https://doi.org/10.1371/journal.pone.0224294

Portillo E, Venuleo M, Janeiro C (2021) Long-term cultivation of a native Arthrospira platensis technical evidence for a viable production of. Processes 9:1–27

Rakhimberdieva MG, Vavilin DV, Vermaas WFJ et al (2007) Phycobilin/chlorophyll excitation equilibration upon carotenoid-induced non-photochemical fluorescence quenching in phycobilisomes of the cyanobacterium Synechocystis sp. PCC 6803. Biochim Biophys Acta Bioenerg 1767:757–765. https://doi.org/10.1016/j.bbabio.2006.12.007

Rosenau S, Wolgast T, Altmann B, Risius A (2023) Consumer preference for altered color of rainbow trout (Oncorhynchus mykiss) fillet induced by spirulina (Arthrospira platensis). Aquaculture 572:739522. https://doi.org/10.1016/J.AQUACULTURE.2023.739522

Saeid A, Chojnacka K (2016) Evaluation of growth yield of Spirulina maxima in photobioreactors. Chem Biochem Eng Q 30:127–136. https://doi.org/10.15255/CABEQ.2015.2269

Safdar W, Shamoon M, Zan X et al (2017a) Growth kinetics, fatty acid composition and metabolic activity changes of Crypthecodinium cohnii under different nitrogen source and concentration. AMB Express 7:85. https://doi.org/10.1186/s13568-017-0384-3

Safdar W, Zan X, Shamoon M et al (2017b) Effects of twenty standard amino acids on biochemical constituents, docosahexaenoic acid production and metabolic activity changes of Crypthecodinium cohnii. Bioresour Technol 238:738–743. https://doi.org/10.1016/j.biortech.2017.04.024

Safdar W, Zan X, Song Y (2017c) Synergistic effects of pH, temperature and agitation on growth kinetics and docosahexaenoic acid production of C. cohnii cultured on different carbon sources. Int J Res Agric Sci 4:94–101

Safdar W, Zan X, Song Y (2017d) Synergistic effect of phosphorus and nitrogen on growth, lipid accumulation and docosahexaenoic acid production in Crypthecodinium cohnii. Int J Agric Innov Res 5:768–775

Saral JS, Satheesh AR, Ranganathan P (2022) Economic and environmental analysis of algal biorefinery for the production of renewable fuels and co-product. Energy Convers Manag X 14:100189. https://doi.org/10.1016/J.ECMX.2022.100189

Saxena R, Rodríguez-Jasso RM, Chávez-Gonzalez ML et al (2022) Strategy development for microalgae Spirulina platensis biomass cultivation in a bubble Photobioreactor to promote high carbohydrate content. Fermentation 8. https://doi.org/10.3390/fermentation8080374

Schreiber U (1979) Cold-induced uncoupling of energy transfer between phycobilins and chlorophyll in Anacystis nidulans. Antagonistic effects of monovalent and divalent cations, and of high and low pH. FEBS Lett 107:4–9. https://doi.org/10.1016/0014-5793(79)80450-0

Shekhar Sarker S, Akter T, Parveen S et al (2023) Microalgae-based green approach for effective chromium removal from tannery effluent: a review. Arab J Chem 105085:105085. https://doi.org/10.1016/J.ARABJC.2023.105085

Sim SJ, Joun J, Hong ME, Patel AK (2019) Split mixotrophy: a novel cultivation strategy to enhance the mixotrophic biomass and lipid yields of Chlorella protothecoides. Bioresour Technol 291:121820. https://doi.org/10.1016/j.biortech.2019.121820

Simpósio XII, Enzimática DEH, Holz LC, et al (2017) Strategy of Spirulina platensis cultivations for high carbohydrate content in raceway pond, pp 1–4

Soni RA, Sudhakar K, Rana RS (2017) Spirulina – from growth to nutritional product: a review. Trends Food Sci Technol 69:157–171. https://doi.org/10.1016/j.tifs.2017.09.010

Sow S, Ranjan S (2021) Cultivation of spirulina: an innovative approach to boost up agricultural productivity. Pharma Innov 10:799–813. https://doi.org/10.22271/tpi.2021.v10.i3k.5889

Susanna D, Dhanapal R, Mahalingam R, Ramamurthy V (2019) Increasing productivity of Spirulina platensis in photobioreactors using artificial neural network modeling. Biotechnol Bioeng 116:2960–2970. https://doi.org/10.1002/bit.27128

Thevarajah B, Nishshanka GKSH, Premaratne M et al (2022) Large-scale production of spirulina-based proteins and c-phycocyanin: a biorefinery approach. Biochem Eng J 185:108541. https://doi.org/10.1016/J.BEJ.2022.108541

Thomas JB, Govindjee (1960) Changes in quantum yield of photosynthesis in the red alga Porphyridium cruentum caused by stepwise reduction in the intensity of light preferentially absorbed by the Phycobilins. Biophys J 1:63–72. https://doi.org/10.1016/S0006-3495(60)86876-2

Tomazic N, Overkamp KE, Wegner H et al (2021) Exchange of a single amino acid residue in the cryptophyte phycobiliprotein lyase GtCPES expands its substrate specificity. Biochim Biophys Acta Bioenerg 1862:148493. https://doi.org/10.1016/j.bbabio.2021.148493

Ugya AY, Meguellati K (2022) Microalgae biomass modelling and optimisation for sustainable biotechnology - a concise review. J Ecol Eng 23:309–318. https://doi.org/10.12911/22998993/150627

Usai A, Theodoropoulos C, Di Caprio F et al (2023) Structured population balances to support microalgae-based processes: review of the state-of-art and perspectives analysis. Comput Struct Biotechnol J 21:1169–1188. https://doi.org/10.1016/J.CSBJ.2023.01.042

Wang M, Morón-Ortiz Á, Zhou J et al (2023) Effects of pressurized liquid extraction with dimethyl sulfoxide on the recovery of carotenoids and other dietary valuable compounds from the microalgae Spirulina, Chlorella and Phaeodactylum tricornutum. Food Chem 405:134885. https://doi.org/10.1016/J.FOODCHEM.2022.134885

Wicker RJ, Daneshvar E, Kumar Gupta A et al (2023a) Hybrid planktonic-biofilm cultivation of a Nordic mixed-species photosynthetic consortium: a pilot study on carbon capture and nutrient removal. Chem Eng J 144585:144585. https://doi.org/10.1016/J.CEJ.2023.144585

Wicker RJ, Kwon E, Khan E et al (2023b) The potential of mixed-species biofilms to address remaining challenges for economically-feasible microalgal biorefineries: a review. Chem Eng J 451:138481. https://doi.org/10.1016/J.CEJ.2022.138481

Wuang SC, Khin MC, Chua PQD, Luo YD (2016) Use of spirulina biomass produced from treatment of aquaculture wastewater as agricultural fertilizers. Algal Res 15:59–64. https://doi.org/10.1016/j.algal.2016.02.009

Yu Y, Hou X, Yu Q et al (2023) A novel two-stage culture strategy to enhance the C-phycocyanin productivity and purity of Arthrospira platensis. LWT 184:115010. https://doi.org/10.1016/J.LWT.2023.115010

Zhang M, Yao L, Maleki E et al (2019) Membrane technologies for microalgal cultivation and dewatering : recent progress and challenges. Algal Res 44:101686. https://doi.org/10.1016/j.algal.2019.101686

Zhang Z, Chen Y, Klausen LH et al (2023) The rational design and development of microalgae-based biohybrid materials for biomedical applications. Engineering. https://doi.org/10.1016/J.ENG.2022.09.016

Zhou J, Wang M, Barba FJ et al (2023) A combined ultrasound + membrane ultrafiltration (USN-UF) for enhancing saccharides separation from spirulina (Arthrospira platensis). Innov Food Sci Emerg Technol 103341:103341. https://doi.org/10.1016/J.IFSET.2023.103341

Zhu B, Shen H, Li Y et al (2020) Large-scale cultivation of spirulina for biological CO2 mitigation in open raceway ponds using purified CO2 from a coal chemical flue gas. Front Bioeng Biotechnol 7:1–8. https://doi.org/10.3389/fbioe.2019.00441

Chapter 8
Cyanobacteria-Based Green Synthesis of Nanoparticles for Industrial Applications

Muhammad Rizwan Javed, Shaista Shafiq, Elsayed Fathi Abd Allah, Mahwish Salman, Naz Perver, Asifa Anwar, and Fatima tul Zahra

Abstract Cyanobacteria are microscopic Gram-negative photoautotrophs present in fresh and marine water. They are mainly used as source of pigments in many industries. Nanoparticles synthesized from cyanobacteria are gaining interest in recent years. There are many physical and chemical methods for synthesizing nanoparticles, but these methods use toxic chemicals and are much more expensive. Green synthesis of nanoparticles from cyanobacteria is an eco-friendly approach synthesizing various nanoparticles with unique physical, chemical, and biological features. Cyanobacterial cell extract contains many biomolecules that help in the synthesis and stabilization of nanoparticles. Such biomolecules could be used to produce food, metabolites, and biodiesel products. They are also used as antibacterial, antifungal, and antiviral agents. In industry, nanoparticles synthesized from certain species of cyanobacteria are used as biofertilizers and biopesticides. They produce exopolysaccharides that decrease surface tension and help in oil recovery. They also help in the removal of dyes from industrial effluents. Moreover, cyanobacterial synthesized nanoparticles along their synthesis, can also accumulate polyhydroxyalkanoates that being biodegradable materials are used in pharmaceutical applications. In this chapter various methods to synthesize nanoparticles from cyanobacteria have been discussed along with their industrial applications.

Keywords Nanoparticles · Green synthesis · Cyanobacteria · Industrial applications

M. R. Javed (✉) · N. Perver · A. Anwar · F. tul Zahra
Department of Bioinformatics and Biotechnology, Government College University Faisalabad, Faisalabad, Pakistan
e-mail: rizwan@gcuf.edu.pk

S. Shafiq
Department of Biotechnology, The University of Faisalabad, Faisalabad, Pakistan

E. F. Abd Allah
Plant Production Department, College of Food and Agricultural Sciences, King Saud University, Riyadh, Saudi Arabia

M. Salman
Department of Biochemistry, Government College University Faisalabad, Faisalabad, Pakistan

M. A. Mehmood et al. (eds.), *Pharmaceutical and Nutraceutical Potential of Cyanobacteria*, https://doi.org/10.1007/978-3-031-45523-0_8

8.1 Introduction

Nanotechnology is playing an important role in every aspect of our daily lives owing to its broader applications. It is a modern field that is based on the use of nanostructured materials to prepare valuable industrial, agricultural, and medical products. One application of nanotechnology is to synthesize nanoparticles, to characterize their basic structure, and to study their chemical and optical properties. This chapter exclusively describes the role of cyanobacteria in the synthesis of nanoparticles and their subsequent industrial significance. Today, many products available in the market that are used in everyday life are prepared using nanotechnology. These products include stain-repellent fabrics, crack-resistant paints, transparent sunscreens, and scratch-proof sunglasses, among others (Contado 2015).

"Nanomaterials" means a material or a particle that has at least one nano-sized (10^{-9} m) dimension. Nanomaterials are different from their corresponding bulkier counterparts, in terms of their chemical, biological, and physical properties (Hamida et al. 2020b). Nanomaterials can exhibit unique thermal, electrical, magnetic, catalytic, optical, and mechanical characteristics owing to their high surface area to volume ratio. These characteristics make nanomaterials or nanoparticles important for diverse applications (Schmid 2011). For example, nanoparticles are effectively used for optical sensing, screening, imaging, and diagnostics of various diseases that affect human health (Baetke et al. 2015).

8.1.1 Classification of Nanoparticles

Nanoparticles (NPs) can be classified based on multiple parameters including dimensions, composition, shape, nature, and origin (Ahmad et al. 2019). For instance, synthetic nanoparticles are engineered using physical and chemical methods, whereas natural NPs are produced mostly after natural activities or disaster events, e.g., volcanic eruptions, soil erosions, and forest fires. Further, fabrication of heavy metals also leads to the production of NPs. Nanoparticles, based on their composition and physical and chemical properties, can also be divided into organic and inorganic NPs. Organic NPs are strictly based on carbon and/or carbon-containing derivatives and include chitosan, N-halamine compounds, and carbon nanoparticles. Inorganic NPs are non-carbon metals or metal oxides such as gold (Au), silver (Ag), platinum (Pt), zinc oxide (ZnO), and titanium oxide (TiO_2) (Asmathunisha and Kathiresan 2013).

Based on the shapes of nanoparticles, they could be spherical, cylindrical, and disc shaped, among others (Fig. 8.1). However, nanoparticles with low and high aspect ratio are classified separately. High aspect ratio nanoparticles are of various shapes like zigzag, belts, and helices, whereas low aspect ratio nanoparticles include cubic, prism, pillar, and oval shaped (Li et al. 2015).

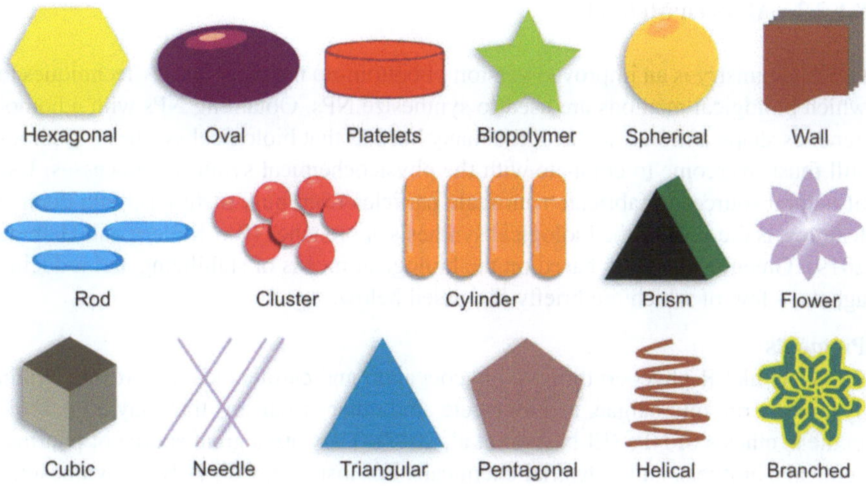

Fig. 8.1 Graphical representation of various shapes of nanoparticles

8.1.2 Methods of Nanoparticle Synthesis

There are various methodologies that are being used for the synthesis of NPs, i.e., conventional methods and modern methods. The conventional methods involve chemical and physical approaches; on the other hand, modern methods are based on biological synthesis that involves living organisms, enzymes, vitamins, etc.

8.1.2.1 Conventional Methods

The chemical methods are based on bottom-up fabrication approach which involves the assembly of atoms to nuclei which leads to the formation of NPs. The main components in chemical methods are metals (main precursor), stabilizing agents, and reducing agents. Some of the commonly used reducing agents are sodium borohydride, sodium citrate, and ascorbate, whereas stabilizing agents include carboxyl methylcellulose, starch, and polyvinyl pyrrolidone. Whereas, physical methods follow top-down approach in which metals are transformed into their nanoforms via physical methods like sputtering, mechanical milling, and laser ablation (Nadagouda and Varma 2006).

The natural synthesis of NPs varies among organisms; even the variations between the strains of the same species have been observed (Husain et al. 2015). Naturally, the synthesis of NPs is initiated on the exposure of microorganisms towards toxic materials either by electrostatic interactions or by secreting extracellular molecules that bind to the material. They are also synthesized intracellularly, which is described ahead in the chapter.

8.1.2.2 Modern Methods

Green chemistry is an improved version of bottom-up nanotechnology techniques in which biological methods are used to synthesize NPs. Obtaining NPs with a homogeneous shape and size is one of the many hurdles that biological synthesis methods still must overcome to compete with the physicochemical synthesis processes. Use of natural sources to fabricate nanoscale particles from bulk materials, with distinct features, is the goal of the biological synthesis approaches (Bin-Meferij and Hamida 2019). Green synthesis is based on the biological means of stabilizing and reducing agents, a few of which are briefly discussed below.

Pigments
Pigments like R-phycoerythrin, C-phycocyanin, and carotenoids are present within cyanobacteria, microalgae, actinomycete, and other organisms that play a vital role in the synthesis of NPs (El-Naggar et al. 2017). They are a great source of pharmaceuticals, biofuels, and industrial chemicals. For instance, CdS NPs are synthesized by the pigment C-phycoerythrin that is extracted from marine cyanobacterium, *Phormidium tenue* (Mubarakali et al. 2012). C-phycocyanin has been utilized to produce silver NPs. These silver NPs have many commercial uses like antimicrobial activity (Patel et al. 2015).

Proteins
Proteins have been studied extensively for their function in NP stability and reduction. *Verticillium* sp. and *Fusarium oxysporum* release cationic proteins with hydrolytic activity, allowing them to decrease and/or cap magnetite NPs (Hamida et al. 2020b). In the bio-fabrication of metal nanoparticles (MNPs), amino acids and proteins act as reducing as well as stabilizing agents. The presence of sulfate, carboxyl, amino, and other groups in the proteins of cyanobacteria enables the bio-reduction of minuscule NPs with a homogeneous particle size distribution (Ali et al. 2011).

Enzymes
Various enzymes such as NADH-dependent enzymes and nitrate reductases have been reported to play a role in the bio-reduction of NPs. Assimilatory nitrate reductases are metalloproteins that catalyze various processes in the nitrogen, carbon, and sulfur cycles using molybdenum ions as a cofactor. Many studies have proven the significance of these enzymes as a powerful reducing agent in the bio-fabrication of MNPs (Ali et al. 2019). For instance, nitrate reductase activity is optimized by varying the cultural conditions to synthesize silver NPs (Khodashenas and Ghorbani 2016). Using enzymes is a cost-efficient and environment-friendly approach as compared to other methods. Enzymes also help to regulate the size and morphology of nanoparticles (Lahiri et al. 2021).

Polysaccharides
Extracellular polysaccharides (EPSs) from cyanobacteria *Nostoc commune* have been used to convert silver nitrate to Ag-NPs. It has been reported that EPSs are

effective as reducing and capping agents. Moreover, encasing Ag-NPs with EPSs increased the stability of the particles by preventing their interaction with oxygen (Ahmad et al. 2019).

8.1.3 *Characterization of Nanoparticles*

Nanoparticles possess various physicochemical properties that make them different from their material in bulk. Therefore, it is important to characterize them to develop products with novel applications to serve mankind. The characterization approaches to investigate the physicochemical properties of NPs include morphological, particle size, surface area, structural, and optical characterization.

The shape of NPs has an impact on their properties. For morphological characterization, some of the main microscopic techniques that are being employed include transmission electron microscopy (TEM) (Mohamad et al. 2018), scanning electron microscopy (SEM), and polarized optical microscopy (POM) (Saeed and Khan 2016).

The SEM and TEM are also used to study particle size and surface area of NPs. Other commonly used characterization techniques include dynamic light scattering (DLS), X-ray diffraction (XRD), atomic force microscopy (AFM), and zeta potential size analyzer (Kestens et al. 2016). Among all the abovementioned techniques, SEM, TEM, XRD, and AFM provide information about particle size, whereas DLS and zeta potential size analyzer are specifically used where the size of NPs needs to be determined at low level. It measures and calculates the zeta potential of nanoparticles. Zeta potential is a physical property which is based on the charge on particles either present in an emulsion or suspension (Clogston and Patri 2011).

The structural properties of NPs describe the nature and composition of bonding materials present in particles or materials. The techniques that are used to study structural characterization include energy dispersive X-ray (EDX) spectroscopy, Brunauer–Emmett–Teller (BET), XRD, X-ray photoelectron spectroscopy (XPS), Fourier transform infrared (IR) spectroscopy, Raman spectroscopy, and zeta potential size analyzer (Emery et al. 2016).

Optical characterization is frequently used in photochemical reactions to study and analyze photocatalytic applications. Their descriptions are based on Beer–Lambert law and basic light concepts (Swinehart 1962). This approach is useful for determining absorption, reflectance, luminescence, and phosphorescence properties of NPs (Mohamad et al. 2018). The instruments that are used for optical characterization of NPs include ultraviolet-visible photoluminescence (UV-Vis PL) and null ellipsometer (Mohamad et al. 2018). A recent study found UV/Vis diffuse reflectance spectrometer (DRS) as an efficient and a complete device that effectively measures optical absorption, transmittance, and reflectance of NPs (Mohamad et al. 2018).

8.2 Cyanobacteria as Useful Bio-Machinery

Cyanobacteria are one of the most important species on the planet. They are photoautotrophic prokaryotes that are abundantly present in luminous environments. Their estimated biomass on Earth is 3×10^{14} g C. They all make chlorophyll a and during photosynthesis release oxygen (Whitton and Potts 2012), thereby significantly contributing to the atmospheric oxygen. Further, most of the cyanobacteria produce pigments such as phycocyanin and phycoerythrin that are responsible for the blue and red colors in cells, respectively. With the advancement of scientific areas, modern methodologies necessitate more precise taxonomic classification criteria and use of diversification techniques for the classification of a complex group of organisms like cyanobacteria.

8.2.1 Classification of Cyanobacteria

Cyanobacteria are traditionally classified into various sections based on their morphology. However, there is a need to revise the criteria for taxonomic categorization of cyanobacteria and to implement modern technologies like bioinformatics tools that will allow preparing the phylogenetic map more precisely while also considering evolutionary marks (Komárek 2014). Table 8.1 shows the biological significance and economic importance of classes of cyanobacteria.

8.2.1.1 Chroococcales

In comparison to the previous notion which contained the coccoid forms with more complex cytology and no baeocyte formation, this order has been decreased drastically. Other characteristics of these species include the ability to reproduce through binary fission and the ability to build colonies that form dense visible masses on damp surfaces such as rocks. As they age, they form colonies that are held together by slimy matrix. *Chroococcales* are composed of mainly two classes that are *Chroococcaceae* and *Entophysalidaceae* (Billi and Caiola 1996).

8.2.1.2 Pleurocapsales

In the order *Pleurocapsales*, there are six genera, each having members from marine, terrestrial, and freshwater habitats: *Dermocarpa*, *Xenococcus*, *Dermocarpella*, *Myxosarcina*, *Chroococcidiopsis*, and *Pleurocapsa*. The analyzed provisional assemblage of the 16S rRNA sequences of one strain each of *Dermocarpa*, *Pleurocapsa*, and *Myxosarcina* revealed that they were phylogenetically related (Caudales et al. 2000). They are unicellular and reproduced by creating small,

Table 8.1 Classification, biological, and economic significance of cyanobacteria

Group	Cell size	Biological significance	Economic importance	References
Chroococcales	<2 μm	• Helps in rice cultivation due to nitrogen-fixing ability • Act as biofertilizer	• Enzyme inhibitor • Cytotoxic • Cell differentiation • Endotoxic • Hepatotoxic	Fenical (1996), Li and Chase (2010)
Pleurocapsales	2–3 μm	• Mat builders • Part of tidal flat communities • Grazer-resistant • Forming calcified stromatolitic structures	• Antifungal • Antimicrobial • Helps in nitrogen fixation • Dye removal	Whitton and Potts (2012)
Oscillatoriales	2–8 μm	• Helps in nitrogen fixation	• Antifeedant • Antifungal • Anti-inflammatory • Antimicrobial • Antimitotic • Antiproliferative • Antiviral/anti-HIV • Cytotoxic	Harada et al. (1993)
Nostocales	~25 cm	• Use as food source in China • Increase the fertility of soil	• Anticancer • Antifungal • Antimalarial • Cardioactive • Hepatotoxic • Antimicrobial	Blunt et al. (2003)
Stigonematales	2 to 450 μm	• Capable of producing hormogonia (motile trichomes)	• Antifungal • Antimicrobial • Anticancer • Antimitotic • Cytotoxic • Herbicidal	Pang et al. (2018), Patterson et al. (1991), Walton and Berry (2016)

spherical baeocytes (a type of reproductive cell that forms in cyanobacteria) through internal multiple fission. They also synthesize pseudo filaments through baeocytes. They develop epiliths on moist rock surfaces and epiphytes on algae (Shalygin et al. 2019).

8.2.1.3 Oscillatoriales

This order includes filamentous taxa (phylogenetically close genus). Earlier the filamentous taxa were characterized as narrow trichomes and parietal thylakoids (*Synechococcales*, *Spirulinales*). Lately it has been redefined after complicated cytology is revealed, as thylakoid arrangements were found to be more irregular, radial, and fasciculated (Komárek 2014). This genus is placed between *Chroococcales* and orders with similar thylakoid architecture. This can further group down to a more monophyletic-higher level taxa after required modifications (Komárek 2014). They consist of filamentous cyanobacteria. However, they don't have akinetes (dormant cells of cyanobacteria that allow them to survive in harsh and starvation conditions), heterocysts, and true branches. They are present in different environments from saline water to freshwater bodies. They are also found in terrestrial environments. They usually form elongated structures that are multicellular (Gómez et al. 2021).

8.2.1.4 Nostocales

This order includes a vast cluster of monophyletic filamentous cyanobacteria bearing prominent cells like heterocytes and akinetes with a variety of thallus. In *Nostocales*, filaments are branched (true or false types), unbranched, and isopolar; in few families, the structure of filaments is heteropolar too (Komárek 2014). They are involved in nitrogen fixation through their thick hyaline protoplasts. Their akinetes help them to survive in unfavorable conditions. They also have false branching in some of their species (Usharani et al. 2012).

8.2.1.5 Stigonematales

Stigonematales are categorized as heterocystous cyanobacteria that fix nitrogen and bear true branched filaments (Walton and Berry 2016). Further, their vegetative cells evolve as spore-like, thick-walled, and non-motile cells.

8.2.2 Significance of Cyanobacteria in NP Synthesis

Recently, biological production of metal nanoparticles, i.e., green synthesis of NPs, is gaining popularity due to its simple methodologies, less time consumption, and environmental friendliness. Compared to the other methodologies, the green synthesis of metallic nanoparticles is better to protect the environment as there are reduced biological threats particularly in biomedical applications and due to reduction in the use of harmful chemicals (Husain et al. 2019). Therefore, it is crucial to develop a reliable and environmentally safe approach to gain benefits of versatile nanotechnology while preparing metallic nanoparticles; fortunately cyanobacteria is one of the best choices available (Husain et al. 2015).

Cyanobacteria can accumulate heavy metals from their surroundings. They also contain bioactive components such as enzymes and pigments that can act as stabilizing and reducing agents. These properties of cyanobacteria make them among the most advantageous natural species for NPs synthesis. The advantages of cyanobacteria-mediated NPs are to achieve various significant characteristics including antibacterial, antifungal, anti-cancerous, and photocatalytic properties (Hamida et al. 2020b).

8.3 Mechanism of NP Synthesis

Cyanobacteria provide efficient bio-systems to prepare NPs either through intracellular or extracellular synthesis pathways (Fig. 8.2).

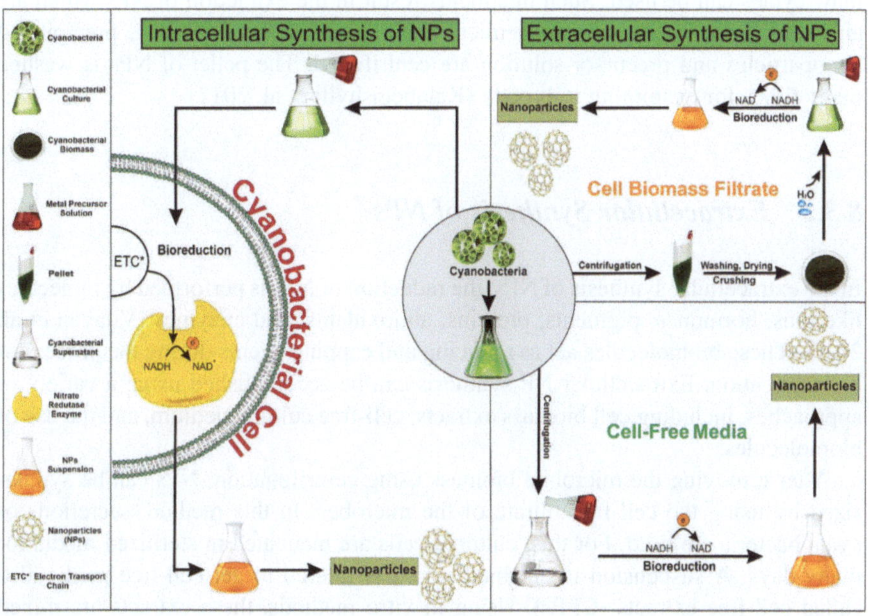

Fig. 8.2 Mechanisms of NP synthesis. **Left**: Schematic diagram of intracellular biosynthesis of NPs using cyanobacteria. Intracellular biosynthesis involves enzyme-mediated synthesis (e.g., nitrate reductase enzyme). **Right**: Schematic diagram of extracellular biosynthesis of NPs using cyanobacteria. Extracellular biosynthesis involves two approaches: cell-free media, in which cell supernatant is used to reduce metals into NPs and cell biomass filtrate, using cyanobacterial biomass which reduce metals into metallic NPs

8.3.1 Intracellular Synthesis of NPs

Intracellular fabrication of nanoparticles involves the reduction of bulk mass into nanoparticles. This is done within the host cell by controlling various biological factors. However, in general, there are three basic steps for the intracellular synthesis of NPs in the laboratory (Lengke et al. 2007):

(i) Culturing of target microbe.
(ii) Interaction of cultured microorganisms with precursor solution and enzymatic reduction of precursor ions into NPs within cells.
(iii) Separation and refinement of the synthesized nanoparticles for characterization using physiochemical techniques.

In cyanobacteria, ions are reduced intracellularly with the help of electrons through electron transport system, using photosynthesizing reactions (that synthesize energy) and sometimes using respiratory electron chain. These reactions use many enzymes like NADH-dependent reductase. The redox reaction occurs in the thylakoid membrane, cytoplasm, and cell membrane of the cell (Saim et al. 2021).

If the synthesis of nanoparticles is intracellular, then for the lysis of cells, different methods like sonication, mechanical disruption, liquid homogenization, and freeze-thaw cycles can be used. Such treatments result in the extraction of NPs which are analyzed further using physiochemical techniques. In the final step, precipitated nanoparticles and precursor solution are centrifuged. The pellet of NPs is washed many times for examining it directly (Kalabegishvili et al. 2012).

8.3.2 Extracellular Synthesis of NPs

In the extracellular synthesis of NPs, the reduction of NPs is performed by molecules like ions, hormones, pigments, proteins, antioxidants, and enzymes (Vijayan et al. 2014). These biomolecules act as reducing and capping agents during the process of bio-fabrication. Extracellular NP synthesis can be accomplished using a variety of approaches, including cell biomass extracts, cell-free culture medium, and the use of biomolecules.

After removing the microbial biomass using centrifugation, NPs can be synthesized by using the cell-free filtrate of the microbes. In this method, secretions of cyanobacteria are used. For this, cultured cells are incubated in sterilized media for many days. A suspension is obtained which is filtered to get cell-free media also called cell-free extracts (CFEs). Using in vitro methods, these extracts are mixed with metal ions to synthesize NPs. Metal ions act as electron donors in the synthesis of NPs. Numerous researchers have shown the importance of biomolecules including phenolics, enzymes, antioxidants, and alkaloids in a cell-free filtrate in regulating the bio-reduction mechanism of NPs (Shaikh et al. 2021).

Several studies describe the cell biomass filtrate synthesis method, which involves centrifuging microbial cultures and washing the cell pellets three times with distilled water to remove any excess metal from the medium. The freeze-dried (in a lyophilizer) or oven dried (at approx. 40 °C) pellets are then crushed with a mortar and pestle. Following that, distilled water is added to the fine powder and then filtered through Whatman filter paper. The filtrate is then mixed with a solution containing metal precursor, and the mixture is incubated at optimized conditions. Active bio-moieties in the cell's filtrate, such as proteins and enzymes, play an important role in the bio-fabrication and stability of NPs in this approach (Bin-Meferij and Hamida 2019).

Another method describes collecting cell biomass at the middle of exponential phase, washing it multiple times to remove components of media, dipping it in water for a set duration, and then utilization of the filtrate for the synthesis of nanoparticles (Ali et al. 2019).

8.4 Applications of Nanoparticles Synthesized by Cyanobacteria

8.4.1 Medical Applications

Nanoparticles prepared through different strains of cyanobacteria are reported to have bioactivities such as antitumor, antibacterial, and antioxidant properties (Table 8.2) (Ahmad et al. 2019). Human cancer is one of the primary causes of death. Cancer is becoming more common because of a variety of causes, including aging, population growth, and lifestyles that predispose people to the disease. New strategies to prevent metastasis and tumor progression have been developed, thanks to advances in nanoscience. Magnetic iron oxide NPs, for example, have improved the cellular imaging, resulting in increased diagnostic quality and timeliness. Several studies on the biological synthesis of NPs via cyanobacteria and their anticancer activity have been reported (Hamida et al. 2020b).

NPs are appealing for medical applications because of their high surface area to volume ratio. Smaller NPs have a larger surface area to interact more with target bacterial cells. Ag-NPs produced by *Synechococcus* sp. showed antibacterial potential against a wide range of bacteria, such as *Escherichia coli, Bacillus subtilis* and *Staphylococcus aureus*, of which *B. subtilis* is the most susceptible against the synthesized Ag-NPs (Sharma et al. 2016). Antimicrobial potential of Ag-NPs produced by *Scenedesmus obliquus*, *Chlorella vulgaris*, and *S. platensis* against *Microcystis aeruginosa* was examined (Bakir et al. 2018). *M. aeruginosa* is characterized as a hazardous microalga which is involved in the production of cancer-causing toxins. Such findings support the application of Ag-NPs as antialgal agents to minimize the adverse effects of *M. aeruginosa*.

Table 8.2 Applications of NPs synthesized by cyanobacteria

Metal oxide nanoparticles	Applications	Species	Mode of synthesis	Salts used	References
Ag-NPs (silver NPs)	• Antibacterial • Antialgal • Anticancer • Antileukemic • Antifungal • Photocatalytic • Dye decolorization	*Arthrospira platensis*	Extra- and intra-cellular	$AgNO_3$	Cicci et al. (2017)
		Aulosira fertilissima	Extra-cellular		Husain et al. (2015)
		Calothrix pulvinate	Extra- and intra-cellular		Brayner et al. (2007)
		Chroococcus sp. *NCCU-207*	Extra-cellular		Husain et al. (2015)
		Cylindrospermopsis sp. 121–1	Extra- and intra-cellular		Patel et al. (2015)
		Desertifilum sp. *IPPAS B*-1220	Extra-cellular		Hamida et al. (2020a)
		Gloeocapsa gelatinosa NCCU-430	Extra-cellular		Husain et al. (2015)
		Hapalosiphon fontinalis NCCU-339	Extra-cellular		Husain et al. (2015)
		Lyngbya sp. 15–2	Extra- and intra-cellular		Patel et al. (2015)
		Leptolyngbya sp. *JSC-1*	Extra-cellular		Zada et al. (2018)
		Microchaete sp. NCCU-342	Extra-cellular		Husain et al. (2019)
		Nostoc carneum	Extra-cellular		El-Naggar et al. (2018)
		Oscillatoria limnetica	Extra-cellular		Hamouda et al. (2019)
		Tolypothrix tenuis	Extra-cellular		Husain et al. (2015)
Au-NPs (gold NPs)	• Antitumor • Antibacterial • Antioxidant • Anticancer	*Anabaena* sp. PCC 7120	Extra- and intra-cellular	$AuHCl_4$	Radtsig et al. (2016)
			Intra-cellular		Parial et al. (2016)

	• Anti-myocardial • Pharmaceutical • Medical purposes	*Coleofasciculus chthonoplastes* BDU 61,001			
		Cyanothece sp.	Intra-cellular		Younis et al. (2019)
		Gloeocapsa sp.	Intra-cellular		Geetha et al. (2014)
		Lyngbya majuscula	Intra-cellular		Bakir et al. (2018)
		Leptolyngbya tenuis BDU 20,391	Intra-cellular		Parial et al. (2016)
		Microcoleus chthonoplastes	Intra-cellular		Parial et al. (2012)
		Nostoc ellipsosporum	Intra-cellular		Asmathunisha et al. (2018)
		Oscillatoria sp.	Intra-cellular		Asmathunisha et al. (2018)
		Phormidium willei	Intra-cellular		Mubarakali et al. (2013)
		Synechocystis sp. PCC 6803	Intra-cellular		Focsan et al. (2011)
		Spirulina platensis	Intra-cellular		Kalabegishvili et al. (2012)
β-FeOOH (Akaganeite)	• "One-pot" hybrid biofilms	*Anabaena flos-aquae* ALCP B24	Intra-cellular	(Fe^{2+}; Fe^{3+})	Dahoumane et al. (2010)
ZnO (zinc oxide)	• Antibacterial • Antibiofilm • Anticancer	*Nostoc* sp. EA03	Extra-cellular	$Zn(CH_3CO_2)_2$	Ebadi et al. (2019)
CdS (cadmium sulfide)	• Bio-labeling	*Phormidium tenue* NTDM05	Extra-cellular	$CdCl_2$ and Na_2S	Mubarakali et al. (2012)
Pt (platinum)	• Redox sensing/signaling for photosynthetic acclimation	*Plectonema boryanum* UTEX 485	Extra- and intra-cellular	$PtCl_4$	Lengke et al. (2007)
CuO (copper oxide)	• Antibacterial	*Spirulina platensis*	Extra-cellular	$Cu(CH_3COO)_2$	Saran et al. (2017)

8.4.2 Agricultural and Food Applications

Cyanobacteria have an important role in agricultural applications. Several cyanotoxins generated from cyanobacteria could be used as biocides due to their varying bioactivities. These biocides limit the growth of bacteria, viruses, fungi, and some invertebrates such as crustaceans and bivalves, as well as some vertebrates such as fish, birds, and mammals. Because of their allelopathic effects, these cyanotoxins have a lot of potential for developing active biological compounds that could be used in crop fields such as insecticides, herbicides, algicides, and fungicides (Teneva et al. 2023). Many problems have been reported because of the use of synthetic pesticides, necessitating the development of biocides with low environmental risk (Pathak et al. 2022).

Nanoparticles synthesized from cyanobacteria are also widely used in the food industry for packaging and preventing food spoilage from pathogens (Aziz et al. 2021). These nanoparticles have shown bacteriostatic and bactericidal effects. The nano system packaging limits supply of oxygen and hence limits proliferation of microbes (Dos Santos et al. 2020). Most widely used nanoparticles for this purpose are silver nanoparticles because of their high stability and toxicity against a wide range of bacteria (Carbone et al. 2016).

8.4.3 Industrial and Environmental Applications

Cyanobacteria produce several bioactive compounds making them more useful for many industrial applications like antifungal, antibacterial, and antiviral. Additionally, their secondary metabolites such as toxins and vitamins are used in pharmaceutical industries. Moreover, some species can accumulate polyhydroxyalkanoates (PHA) intracellularly that can be used to produce bioplastics. Additionally, applications of cyanobacteria include uses in biofertilizers, biopesticides, and nutraceuticals (Gupta et al. 2022).

Cyanobacterial nanoparticles produce many exopolysaccharides (EPSs) that can be helpful in oil recovery. These nanoparticles decrease surface tension and increase solubility of oil (Abed et al. 2009). Moreover, EPS is widely used to remove dyes from industrial effluents. The most used strains to produce EPS are *Aphanocapsa halophytia*, *Cyanothece* sp., and *Anabaena* sp. (Carmichael and Gorham 1980).

Polyhydroxyalkanoates (PHA) are crystalline thermoplastic compounds that have properties like polypropylene. They have many applications in biopharmaceuticals and biomedical area being an ideal biodegradable material (Sudesh 2004), e.g., *S. platensis* can accumulate PHA under mixotrophic conditions of growth; therefore, nanoparticles that are formed intracellularly also contain concentrations of PHA (de Philippis et al. 1992).

Dyes are synthetic organic compounds that are released from food, cosmetics, textile, paper, pharmaceutical, and leather industries (Sami and Fatma 2019).

Unfortunately, the release of these dyes contributes to environmental pollution. For instance, azo dyes are highly carcinogenic when discharged into environment without being treated (Kusic et al. 2011). Controlling dye concentration through conventional biodegradation methods is found unsuccessful because these dyes are resistant to microbial degradation. Nanoparticles synthesized from cyanobacteria are involved in dye decolorization and degradation, for example, acid red and sky-blue dyes are decolorized up to $>80\%$ by *Gloeocapsa pleurocapsoides* and *Phormidium ceylanicum* (Parikh and Madamwar 2005).

8.5 Conclusion and Prospects

Green nanotechnology is recognized as one of the most promising fields of modern science. This approach allows for the environmentally safe and low-cost synthesis of high yield NPs. Plants, algae, cyanobacteria, fungi, and bacteria are among the natural resources used to biosynthesize NPs. However, several areas still need to be investigated further; for example, many cyanobacterial strains are yet unknown. More research is needed to learn more about these cyanobacteria's biocatalytic skills and their ability to bio-transform precursor metals into nanoforms. The use of cyanobacteria as bio-fabrication machinery for the synthesis of NPs has enormous value as it allows the safe, eco-friendly, sustainable, energy-efficient, and economical synthesis. It can produce novel NPs of varying shapes and sizes. Cyanobacteria-mediated NPs exhibit a variety of biological, physical, and chemical properties that allow for a wide range of applications. Numerous investigations have demonstrated the antibacterial, anticancer, and photocatalytic potential of such NPs. Some of the cyanobacterial strains including *Desertifilum* sp. and *Nostoc* sp. have been characterized to produce Ag-NPs with substantial anticancer potential against a variety of cancer cell lines. Moreover, Au-NPs and Ag-NPs synthesized from *Anabaena* sp. and *Spirulina* sp. have a wide range of inhibitory abilities against Gram-negative and Gram-positive bacteria. However, preclinical toxicology, safety, and therapeutic assessment of cyanobacterial NPs are recommended for their successful implementation. The development and improvement of cyanobacterial-based biosynthetic processes may result in the identification of novel biogenic NPs with special features to meet a range of applications. Incorporating them into medical therapies may improve effectiveness and reduce negative effects. Although numerous theories have been proposed, the exact mechanism through which the living organisms synthesize NPs has yet to be determined. Therefore, it will be required to further investigate microbial synthesis mechanisms of NPs and to explore the reasons underlying the production of NPs with different morphologies, even from the various strains of the same species. Moreover, during the green synthesis of NPs, control of the biological and physicochemical characteristics of NPs will be possible by understanding the impact of biotic and abiotic factors (including metabolic status, growing conditions, electrostatic force, and role of biomolecules such as proteins, pigments, enzymes). The increasing demand for greener and safer products promotes the research into green synthesis of NPs in a variety of fields.

Acknowledgments Authors are grateful to Higher Education Commission (HEC) of Pakistan for providing financial support under the projects NRPU-5590 and NRPU-16342.

References

Abed RMM, Dobretsov S, Sudesh K (2009) Applications of cyanobacteria in biotechnology. J Appl Microbiol 106:1–12

Ahmad S, Munir S, Zeb N, Ullah A, Khan B, Ali J, Bilal M, Omer M, Alamzeb M, Salman SM (2019) Green nanotechnology: a review on green synthesis of silver nanoparticles - an ecofriendly approach. Int J Nanomedicine 14:5087

Ali DM, Sasikala M, Gunasekaran M, Thajuddin N (2011) Biosynthesis and characterization of silver nanoparticles using marine cyanobacterium, Oscillatoria willei NTDM01. Dig J Nanomater Biostruct 6:385–390

Ali J, Ali N, Wang L, Waseem H, Pan G (2019) Revisiting the mechanistic pathways for bacterial mediated synthesis of noble metal nanoparticles. J Microbiol Methods 159:18–25

Asmathunisha N, Kathiresan K (2013) A review on biosynthesis of nanoparticles by marine organisms. Colloids Surf B Biointerfaces 103:283–287

Asmathunisha N, Anburaj R, Kathiresan K (2018) Synthesis of silver and gold nanoparticles by mangrove-derived cyanobacteria. J Adv Nanomater 3:1–7

Aziz MN, Zaki A, Ahamad I, Fatma T (2021) Silver nanoparticle synthesis from Cyanobacteria: environmental and biomedical applications. In: Patel JK, Pathak YV (eds) Emerging technologies for nanoparticle manufacturing. Springer, Cham

Baetke SC, Lammers T, Kiessling F (2015) Applications of nanoparticles for diagnosis and therapy of cancer. Br J Radiol 88:20150207

Bakir EM, Younis NS, Mohamed ME, Semary EL, N. A. (2018) Cyanobacteria as nanogold factories: chemical and anti-myocardial infarction properties of gold nanoparticles synthesized by *Lyngbya majuscula*. Mar Drugs 16:217

Billi D, Caiola MG (1996) Effects of nitrogen limitation and starvation on *Chroococcidiopsis sp.* (Chroococcales). New Phytol 133:563–571

Bin-Meferij MM, Hamida RS (2019) Biofabrication and antitumor activity of silver nanoparticles utilizing novel *Nostoc sp*. Bahar M. Int J Nanomed 14:9019

Blunt JW, Copp BR, Munro MH, Northcote PT, Prinsep MR (2003) Marine natural products. Nat Prod Rep 20:1–48

Brayner R, Barberousse H, Hemadi M, Djedjat C, Yéprémian C, Coradin T, Livage J, Fiévet F, Couté A (2007) Cyanobacteria as bioreactors for the synthesis of Au, Ag, Pd, and Pt nanoparticles via an enzyme-mediated route. J Nanosci Nanotechnol. 7(8):2696–2708

Carbone M, Donia DT, Sabbatella G, Antiochia R (2016) Silver nanoparticles in polymeric matrices for fresh food packaging. J King Saud Univ Sci 28:273–279

Carmichael W, Gorham P (1980) Freshwater cyanophyte toxins. Algae biomass. Elsevier, New York

Caudales R, Wells JM, Butterfield JE (2000) Cellular fatty acid composition of cyanobacteria assigned to subsection II, order *Pleurocapsales*. Int J Syst Evol Microbiol 50:1029–1034

Cicci A, Sed G, Tirillò J, Stoller M, Bravi M (2017) Production and characterization of silver nanoparticles in cultures of the cyanobacterium *A. platensis* (*Spirulina*). Chem Eng Trans 57: 1405–1410

Clogston JD, Patri AK (2011) Zeta potential measurement. Characterization of nanoparticles intended for drug delivery. Springer

Contado C (2015) Nanomaterials in consumer products: a challenging analytical problem. Front Chem 3

Dahoumane SA, Djediat C, Yéprémian C, Couté A, Fiévet F, Brayner R (2010) Design of magnetic akaganeite-cyanobacteria hybrid biofilms. Thin Solid Films 518:5432–5436

de Philippis R, Ena A, Guastini M, Sili C, Vincenzini M (1992) Factors affecting poly-β-hydroxybutyrate accumulation in cyanobacteria and in purple non-sulfur bacteria. FEMS Microbiol Rev 9:187–194

Dos Santos CA, Ingle AP, Rai M (2020) The emerging role of metallic nanoparticles in food. Appl Microbiol Biotechnol 104:2373–2383

Ebadi M, Zolfaghari MR, Aghaei SS, Zargar M, Shafiei M, Zahiri HS, Noghabi KA (2019) A bio-inspired strategy for the synthesis of zinc oxide nanoparticles (ZnO NPs) using the cell extract of cyanobacterium *Nostoc sp.* EA03: from biological function to toxicity evaluation. RSC Adv 9:23508–23525

El-Naggar NE-A, Hussein MH, El-Sawah AA (2017) Bio-fabrication of silver nanoparticles by phycocyanin, characterization, in vitro anticancer activity against breast cancer cell line and in vivo cytotoxicity. Sci Rep 7:1–20

El-Naggar NE-A, Hussein MH, El-Sawah AA (2018) Phycobiliprotein-mediated synthesis of biogenic silver nanoparticles, characterization, *in vitro* and *in vivo* assessment of anticancer activities. Sci Rep 8:1–20

Emery AA, Saal JE, Kirklin S, Hegde VI, Wolverton C (2016) High-throughput computational screening of perovskites for thermochemical water splitting applications. Chem Mater 28:5621–5634

Fenical W (1996) Marine biodiversity and the medicine cabinet the status of new drugs from marine organisms. Oceanography 9:23–27

Focsan M, Ardelean II, Craciun C, Astilean S (2011) Interplay between gold nanoparticle biosynthesis and metabolic activity of cyanobacterium *Synechocystis sp.* PCC 6803. Nanotechnology 22:485101

Geetha S, Sathakkathulzariya J, Aarthi R, Blessie H (2014) Green synthesis of gold nanoparticle using marine cyanobacteria *Gloeocapsa sp* and the antitumor potential. J Chem Pharm Sci 4: 172–174

Gómez C, Guzmán-Carrasco A, Lafarga T, Acién-Fernández FG (2021) Optimization of a new culture medium for the large-scale production of protein-rich *Arthrospira platensis* (oscillatoriales, cyanophyceae). J Phycol 57:636–644

Gupta V, Verma N, Choudhary K (2022) Cyanobacterial nanoparticles: application in agriculture and allied sectors. In: Cyanobacterial lifestyle and its applications in biotechnology. Springer

Hamida RS, Abdelmeguid NE, Ali MA, Bin-Meferij MM, Khalil MI (2020a) Synthesis of silver nanoparticles using a novel cyanobacteria *Desertifilum sp.* extract: their antibacterial and cytotoxicity effects. Int J Nanomedicine 15:49–63

Hamida RS, Ali MA, Redhwan A, Bin-Meferij MM (2020b) Cyanobacteria - a promising platform in green nanotechnology: a review on nanoparticles fabrication and their prospective applications. Int J Nanomedicine 15:6033–6066

Hamouda RA, Hussein MH, Abo-Elmagd RA, Bawazir SS (2019) Synthesis and biological characterization of silver nanoparticles derived from the cyanobacterium *Oscillatoria limnetica*. Sci Rep 9:1–17

Harada K-I, Mayumi T, Shimada T, Suzuki M, Kondo F, Watanabe MF (1993) Occurrence of four depsipeptides, aeruginopeptins, together with microcystins from toxic cyanobacteria. Tetrahedron Lett 34:6091–6094

Husain S, Sardar M, Fatma T (2015) Screening of cyanobacterial extracts for synthesis of silver nanoparticles. World J Microbiol Biotechnol 31:1279–1283

Husain S, Afreen S, Yasin D, Afzal B, Fatma T (2019) Cyanobacteria as a bioreactor for synthesis of silver nanoparticles-an effect of different reaction conditions on the size of nanoparticles and their dye decolorization ability. J Microbiol Methods 162:77–82

Kalabegishvili T, Kirkesali E, Rcheulishvili A (2012) Synthesis of gold nanoparticles by blue-green algae *Spirulina platensis*. Frank Laboratory of Neutron Physics

Kestens V, Roebben G, Herrmann J, Jämting Å, Coleman V, Minelli C, Clifford C, De Temmerman P-J, Mast J, Junjie L (2016) Challenges in the size analysis of a silica nanoparticle mixture as candidate certified reference material. J Nanopart Res 18:1–22

Khodashenas B, Ghorbani HR (2016) Optimisation of nitrate reductase enzyme activity to synthesise silver nanoparticles. IET Nanobiotechnol 10:158–161

Komárek J (2014) Modern classification of cyanobacteria. In: Cyanobacteria: an economic perspective. Wiley, pp 21–39

Kusic H, Juretic D, Koprivanac N, Marin V, Božić AL (2011) Photooxidation processes for an azo dye in aqueous media: modeling of degradation kinetic and ecological parameters evaluation. J Hazard Mater 185:1558–1568

Lahiri D, Nag M, Sheikh HI, Sarkar T, Edinur HA, Pati S, Ray RR (2021) Microbiologically-synthesized nanoparticles and their role in silencing the biofilm signaling cascade. Front Microbiol 12:636588

Lengke MF, Fleet ME, Southam G (2007) Synthesis of palladium nanoparticles by reaction of filamentous cyanobacterial biomass with a palladium(II) chloride complex. Langmuir 23:8982–8987

Li J, Chase HA (2010) Applications of membrane techniques for purification of natural products. Biotechnol Lett 32:601–608

Li M-C, Wu Q, Song K, Lee S, Qing Y, WU, Y. (2015) Cellulose nanoparticles: structure–morphology–rheology relationships. ACS Sustain Chem Eng 3:821–832

Mohamad AT, Kaur J, Sidik NAC, Rahman S (2018) Nanoparticles: a review on their synthesis, characterization and physicochemical properties for energy technology industry. J Adv Res Fluid Mech Therm Sci 46:1–10

Mubarakali D, Gopinath V, Rameshbabu N, Thajuddin N (2012) Synthesis and characterization of CdS nanoparticles using C-phycoerythrin from the marine cyanobacteria. Mater Lett 74:8–11

Mubarakali D, Arunkumar J, Nag KH, Sheiksyedishack K, Baldev E, Pandiaraj D, Thajuddin N (2013) Gold nanoparticles from pro and eukaryotic photosynthetic microorganisms—comparative studies on synthesis and its application on biolabelling. Colloids Surf B Biointerfaces 103: 166–173

Nadagouda MN, Varma RS (2006) Green and controlled synthesis of gold and platinum nanomaterials using vitamin B2: density-assisted self-assembly of nanospheres, wires and rods. Green Chem 8:516–518

Pang K, Tang Q, Chen L, Wan B, Niu C, Yuan X, Xiao S (2018) Nitrogen-fixing heterocystous cyanobacteria in the Tonian period. Curr Biol 28:616–622.e1

Parial D, Patra HK, Dasgupta AK, Pal R (2012) Screening of different algae for green synthesis of gold nanoparticles. Eur J Phycol 47:22–29

Parial D, Gopal PK, Paul S, Pal R (2016) Gold(III) bioreduction by cyanobacteria with special reference to in vitro biosafety assay of gold nanoparticles. J Appl Phycol 28:3395–3406

Parikh A, Madamwar D (2005) Textile dye decolorization using cyanobacteria. Biotechnol Lett 27: 323–326

Patel V, Berthold DE, Puranik P, Gantar M (2015) Screening of cyanobacteria and microalgae for their ability to synthesize silver nanoparticles with antibacterial activity. Biotechnol Rep 5:112–119

Pathak VM, Verma VK, Rawat BS, Kaur B, Babu N, Sharma A, Dewali S, Yadav M, Kumari R, Singh S, Mohapatra A, Pandey V, Rana N, Cunill JM (2022) Current status of pesticide effects on environment, human health and it's eco-friendly management as bioremediation: a comprehensive review. Front Microbiol 13:962619

Patterson GM, Baldwin CL, Bolis CM, Caplan FR, Karuso H, Larsen LK, Levine IA, Moore RE, Nelson CS, Tschappat KD (1991) Antineoplastic activity of cultured blue-green algae (cyanophyta). J Phycol 27:530–536

Radtsig M, Koksharova O, Nadtochenko V, Khmel I (2016) Production of gold nanoparticles by biogenesis using bacteria. Microbiology 85:63–70

Saeed K, Khan I (2016) Preparation and characterization of single-walled carbon nanotube/nylon 6,6 nanocomposites. Instrum Sci Technol 44:435–444

Saim AK, Kumah FN, Oppong MN (2021) Extracellular and intracellular synthesis of gold and silver nanoparticles by living plants: a review. Nanotechnol Environ Eng 6:1–11

Sami N, Fatma T (2019) Studies on estrone biodegradation potential of cyanobacterial species. Biocatal Agric Biotechnol 17:576–582

Saran S, Sharma G, Kumar M, ALI, M. (2017) Biosynthesis of copper oxide nanoparticles using cyanobacteria *spirulina platensis* and its antibacterial activity. Int J Pharmaceut Sci Res 8:3887–3892

Schmid G (2011) Nanoparticles: from theory to application. John Wiley & Sons

Shaikh WA, Chakraborty S, Owens G, Islam RU (2021) A review of the phytochemical mediated synthesis of AgNP (silver nanoparticle): the wonder particle of the past decade. Appl Nanosci 11:2625–2660

Shalygin S, Kavulic KJ, Pietrasiak N, Bohunická M, Vaccarino MA, Chesarino NM, Johansen JR (2019) Neotypification of *Pleurocapsa fuliginosa* and epitypification of *P. minor (Pleurocapsales)*: resolving a polyphyletic cyanobacterial genus. Phytotaxa 392:245

Sharma A, Sharma S, Sharma K, Chetri SP, Vashishtha A, Singh P, Kumar R, Rathi B, Agrawal V (2016) Algae as crucial organisms in advancing nanotechnology: a systematic review. J Appl Phycol 28:1759–1774

Sudesh K (2004) Microbial polyhydroxyalkanoates (PHAs): an emerging biomaterial for tissue engineering and therapeutic applications. Int Med J Malaysia 59:55–56

Swinehart DF (1962) The beer-lambert law. J Chem Educ 39:333

Teneva I, Velikova V, Belkinova D, Moten D, Dzhambazov B (2023) Allelopathic potential of the cyanotoxins microcystin-LR and cylindrospermopsin on green algae. Plan Theory 12:1403

Usharani G, Saranraj P, Kanchana D (2012) *Spirulina* cultivation: a review. Int J Pharm Biol Arch 3:1327–1341

Vijayan SR, Santhiyagu P, Singamuthu M, Kumari Ahila N, Jayaraman R, Ethiraj K (2014) Synthesis and characterization of silver and gold nanoparticles using aqueous extract of seaweed, *Turbinaria conoides*, and their antimicrofouling activity. Sci World J 2014:938272

Walton K, Berry JP (2016) Indole alkaloids of the *Stigonematales* (Cyanophyta): chemical diversity, biosynthesis and biological activity. Mar Drugs 14:73

Whitton BA, Potts M (2012) Introduction to the cyanobacteria. In: Ecology of cyanobacteria II. Springer, New York

Younis NS, Bakir EM, Mohamed ME, El Semary NA (2019) Cyanobacteria as nanogold factories II: chemical reactivity and anti-myocardial infraction properties of customized gold nanoparticles biosynthesized by *Cyanothece sp.* Mar Drugs 17:402

Zada S, Ahmad A, Khan S, Yu X, Chang K, Iqbal A, Ahmad A, Ullah S, Raza M, Khan A (2018) Biogenic synthesis of silver nanoparticles using extracts of *Leptolyngbya JSC-1* that induce apoptosis in HeLa cell line and exterminate pathogenic bacteria. Artif Cells Nanomed Biotechnol 46:S471–S480

Chapter 9
Cyanobacterial Bioactive Compounds: Synthesis, Extraction, and Applications

Fahad Khan, Azka Akhlaq, Muhammad Hidayat Rasool, and Sirasit Srinuanpan

Abstract Cyanobacteria, as ancient photosynthetic prokaryotes, have garnered attention due to their ability to produce an array of biologically active metabolites, including alkaloids, terpenoids, polysaccharides, pigments, cyclic peptides, phenols, lipids, and vitamins. The diverse morphological, physiological, and genetic traits of cyanobacteria contribute to the synthesis of these compounds, which possess promising medicinal properties, excellent abilities for bioremediation, and potential to be used as nutraceuticals, biofertilizers, cosmetics, nanomaterials, and biofuels. This chapter provides a deeper insight into the extraction techniques employed to obtain these bioactive compounds from cyanobacteria, encompassing both traditional and innovative methods. Furthermore, the synthesis pathways and genetic manipulation of cyanobacteria are explored, focusing on enhancing the quality and quantity of bioactive compounds through genetic engineering approaches. The applications of cyanobacterial bioactive compounds are extensive, with a particular emphasis on their utilization in drug discovery and development. The potential therapeutic uses, such as antiviral, anti-tumor, antibacterial, anti-HIV, and food additives, are discussed, showcasing the established roles of cyanobacterial compounds in various fields. Additionally, the chapter highlights the emerging trends in the application of cyanobacterial compounds in nanobiotechnology, with a focus on their incorporation into nanoconjugates and nanomaterials for advanced drug delivery systems and

F. Khan
Tasmanian Institute of Agriculture, University of Tasmania, Hobart, TAS, Australia

A. Akhlaq
School of Chemistry, University of the Punjab, Lahore, Pakistan

M. H. Rasool
Institute of Microbiology, Government College University Faisalabad, Faisalabad, Pakistan

S. Srinuanpan (✉)
Biorefinery and Bioprocess Engineering Research Cluster, Chiang Mai University, Chiang Mai, Thailand

Center of Excellence in Microbial Diversity and Sustainable Utilization, Chiang Mai University, Chiang Mai, Thailand
e-mail: sirasit.s@cmu.ac.th

M. A. Mehmood et al. (eds.), *Pharmaceutical and Nutraceutical Potential of Cyanobacteria*, https://doi.org/10.1007/978-3-031-45523-0_9

biomedical applications. This comprehensive chapter provides valuable insights into the synthesis, extraction, and applications of cyanobacterial bioactive compounds, contributing to the growing body of knowledge in the field of natural product research and pharmaceutical development.

Keywords Cyanobacteria · Bioactive compounds · Synthesis · Extraction techniques

9.1 Introduction

Cyanobacteria are a specialized group of kingdom algae that comprises almost 2000 species and 150 genera, and they appeared almost 3.5 billion years ago (Vincent 2009). They are known to be oxygenic autotrophs and gram-negative, and they have evolved to play their role in biotic and geochemical transitions also including the role in creating oxygenic photosynthesis. Algae are the most fundamental producers of this planet which are efficient in generating biomass by using inexpensive solar energy, nutrients from water in diverse environmental conditions, and atmospheric CO_2. Therefore, they are critically important for the sustainability of the ecosystem on Earth, and the most significant impact on climate performed by them is their responsibility to reduce the greenhouse effect as they capture atmospheric CO_2 and deposit it in the bottom layers of oceans (Falkowski 2002). Most commonly, they are associated with toxic algal blooms with their increased risks to animals and human health due to their potential ability to produce a variety of tumor-promoting, hepatotoxic, and neurotoxic secondary metabolites. Besides these, this group of marine organisms are promising sources of many biologically active compounds by their primary or secondary metabolism. Various cyanobacterial species produce a range of secondary metabolites particularly belonging to genera of *Calothrix*, *Synechocystis*, *Oscillatoria*, *Lyngbya*, etc. Their versatility in adaptation to inhabit a diverse range of environments, variability of chemical defense systems against various macro-grazers, and morphology along with some other factors are involved in producing various natural products. The natural products produced by them are grouped according to biosynthetic origins such as isoprenoids, cyano-peptides, polyketides, alkaloids, and other metabolites. Intensive research is being done to investigate the toxicity and use of these compounds for increased biotechnological interests and pharmaceutical applications (Raja et al. 2016).

The composition of cyanobacterial compounds is 9% amides, 40% lipopeptides, 5.6% amino acids, 4.2% macrolides, and 4.2% fatty acids (Ahmad et al. 2020). The most active component of these compounds is lipopeptides which act as antitumor, antibiotic, antimalarial, cytotoxic, multidrug resistance reversing agents, herbicides, antiviral, immunosuppressive agents, anti-mycotic, and antifeedant. Besides their role in producing important pharmaceutical compounds, some species of blue-green algae are also distributed commercially as the dietary supplements of organic algae (Burja et al. 2001). The species include *Chlorella* sp. *Aphanizomenon flos-aquae*, and *Spirulina* sp. and contain significant amounts of minerals, vitamins, lipids, proteins, chlorophyll, and other unique pigments, and they may have potent health-enhancing probiotic compounds. *Spirulina* sp. is a rich source of vitamin B (particularly Vitamin B12), minerals (zinc, iron, potassium, selenium, calcium, manganese, magnesium), and some carotenoids. It is proven an important source of gamma linoleic acid (GLA) which is an essential fatty acid that helps in stimulating hair and skin growth, regulates metabolism, and maintains the reproductive system and bone health. Minerals and vitamins obtained have antioxidant properties which have the potential to eliminate toxins and actively fight against diseases (Raja et al. 2016).

This group of immense importance is not only useful for pharmaceuticals but also a source of food in aquaculture to feed fish and other creatures of economic importance like crustaceans and molluscs. They are also found important in treating wastewater in industrial processes to remove contaminants and biological detoxification, in the generation of biofertilizers, biofuels, synthesis of cosmetics, and biopolymers. Out of several million existing species of microalgae, most of the species are unexplored. It is expected that the unexplored species are of great economic importance and may act as a treasure house to identify or extract new bioactive molecules. However, recently the emphasis is on the cultivation of these small creatures at a large industrial scale to generate biomolecules or a variety of bioactive compounds commercially (Hassan et al. 2022). This chapter includes all the bioactive compounds obtained from cyanobacteria with their potential uses in different lifesaving and lifestyle drugs. Different techniques for the extraction of bioactive compounds and the process of synthesis for these compounds at the genetic level, are discussed in this chapter.

9.2 Bioactive Compounds Produced by Cyanobacteria

9.2.1 Alkaloids

Cyanobacterial alkaloids include these main groups, e.g., ion channel blocker saxitoxins, neuromuscular transmission blocker anatoxins, β-methylamino L-alanine (the degenerated amino acid), guanidine alkaloid cylindrospermopsins as a protein synthesis inhibitor, and indole alkaloids of cyanobacteria for cytotoxic, antiviral, and antifungal activity. The isonitrile family contains indole alkaloids such as ambiguine isonitrile, wetwitindolinones, fischerindoles, and hapalindoles that have been isolated from cyanobacteria having branched filaments and belong to the genera of *Hapalosiphon*, *Westiellopsis*, *Fischerella*, *Westiella*, etc. They all contain a polycyclic carbon skeleton which is derived from geraniol pyrophosphate and L-tryptophan due to which they possess various biological activities including cytotoxic, antiviral, and antibacterial activities (Vasas et al. 2010). Hapalindolinone A obtained from *Fischerella* works as vasopressin antagonists in the prevention or treatment of disease states like congestive heart failure, edema, hyponatremia, and hypertension. Synthetic products and their chemists are captivated by the alkaloid family of cyanobacteria worldwide where the various applications of welwitindolinones have been observed (Baran and Richter 2005).

9.2.2 Terpenoids

Terpenoids being one of the groups in natural products have a variety of commercial applications. Among different hosts of microbes, cyanobacteria have the potential to produce terpenoids sustainably by using CO_2 and light. Some species of cyanobacteria and their enzymes such as *Synechococcus* sp. PCC 7002, *Synechococcus elongatus* PCC 7942, and *Synechocystis* sp. PCC 6803 have been evolved as a promising host in recent years to produce terpenoids. These unicellular bacteria are the natural producers of terpenoids as an important component of their photosynthetic machinery and include hopanoids and carotenoids along with the phytol tail of chlorophylls. Terpenoids possess extensive applications in cosmetics, pharmaceuticals, flavorings, colorants, agrichemicals, disinfectants, and fragrances. Due to all these applications, terpenoids are in demand, but unfortunately, the amount of terpenoids obtained from plants and cyanobacteria is relatively low, and they are difficult to be artificially synthesized due to their complex chemical structure and cost. So, scientists have endeavored to engineer cyanobacteria metabolically using synthetic biology approach to increase the content of terpenoids in them. Cyanobacteria no doubt give the advantage of easy modification through genetic engineering and producing higher densities in photobioreactors, with efficient and simpler extraction of terpenoids or other products and easy purification procedures (Pattanaik and Lindberg 2015).

9.2.3 Polysaccharides

Polysaccharides are present in all organisms where they serve as the source of energy and carbon; they are excreted in both stressful physiological processes and normal processes. They are used as gelling and thickening agents; in addition they have antibacterial, immunomodulatory, antimutagenic, radioprotective, antioxidative, anticancer, antiulcer, anticoagulant, and anti-inflammatory activities. Cyanobacteria or microalgae synthesize polysaccharides that range from ~0.5 g/L to 20 g/L. They produce phyco-colloids, a unique type of polysaccharides prepared by various species of seaweeds. Among different phyco-colloids, alginates, agar, and carrageenan are of immense importance due to multifunctional use. The significant roles they play being antitumoral, antiviral, antioxidant, and anticoagulant are well documented. Alginate also known as algin or alginic acid made up of linear polysaccharides is found in algae and cyanobacteria (Singh et al. 2017). They are extracted from brown algae, e.g., Laminaria, and are extensively used in printer's ink, insecticides, paints, and pharmaceuticals. Moreover, they are used in the textile industry, as gelling agents for the sizing of cotton yarn, and are equally important in pharmaceutical and food industries. Two polysaccharides obtained from broken cell walls of boiled algal species are combined to form agar which is well-known due to its applications. It is used in the food industry as a thickening, stabilizing, or emulsifying agent. Moreover, it also has microbiological applications where it serves as a growth medium for fungi and bacteria in petri dishes due to its property of solidification. Carrageenan, a high molecular weight polysaccharide produced by red algae, is important for the dairy industry, air freshener gels, and toothpaste and in the processing of meat (Mazard et al. 2016).

9.2.4 Pigments

Cyanobacteria are the wider class of photosynthetic microorganisms along with a highly developed system that harvests light and is composed of pigments. The variety of these pigments (their size, arrangement, and cellular localization) makes cyanobacteria relevant to survive in the extreme niches on Earth. The three main classes of cyanobacterial photosynthetic pigments are carotenoids, chlorophylls, and phycobiliproteins. These pigments have immense biotechnological applications due to their bioactive features (e.g., antiviral, antitumor, antioxidant, etc.) and are used in cosmetics, pharmaceuticals, feed, etc. Their natural bright color is highly appealing to the textiles industry and to food colorants (Assunção et al. 2022). Chlorophyll is the most fundamental pigment which is responsible for oxygen-based photosynthetic activity, and it is present in all photosynthetic cyanobacteria and microalgae. Carotenoids include primary and secondary pigments; their profile is changed in different species where the common and commercialized carotenoids include β-carotene, astaxanthin, and lutein. Furthermore, phycobiliproteins are the specialized class of pigments present in red algae and cyanobacteria and are the major light absorbers among all organisms, and the most common pigments of this class are phycoerythrin and phycocyanin. The demand for natural colorants rather than synthetic ones makes them peculiar as the sustainable source of high-value-added products at a large scale by cyanobacteria (Raja et al. 2016). Phycobiliproteins have gained importance in developing phycoflour probes used in immunodiagnostics. C-phycocyanin is constituting 15% of the dry weight of blue-green algal biomass and contributes toward hepatoprotective, anti-inflammatory, antioxidant, and neuroprotective roles.

9.2.5 Cyclic Peptides

An extensive array of biologically active and structurally novel depsipeptides and peptides are found in cyanobacteria. This is a broad class that contains lipophilic cyclic peptides, cyclic hexapeptides, cyclic depsipeptides, and cyclic undecapeptides. Secondary metabolites of cyanobacteria are identified and isolated from this class. *Lissoclinum patella* ascidians have been reported to produce biologically active cyclic peptides. They are potentially important in pharmaceutical leads and have multidrug reversing activities and antimalarial and antitumor activities. Cyanobactins is a collective proposed name for cyclic peptides that have isoprenoid amino acid derivatives or heterocyclized amino acids. They were initially known to contain thiazolines and oxazolines with their oxidized derivatives thiazoles and oxazoles. Over 100 cyanobactins are derived from the symbiotic association between ascidians and cyanobacteria (Sivonen et al. 2010). Cryptophycin is a family of 16-membered macrolides which are antimitotic depsipeptides that have cytotoxic activity even against multidrug-resistant tumor cells. Peptides are produced ribosomal synthesis pathway and non-ribosomal synthesis pathways in cyanobacteria. The most common cyclic peptides are microcystins, the cyclic heptapeptides in association with various cyanobacterial poisonous blooms that are found in eutrophic freshwater lakes. Many important classes of toxins for cancer cells that present apoptotic activities are characterized by marine cyanobacteria, e.g., Lyngbyabellins K–N, Grassypeptolides A–C, Veraguamides, Viequeamide A, Tasiamides C–E, Bouillonamide, Apratoxin D, Hantupeptin A, and Lagunamide C are the antineoplastic agents originated from cyanobacterial metabolites (Raja et al. 2016).

9.2.6 *Phenols and Fatty Acids*

Polyphenols are included in the group of secondary metabolites like flavonoids (chalcones, flavanones, flavones, flavanols, flavanonols, and flavan-3-ols), tannins, lignin, and phenolic acids. Phenolics are characterized as the stress compounds that take part in defense mechanisms when encountered by biotic stresses like settlements of bacteria, grazing, and abiotic stresses like metal toxicity and UV irradiation. Moreover, a cyanobacterium, *Microcystis aeruginosa*, has been found to pose inhibitory effects on the growth in the presence of some polyphenols such as catechin, gallic acid, and ellagic acid. The main compounds of phlorotannins (fuhalols, fucols, fucophlorethols, phlorethols, sulfated, and halogenated phlorotannins) are of great potential under oxidative stress and are responsible for curing diseases particularly caused by free radicals. Researchers have found many medical and therapeutic applications of phenolic compounds extracted from microalgae or cyanobacteria (Singh et al. 2017).

Among different metabolites, fatty acids are important due to their role in metabolic processes. Algae and cyanobacteria contain important fatty acids like arachidonic acid, linolenic acid, and linoleic acid which are the prerequisites for healthy growth. Alcohols and fatty acids are the main ingredients of lipids, and due to their configuration, a variety of glycolipids, waxes, phospholipids, and fats is found. Several species of microalgae have capacity to accumulate high amount of lipids that can serve as a good source of oil yields; the average lipid content varies from 1% to 70% of dry weight and reach up to the highest amount of 90% of dry weight. Among a diverse variety of fatty acids, PUFAs (polyunsaturated fatty acids) are of great concern due to increasing global market demand due to extensive health benefits. Moreover, they are involved in the regulation of a range of cellular processes like membrane fluidity, transport of electrons and oxygen, and heat adaptation. The presence of two or more double bonds in PUFAs makes them valuable from a nutraceutical point of view (Huang and Zimba 2019).

9.2.7 Vitamins

Living organisms utilize various kinds of vitamins which include lipid-soluble vitamins (vitamins K, D, A, and E) and other water-soluble vitamins (Vitamin B1, B2, B3, B5, B6, B7, B9, B12, and Vitamin C). Cyanobacteria are important in producing vitamins, K, E, C, B, and A in response to stress conditions in their environment. Not all the vitamins can be produced by plants as they are scarce in plants like vitamins K, B, and D, while these vitamins are actively generated in cyanobacteria. Cyanobacterial products are fascinating and gained importance due to their higher content of amino acids, proteins, minerals, and vitamins. The products of *Spirulina* sp. are promoted due to their content of vitamin B12, micronutrients, and proteins. *Anabaena cylindrica* is considered an exceptional source of vitamin C that poses an antioxidant activity in the defense system. A cyanobacterial species *Arthrospira* sp. is responsible to generate vitamin D in response to the stress of UV radiation to prevent cell membrane destruction. A significant amount of vitamin K1 is observed to be produced by the marine creature *Anabaena cylindrical* when compared to its content in spinach and parsley. *Spirulina* is also a rich source of Vitamin E and β-carotene due to which it has medicinal importance. Cyanobacteria are denoted as one of the most capable beings for many biotechnological applications that ensure the availability of eco-sustainable production of natural bioactive metabolites. They help in fulfilling the vitamin deficiency in organisms that they cannot gain through their food and those which are difficult to synthesize chemically. They are a rich, easy, sustainable, and cost-effective source of vitamins among all other producers (Żymańczyk-Duda et al. 2022).

9.3 Extraction Techniques for Cyanobacterial Bioactive Compounds

Various metabolites are present in algal biomass including proteins, carbohydrates, carotenoids, phenolics, lipids, and so on. The extraction strategy is selected based on the chemical nature of the metabolite to be extracted. In recent years, bioactive metabolites of algae are being extracted with the use of a combination of green extraction methods that include pressurized liquid extraction (PLE), microwave-assisted extraction method, and supercritical fluid extraction (SFE) rather than using conventional techniques which are laborious and time-consuming and use large quantities of organic acids (Hassan et al. 2022). Here, we have classified the extraction methods into two types, traditional and modern techniques, and summarized the comparison of these techniques in Table 9.1.

9.3.1 Traditional Extraction Methods

Literature evidence showed that several conventional extraction techniques are well exploited to extract bioactive compounds from different materials including marine microalgae. Existing classical or conventional techniques include hot continuous extraction, maceration, infusion, percolation, decoction, Soxhlet extraction, and hydro-distillation which is further divided into three main categories, i.e., direct steam distillation, water and steam distillation, and water distillation. Many abovementioned techniques are strongly dependent upon potentially influencing parameters, e.g., concentration, sample size, extracting power of solvents, etc. The Soxhlet extraction is a solvent-based technique that requires pre-digestion by using acids and is a time-consuming process due to which its applications as modern extraction techniques are limited.

Moreover, crude extracts obtained in many cases are further subjected to either purification and fractionation or anyone by using solvent partition and fractionation. In addition to this, some processes are multi-stepped like in maceration, while others require large water consumption and energy consumption as in hydro distillation; these all possess serious concerns. Some other major limitations or drawbacks of the traditional extraction technique are prolonged procedures, the need for highly pure solvents and evaporation of higher amounts of these solvents, selectivity, low extraction yield, and thermal decomposition of thermal sensitive ingredients (Sosa-Hernández et al. 2018).

9.3.2 Modern Extraction Techniques

To tackle the limitation and gaps of conventional techniques, new promising and alternative non-conventional techniques are proposed and reported in recent literature. The modern techniques include EAE, PLF, MAE, and SFE among others. Many of the extraction techniques are considered green in nature as they meet all the standards set by US Environmental Protection Agency (EPA). As compared to traditional techniques, the modern ones have advantages like eco-friendly processing conditions, safer auxiliary solvents, almost no or less use of hazardous chemicals or substances, reduction in derivative formation, facile preparatory steps, use of renewable feedstocks, higher or greater efficiency, protection and de-protection steps are

Table 9.1 Comparison of conventional and modern extraction techniques for cyanobacterial bioactive compunds.

Conventional extraction techniques						
Names of techniques	Type of method	Selection of bioactive compound	Strain used	Advantages	Limitations	References
Hot continuous extraction (Soxhlet extraction)	Thermochemical	Lipids or oils	*Chlorella* sp., *Spirulina* sp.,	Automatic continuous method, higher extraction efficiency, requires less time, and solvent consumption than other traditional techniques	Extraction time and high temperature can result in thermal degradation	Aravind et al. (2021)
Folch extraction	Chemical based	Lipids	*Micractinium* sp., *Hindakia* sp., and *Scenedesmus* sp.	Do not require high temperature and high pressure, reliable method for higher lipid extraction, based on green solvents, efficient in terms of selectivity	Higher proportion of solvent is required to extract higher proportion of lipids, optimization is required	Onay et al. (2016)
Solid-liquid extraction	Chemical based	Toxins: β-N-methylamino-L-alanine (BMAA) microcystin-arginine-arginine (MC-RR) and microcystin-leucine-arginine (MC-LR)	Blue-green algae *Spirulina* sp.	Ease of operation, low processing cost	Low extraction efficiency, environmental pollution, production of solvent residues, low selectivity	Aparicio-Muriana et al. (2023)
Hydro distillation	Physicochemical	Volatile organic compounds (VOCs)	*Synechococcus* sp. Strain GFB01 cyanobacterium	Suitable for volatile substances, higher extract yield	Some natural compounds are decomposed in steam distillation and water distillation, longer extraction time	do Amaral et al. (2020)

(continued)

Table 9.1 (continued)

Names of techniques	Type of method	Selection of bioactive compound	Strain used	Advantages	Limitations	References
Conventional extraction techniques						
Infusion	Chemical based	*Spirulina platensis*	Pigments: Chlorophyll	Inexpensive and rapid process, lower solvent consumption, improved yield and quality of extract	Water used in it is not a better solvent for many substances	Minchev et al. (2020)
Maceration	Chemical based	Phytochemicals: Quinones, tannins, polyphenols, flavonoids C-Phycocyanin (C-PC)	*Spirulina platensis*	Can be used for thermolabile compounds, soften the material, low cost, low energy consumption	Long extraction time, low extraction efficiency, low extraction yield	Kamble et al. (2013)
Decoction	Heat based	Hepatotoxins	*Microcystis* PCC 7813	Simple, easy inexpensive process with improved quality and quantity of extract, can be used for harsh material as well, volatile substances can easily be extracted from the sample by boiling	Extracts contain a large number of water-soluble impurities and can be used for volatile and thermolabile compounds, hydrolysis efficiency is highly affected by the amount of sample, temperature and pH, high temperature can deactivate the activity of compounds	Metcalf and Codd (2000)
Percolation	Chemical based	Crude C Phycocyanin (C-PC)	*Spirulina* sp.	Efficient than maceration, optimization can make it a suitable method	Saturated solvent is required to be replaced by fresh	Morais Junior et al. (2020)

Reflux extraction	Chemical based	Ionic liquids, fatty acids	Cyanobacterial species	More efficient than maceration and percolation, less extraction time, less solvent consumption	Cannot be used for thermolabile natural products	Ventura et al. (2017)
Bead milling	Mechanical method	Nucleic acids, proteins, fatty acids, etc.	*Chlorella* sp., *Botryococcus* sp., *Chlorococcum* sp., *Scenedesmus* sp., *Nannochloropsis* sp., *Tolypothrix* sp., *Nostoc* sp.	Moderate conditions can reduce energy consumption, fast extraction, efficient cell disruption, effectively extracts the small molecules present in cell, can be used with or without liquid, easy maintenance	Maintenance cost is high, wear occurs due to beads, noisy process, more suitable for hard materials, chances of contamination are more	Hattab and Ghaly (2015)
High-speed homogenization (HSH)	Mechanical process	Lipids, polysaccharides	*Nannochloropsis* sp., *Arthrospira platensis*	Occurs at room temperature, neutral pH, shorter time, HPH with bead milling is an efficient process, extracts lipids efficiently, provides good yield of extracts	Parameters for operating pressure are always required to maintain according to cell wall, energy consumption is very high, non-selectivity, undesirable cell debris is formed, limitations to break hard cell wall	Morais Junior et al. (2020)
Thermolysis, autoclaving, steam explosion	Thermal processes	Lipids, carbohydrates, monomeric sugars, etc.	*Chlorella sorokiniana*, *Phaeodactylum tricornutum*, *Nannochloropsis gaditana*	Low maintenance cost, low temperature can be used to avoid degradation of bio products, combined with Bligh and Dyer method can increase lipid extraction	Physical disruption with low efficiency, high energy consumption, generates large amount of undesirable cell debris, limited to thermal resistant substances	Morais Junior et al. (2020)

(continued)

Table 9.1 (continued)

Conventional extraction techniques						
Names of techniques	Type of method	Selection of bioactive compound	Strain used	Advantages	Limitations	References
				efficiency, promote carbohydrate hydrolysis to convert in monomeric sugars suitable for photosynthesis		
Chemical methods	Chemicals based	Fatty acids, lipids	*Chlorella saccharophila*, *Nannochloropsis salina*	Simple technique, energy efficient, occurs at low temperature, low pressure	Chemical use can raise economic and environmental concerns, target product is chemically contaminated, uses large amount of chemicals	Morais Junior et al. (2020)
Horn sonication	Physical method	Lipids, fatty acids	*Nannochloropsis gaditana*, *Crypthecodinium cohnii*	Efficient disruption method, non-toxic process, low maintenance cost, rapid process,	Expensive, high-energy consumption, localized cavitation	Hattab and Ghaly (2015)
Osmotic shock	Chemical based	Lipids, unsaturated fatty acids, proteins	*Chlorella sorokiniana*, *Chlorella vulgaris*, *Haematococcus pluvialis*	Lower energy consumption, efficient extraction process, simple process, quality of biomass obtained is good	Economically unfeasible at large scale, takes longer time, efficiency for cell disruption is not enough, produces wastewater, not eco-friendly	Morais Junior et al. (2020)
Ultracentrifugation	Physical method	C-phycocyanin, enzymes, proteins	–	Removes impurities and cell debris, separates each required component according to density	Outdated procedure, energy-intensive process, requires maintenance	Morais Junior et al. (2020)

Freeze drying	Thermo-chemical process	Lipids, fatty acids	*Nannochloropsis* sp. *P. tricornutum*, *Scenedesmus* sp., *Dunaliella tertiolecta*	Single step process for drying and extraction, don't affect the activity of bioactive compounds,	Inefficient cell disruption, time consuming, difficult to maintain, high energy consumption, long-term storage requires airtight containers, water required for reconstitution	Hattab and Ghaly (2015)
Micro fluidizer	Physical method	Lipids, fatty acids	*Tetraselmis suecica*, *Chlorella* sp., *Nannochloropsis* sp.	Disruption of cells occurs at room temperature, efficient in extracting lipids, easy cleaning	Requires high energy to use and maintain, not good for protein extraction	Spiden et al. (2013)
Modern extraction techniques						
Pressurized liquid extraction (PLE)	Chemical based	PUFA especially eicosapentaenoic acid [EPA]	*N. oculata*	User-friendly, rapid, and safe extraction, can increase efficiency under high temperatures and pressure	High cost for investment, for analytical use only	Pieber et al. (2012)
Enzyme-assisted extraction (EAE)	Biological process	Peptides, amino acids and proteins, lipids	*Chlorella*, *Chlamydomonas reinhardtii*	Higher extraction efficiency due to enzymatic hydrolytic action, increased extract yield, selective enzymes are used for cell wall breakdown and safe extract collection, can extract targeted bio compound	Enzymes require maintained conditions to act and are expensive, low production capacity	Safi et al. (2014), Sierra et al. (2017)

(continued)

Table 9.1 (continued)

Conventional extraction techniques						
Names of techniques	Type of method	Selection of bioactive compound	Strain used	Advantages	Limitations	References
Microwave-assisted extraction (MAE)	Electromagnetic	Lipids	*Nannochloropsis* sp., *Tetraselmis* sp., *Chlorella* PY-ZU1	Moderate investment, no hazardous fumes, loss of volatile compounds can be avoided, airborne contamination can be avoided	Lipids may oxidize due to energy; vessels require cooling to avoid loss of volatile substances	Teo and Idris (2014)
Ultrasonic microwave-assisted extraction (UMAE)	Electromagnetic	Proteins	–	Accelerates substance dissolution, shortens extraction time	Heat of system increases due to which thermolabile compounds are not suitable to extract	Morais Junior et al. (2020)
Liquefied gas extraction (LGE)	Chemical based	Lipids	*Arthrospira platensis* Bacteria	Liquefied and compressed gas has ability to dissolve natural substances, carried out at room temperature, minimal energy consumption, residual amount of solvent in extract is negligible	Hazardous due to flammable vapors	Hoshino et al. (2017)
Ultrasound-assisted extraction (UAE)	Electromagnetic	Polysaccharides	*Chlorella pyrenoidosa*	Consumes smaller fossil energy, low investment cost	Active constituents are deleteriously affected by ultrasound energy, larger volume of solvent is required for filtration step, repeated extractions are required	Shi et al. (2007)

Supercritical fluid extraction (SFE)	Chemical based	Triglycerides, tocopherols, and fatty acids	*I. galbana*, *C. cohnii*, *N. oculata*, and *T. suecica*	Pure extract, maintains quality of product, low operation temperature, free of organic salts and heavy metals, fast and high yield, effective	Expensive, complex equipment, use of elevated pressure, not suitable for polar substance, high power consumption	Hassan et al. (2022)
Pulsed electric field-assisted extraction technique	Electrical	Lipids	*Ankistrodesmus falcatus*, *Chlorella pyrenoidosa*	No addition of chemical and heat, drying or dewatering step is not required, reduced operational cost, maintains quality of product, less time consuming	Optimization of parameters and process conditions are required to maintain	Zbinden et al. (2013)
Subcritical water extraction (SCWE)	Thermal process	Microalgal carbohydrates	*D. salina* and *H. pluvialis*	Ecofriendly, can extract a variety of biological compounds, use of high pressure and temperature can change the dielectric constants and polarity of solvents to maintain biological activities of products	High pressure and high temperature require attention for safe operation, under high temperature organic degradation may occur	Santoyo et al. (2009)
Solvent extraction	Chemicals based	Hydrophilic proteins, phycobiliproteins, phycocyanobilin (PCB), phycoerythrobilin (PEB), phycobiliviolin (PVB)	Microalgae, macroalgae and cyanobacteria *Spirulina* sp., *Arthrospira platensis*	Several extractions can be done in parallel, requires little training, more sample mass is extracted than others	Larger amount of extractants is wasted, flammable, hazardous organic solvents are used, requires clean up step	Kovaleski et al. (2022)

(continued)

Table 9.1 (continued)

Conventional extraction techniques						
Names of techniques	Type of method	Selection of bioactive compound	Strain used	Advantages	Limitations	References
Ionic liquid extraction (ILE)	Electromagnetic	Lipids	*Galdieria sulphuraria*, *Chlorella sorokiniana*, *Nannochloropsis salina*	Low volatility, greater thermal stability, low or negligible vapor pressure, can extract a variety of bioactive compounds	High preparation cost	Pan et al. (2016)
Ultrasound-assisted extraction	Electromagnetic	Phycobiliproteins, phycocyanin, phenolic compounds	*Spirulina* sp.	Uniform distribution of energy, low cost,-efficient process, simple operating process, increased extraction performance, faster kinetics, and rate of extraction	Effect of ultrasound waves varies at different positions of container, optimization is required	Kovaleski et al. (2022)

avoided, efficient use of water and energy resources, overall cost-effective ratio, prevention of degradation, etc. (Michalak and Chojnacka 2014).

Supercritical fluid extraction (SFE) is a technique that addresses many drawbacks of conventional techniques like long extraction time, waste generation, high energy consumption, etc. which were avoided in this technique. It shows great extraction selectivity, low degradability of extracted useful compounds, less extraction time, requiring minimal solvents, chemical inertness, non-toxic nature, cost-effectiveness, and non-flammability which are some key features of this technique. Microwave-assisted extraction (MAE) is a commendable technique that provides higher titer yield of value-added quality extracts of biological compounds of great economic and industrial interests. Pressurized liquid extraction (PLE) is the best technique to extract polar compounds; it is also known as enhanced solvent extraction (ESE). Like other green extraction techniques, this method also has such advantages that can successfully extract bioactive compounds with other benefits. Enzyme-assisted extraction (EAE) has received much attention over other techniques due to its greater efficiency and process recyclability. The range of enzymes used as catalysts in this process for pre-treatments, breakdown, and hydrolysis makes it an organic and green process (Pagels et al. 2021).

9.4 Synthesis Pathways and Genetic Manipulation of Cyanobacteria

Owing to the source-based variation, extreme diversity, and physiochemical properties, the precise classification of bioactive compounds is not established. It is difficult due to certain similarities and dissimilarities between chemically interrelated compounds and functions. Besides other different perspectives, they are also classified based on synthetic pathways and three categories of bioactive compounds, where the major biosynthetic pathways are the mevalonic acid pathway, non-mevalonate pathway, malonic acid pathway, and shikimic acid pathway. Natural products of cyanobacteria are classified as alkaloids, UV-absorbing compounds, polyketides, peptides, fatty acids, and terpenes. Synthesis of peptides in cyanobacteria occurs through non-ribosomal and ribosomal pathways. Synthesis of fatty acid occurs in two different pathways: type I is formed by the large multifactorial proteins and type II is formed by discrete proteins that act together (Joutsen and Keiko 2015).

Isoprenoids and terpenoids possess great structural diversity within a family essential for living organisms. In cyanobacteria, this group of bioactive compounds is synthesized by methylerythritol-phosphate (MEP) pathway with the use of pyruvate and glyceraldehyde 3-phosphate produced by the process of photosynthesis as substrates. As cyanobacteria have the potential to produce a range of products through their natural synthetic pathways so based on current findings, scientists, researchers, and industrialists are engaged in establishing the frameworks to obtain higher quantities of valuable products from these potential microorganisms. They are using cyanobacteria and existing knowledge to create powerful systems to use cyanobacteria in a range of settings including sustainable agricultural practices, biofuels, therapeutic procedures, environmental impacts, and other relevant by-products.

9.5 Engineering Cyanobacteria for Enhanced Bioactive Compound Production

Complete genome sequencing was begun with the model photosynthetic organism *Synechocystis* sp. PCC 6803, and almost 200 genomes of cyanobacteria have been sequenced. Cyanobacteria are more amendable to genetic manipulation. Bioengineering applications of cyanobacterial species have received significant importance during the last few decades. Multi-omics approaches and modern gene editing techniques have provided new insight into the methods of acquiring genetic material from cyanobacterial species. The reliable technique is a cis genetic alteration in cyanobacteria by genome editing as many strains are susceptible to conformational change and cellular and molecular modifications for genetic insertion, deletion, and mutation. They are modified using the plasmid carrying desired gene flanking regions similar to desired genomic loci and a marker gene. This method used *Synechococcus elongatus* PCC 7942 and *Synechocystis* sp. PCC 6803 for expressing heterologous genes in synthetic pathways for improved production. They are a diverse class of natural CRISPR systems where *Coleofasciculus chthonoplastes* PCC 7420 has the highest number of CRISPR loci in the genome. CRISPR-based genome editing technique is an emerging method for the effective engineering of desired products. Genetically engineered cyanobacteria can efficiently convert photo energy in enhanced form by using water and sunlight only without additional organic carbon. This offers the potential development of an eco-friendly, clean, renewable, alternative source of energy. A carbon compound poly-β-hydroxybutyrate found in biodegradable polymers is obtained from cyanobacteria. *Synechocystis* sp. PCC 6803 was investigated to check the effect of genetically engineered and nanoparticle research for increased production of this polymer. It was studied to explore more and for excess production of this value-added product. *Alcaligenes eutrophus*, a recombinant PHA-negative mutant, was transformed with a vector containing pha (exogenous) phaECSyn or hybrid phaECAB gene from *A. eutrophus* or *Synechocystis* sp. PCC 6803. The production of PHA and PHB was investigated in both engineered and non-engineered cyanobacteria. The amount of PHB was observed high in the dry cell weight (Govindasamy et al. 2022).

HpTG and GroEL chaperons are upregulated by *Nostoc entophytum* ISC32 in hydrocarbon-contaminated soil in response to Cd. The remarkable Cd absorption capacity is due to chaperones and the powerful antioxidant system of this strain.

Cyanobacterial bioreporters are an important tool to assess the toxicity, bioavailability, and ROS production from metallic nanoparticles. These cyanobacterial bioreporters are recombinant as the reporter genes (luxCDABE) of *Photorhabdus luminescens* are linked with the promoters of oxidative stress, metal-sensitive, and toxic genes to make up a bioreporter. Genetically engineered nanosized cyanobacteria modify the nutritional status and adaptive immunological response of transporting proteins and enzymes and hence proved aquaculture's next frontier. The nano-formulations of genetically engineered cyanobacterial high value therapeutic products make them more beneficial. Another use of cyanobacteria in biotechnology records its role in gene delivery for crop improvement through cyanobacterial-mediated nanoparticles. Some cyanobacterial strains are also improved for biofuel production (Kothari et al. 2023).

9.6 Metabolic Engineering Approaches

Metabolic engineering is defined as the practice of optimization of regulatory and genetic processes to increase the production of required metabolic substances. The technology has been used in bacteria as well as cyanobacteria to expand the product list and production efficiency. Traditional approaches for improved production were based on the selection and targeting of the individual gene and mutation, or random mutagenesis took considerable time for design and implementation. In recent years systems metabolic engineering has now emerged as a new method to solve issues. It is based on using mathematical models to predict and simulate the behaviors emerging in complex systems and is used extensively to improve microbial production. In parallel, the principle of synthetic biology has promoted a bottom-up approach to design biological systems, recombining the defined parts and modules of existing systems restructured to build de novo pathways. It is based on the construction of intricate biological systems by using well-characterized interchangeable and standardized biological parts or modules.

These modules collectively form toolboxes of components used to modify organisms and include characterized biological parts (promoters, riboswitches, ribosomal binding sites (RBS), and terminator libraries) along with standardized assembly methods, manipulated genetic components, and predictive models to facilitate the optimization of pathways. Some important approaches and methods used for cyanobacteria to improve production include (i) engineered promoters for enhanced protein expression; (ii) optimization of ribosome binding site for various applications in biotechnology; (iii) riboswitches are used as tools for the robust control of expression of genes; (iv) use of reporter proteins for easy quantitation of gene expression; (v) modular vector system to engineer cyanobacteria; (vi) markerless selection is used as a tool to facilitate the engineering of cyanobacteria; (vii) predicting and engineering metabolism of cyanobacteria via genome-scale models; (viii) use of tools to improve the reconstruction and quality of GSM and (ix) flux balance analysis for metabolic engineering (Santos-Merino et al. 2019).

9.7 Applications of Cyanobacterial Bioactive Compounds

Cyanobacteria are considered the most important group among microorganisms on this planet. They were the early settlers in many oceanic regions and fulfill ecological functions in barren parts of the world's oceans being the major contributors to global nitrogen and carbon budgets. Recently they have caught significant attention due to their applications in various areas of biotechnology. Due to their industrial applications, they are parts of different studies relating to coloring dyes, fertilizers, food additives, biofuels, water treatment, hydrogen production, forestry, cosmetics, bioplastic production, animal feed, energy resource (biogas, biohydrogen, biodiesel, bioethanol, biofuel production), and nanobiotechnological applications. All the applications suggest that the bioactive compounds obtained from cyanobacteria are the treasure for biotechnological studies and ultimately for human beings. Figure 9.1 highlights the important applications of cyanobacterial bioactive compounds.

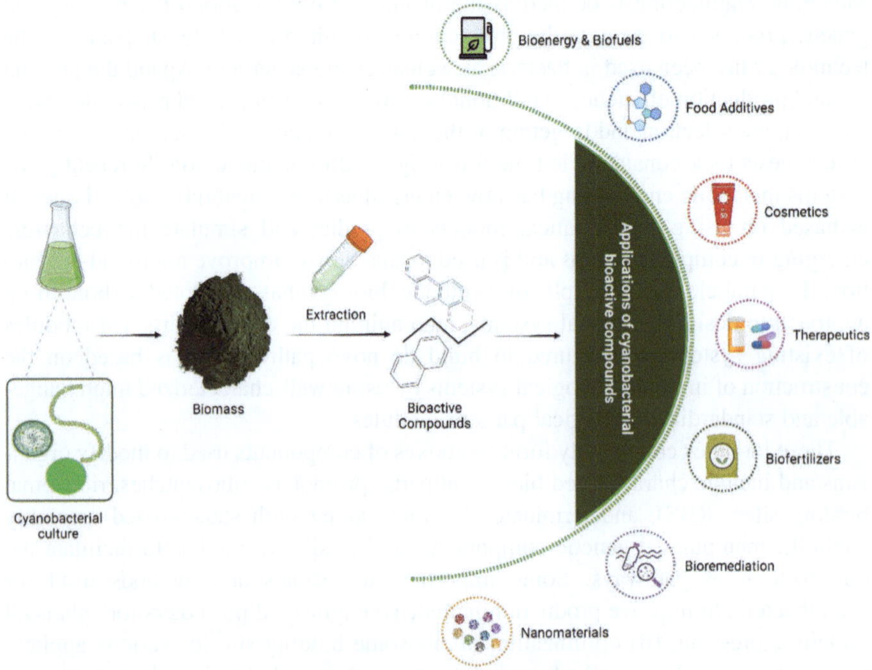

Fig. 9.1 Diverse applications of bioactive compounds from cyanobacteria

9.7.1 Therapeutic Applications

The bioactive compounds of cyanobacteria have anticancer and anti-inflammatory properties and some enzymes, and antibiotics have also been isolated from them which suggests that they are remarkable creatures along with various important biotechnological applications. It is indicated that bioactive compounds obtained from cyanobacteria have a contradiction with synthetic drugs in relevance to their composition and arrangements of atoms or radicals. They are potent in inhibiting the interaction between proteins which results in effective immune response, transduction of signals, mitosis, and apoptosis without posing any harmful impacts on living organisms. Cyanobacteria can also be a potential source of anti-hemolytic or hemolytic effect on erythrocytes in the human body. Stypoldione is a marine natural product that has an anti-quinone functional group to inhibit cell division and other biological processes. Hence, bioactive compounds demonstrate a broad spectrum of activities like protease inhibition, antiviral, antibacterial, antifungal, anti-inflammatory, immunosuppressive, antitumor effects, and antioxidant properties. The drugs obtained from microalgae are cost-effective as the cultivation of cyanobacteria only requires micronutrients which are not costly. Cyanobacteria are rich and potential sources of omega-3 fatty acids, e.g., docosahexaenoic acid (DHA) and eicosapentaenoic acid (EPA) that are known to prevent cardiovascular diseases (Sitther et al. 2022).

9.7.2 Bioremediation

Cyanobacteria are responsible for the bioremediation of wastewater through different processes and mechanisms. It includes various pathways to the transfer of hazardous substances in different mediums. Bioremediation applications of microorganisms are due to their ubiquity, larger surface-to-volume cell ratio, higher rate of reproduction, and small size. Cyanobacterial mechanisms of interactions with inorganic pollutants include bioaccumulation and biosorption (efficient biosorbents to eliminate heavy metals from the environment). They have the capacity to oxidize complex organic compounds present in the environment and are potent degraders and accumulators of different pollutants like crude oils, catechols, naphthalene, pesticides, phenanthrene, xenobiotics, and heavy metals. Biodegradative and biosorptive capacity of some species like *Synechococcus* sp., *Nodularia* sp., *Cyanothece* sp., *Oscillatoria* sp., and *Nostoc* sp. have dominated the effluents from industries. Their spectrum and capacity of activity increase in combination with some bacteria for instance *Oscillatoria* (cyanobacteria) in the presence of gammaproteobacteria, can efficiently degrade pristine, n-octadecane, phenanthrene, and dibenzothiophene. Many species of cyanobacteria can treat industrial effluents and remove metals like nickel and zinc and other pollutants like phosphates and calcium ions, raise pH, and reduce biological oxygen demand (BOD) and chemical oxygen demand (COD). They are also efficient in clearing oil spills as the coastal areas are highly threatened by crude oil pollution that is in turn a threat to the biodiversity and sustainability of the Earth (Priyanka et al. 2020).

9.7.3 Food Additives

Among various metabolites procured from microalgae or cyanobacteria, some fatty acids particularly PUFAs are of great consideration due to their nutritional benefits. Cyanobacteria produce a large amount of PUFAs and so contribute to manufacturing oils and fats at commercial levels. PUFAs especially docosahexaenoic acid (DHA) and eicosapentaenoic acid (EPA) are important in preventing cardiovascular diseases. Microalgal fatty acids are consisting of triacyl-glycerides (TAG) which is a class of lipids and is the best option to replace the function of vegetable oils currently in use. Moreover, *Arthrospira platensis*, a cyanobacterial species, is a rich source of protein feed supplements and fatty acids. Also, the frequent use of *Spirulina* in the human diet can encourage the health of people who suffers immunosuppression, malnutrition, and compromised hepatic and neural functions. Carrageenan obtained from cyanobacteria *Kappaphycus alvarezii* is supplied in the food industry. It is used as a stabilizer or emulsifier due to its suspension forming and thickening ability and is used in numerous foods, e.g., ice creams, jellies, chocolate milk, evaporated milk, dessert gels, meat products, pet foods, salad dressings, puddings, etc. β-Carotene obtained from cyanobacteria is used as a food additive to enhance coloration, in food products like baked goods, margarine, dairy products, fruit juices, canned goods, cheese, and confectionery. It also captures a vast market of microalgal food colorants (Udayan et al. 2017).

9.7.4 Biofertilizers

The food production scenarios require a green revolution in order to cope with demand which requires green technology to make an eco-friendly environment. Cyanobacteria are emerging for sustainable development of agriculture as they have applications in agriculture, e.g., improving fertility of the soil, biocontrol, reclamation of waste land, biofertilizers, improvement of crop productivity, etc. *Diazotrophs* are useful cyanobacteria in generating eco-friendly biofertilizers which are available at a low cost easily. They can efficiently control the nitrogen deficiency in plants, improve water holding capacity, add vitamin B12, and improve the aeration of soil. Cyanobacteria that can fix nitrogen efficiently are *Aulosira fertilissima*, *Scytonema* sp., *Calothrix* sp., *Anabaena variabilis*, *Nostoc linckia*, and *Tolypothrix* sp., which are present in the cultivation area of rice crops. The major action of cyanobacteria includes the excretion of phytohormones (gibberellins, auxins, etc.), amino acids, and vitamins. They decrease soil salinity, control the growth of weeds, efficiently absorb heavy metals on the surface of microbes (bioremediation), make soil porous, produce adhesive substances, after their decomposition and death increase the biomass of soil, and excrete organic acids to ensure availability of phosphates in soil (Chittora et al. 2020). Azolla-Anabaena lives in symbiotic association for nutrient enrichment and nitrogen fixation in rice fields. They also exhibit lysis of lignin present in the cell wall to release phenolic compounds that in turn induce sporulation. Applications of these biofertilizers are reported in lettuce, sugarcane, chilli, maize, radish, tomato, cotton, barley oats, etc. (Thajuddin and Subramanian 2005).

9.7.5 Cosmetics

Cyanobacteria are involved in producing toxic metabolites which are allergens and pose negative effects on human skin and its health (Meléndez-Martínez et al. 2019). Despite the inherent poisoning and cytotoxicity associated with certain cyanobacterial genera, they are also explored for anticancer activities. Some recent studies have shown that cyanobacterial compounds like astaxanthin, carotenoids, phytofluene, and phytoene play healing and antiaging roles for skin appearance and health; therefore, they are used in cosmetics. Non-melanoma skin cancer (NMSC) is increased in the last two decades, so healthcare specialists recommend sunscreens. Cyanobacterial products can prevent damage from UV radiation and skin protection due to which these products are used in cosmetics and sunscreens. Skin bleaching is common all over Asia and is considered a parameter of beauty. Tyrosine kinase inhibitors from cyanobacteria can perform well for this purpose as they can catalyze the rate-limiting step of pigmentation. Chemical-based products are not only costly but also harmful and toxic to the skin and cause aging. Compared to synthetic products, natural or herbal products are of low cost, biodegradable, mild, and safe with low side effects, but unfortunately, they remained in the smaller fraction of the market. Besides cosmetics cyanobacterial products are also important in cosmeceuticals where cosmetic products have active ingredients that pose pharmaceutical therapeutic benefits. Cyanobacteria have a variety of bioactive molecules for health defense, which includes phenols, steroids, vitamins, saponins, terpenes, tannins, phenols, and flavonoids' pigments (phycobiliproteins, β-carotene, and c-phycoerythrin) (Khalifa et al. 2021). Herbal products improve skin appearance and enhance gloss, preventing aging and chronic inflammation.

9.7.6 Bioenergy and Biofuels

Global warming and energy crisis are the main burning problems for human beings in the current scenario which occurred due to disturbed equilibrium in industrialization, population growth, and availability of fossil fuels. In this scenario, cyanobacteria have emerged as a producer of biofuels with their synthesis and provision at minimum cost in an eco-friendly way (Li et al. 2008). Cyanobacteria are capable to convert 10% of solar energy into biomass, while other energy crops and algae have the capability to convert only 1% and 5% of solar energy into biomass. Microalgae produce oil of almost 16–68% of dry weight where the yield is recorded to 136,900 L/ha compared to other plant crops that produce from 172 to 5950 L/ha. Several metabolites produced during the Calvin cycle of microalgae and cyanobacteria are important components of biofuels. Fermentation is a process in which a large amount of carbohydrates is converted into bioethanol butyrate into propionate and fatty acids into acetate at the commercial level. Furthermore, some more green algae like *Nannochloropsis*, *Botryococcus*, *Scenedesmus*, *Dunaliella*, *Chlamydomonas*, and *Chlorella* are important in providing raw materials for biodiesel production. High lipid content, no seasonal limitation on culturing, high growth rate, and resistant nature under different conditions of environmental stimuli make cyanobacteria perfect and promising for biodiesel production at low costs. The lipid profile of microalgae is checked as it is important for the excellency of biodiesel and can be used in the heating of power engines and the efficient combustion process (Heimann 2016).

9.7.7 Nanobiotechnological Applications

Nanotechnology and nanoscience as a modern discipline have a variety of applications in fundamental science. It has roles in many areas from health and life sciences with a main focus on ecofriendly techniques. This field has encouraged the new way to prevent naturally occurring hydrophobic marine medicines with their low water solubility by using highly complex eukaryotes and simple bacteria. Nanotechnology has proved the fastest industrial and medical platform that can be implemented using desirable methods to improve bioavailability, solubility, and stability (Mohanpuria et al. 2008). Marine cyanobacteria have many applications in the field of nanobiotechnology; first is their direct use to produce nanoparticles of various metals and second is through nanotechnological processing of bioactive metabolite obtained from them to produce medicines. Many cyanobacterial species *Plectonema boryanum*, *Cylindrospermopsis* sp., *Anabaena* sp., *Lyngbya* sp., *Synechocystis* sp., and *Synechococcus* sp. are known to be incorporated in producing silver nanoparticles. Some other genera of cyanobacteria *Leptolyngbya*, *Calothrix,* and *Anabaena* have applications in modifying the nanoparticle shape of platinum, palladium, gold, and silver. Such nanoparticles of metals have antimicrobial impacts against many bacteria including *Micrococcus luteus*, *Bacillus megaterium*, *Bacillus subtilis*, *Staphylococcus aureus*, *Escherichia coli*, and *Pseudomonas aeruginosa*. Nanoparticles of silver have great importance in areas particularly in cosmetics and implantable and impregnating medical devices, along with their antimicrobial effectiveness. Nano-formulated antioxidant, anti-inflammatory, and antiaging medicines or creams have been made from secondary metabolites of cyanobacteria.

Nano-formulated anticancer agents with simplified delivery can be used in many cancer states (Khalifa et al. 2021).

9.8 Conclusion and Prospects

The biologically active compounds obtained from algae and cyanobacteria are under study, and a diverse range of compounds has been explored from the beginning. Cyanobacteria and microalgae are rich sources of various bioactive compounds including fibers, halogenated compounds, antioxidants, steroids, terpenoids, polysaccharides, proteins, vitamins, essential lipids, free fatty acids, polyketides, lectins, pigments, alkaloids, and other metabolites. Due to the presence of such a variety of secondary metabolites, cyanobacterial products have potential health benefits and a sustainable source of pharmaceuticals, food, and feed products. They possess anticancer, antiviral, antibacterial, antifungal, antioxidant, and other medical, health, and pharmaceutical benefits. Other than the medical and pharmaceutical benefits, they are equally important for sustainable agriculture and maintaining the environment. Their applications also include biofertilizer production and bioremediation to save the planet from harmful and toxic substances. Various easy, economical, eco-friendly, cost-effective, and resource-saving complex extraction techniques for the bioactive metabolites are also discussed in this chapter. Furthermore, a complete profile of secondary metabolites of cyanobacteria is required to make them more useful for human beings and for the planet we live in. Some more focus is also required to convert present-day technology to exploit cyanobacteria and algae into green technology. The strategy must be designed to disseminate the commercialization of more products at small and large scales. The production rate of various metabolites must be increased by incorporating biotechnological techniques for accomplishing the future food, feed, nutrients, drugs, pharmaceuticals, nutraceuticals, and cosmetical products. The application of bioengineering techniques can increase the production rate more than plants in the future. This will no doubt reveal more novel secondary metabolites in cyanobacteria and microalgae due to which future research and its commitments will be more promising.

References

Ahmad IZ, Parvez S, Tabassum H (2020) Cyanobacterial peptides with respect to anticancer activity: structural and functional perspective. Stud Nat Prod Chem 67:345–388. https://doi.org/10.1016/B978-0-12-819483-6.00010-2

Aparicio-Muriana MDM, Lara FJ, Del Olmo-Iruela M, García-Campaña AM (2023) Determination of multiclass cyanotoxins in blue-green algae (BGA) dietary supplements using hydrophilic interaction liquid chromatography-tandem mass spectrometry. Toxins 15(2). https://doi.org/10.3390/TOXINS15020127

Aravind S, Barik D, Ragupathi P, Vignesh G (2021) Investigation on algae oil extraction from algae *Spirogyra* by Soxhlet extraction method. Mater Tod Proc 43:308–313. https://doi.org/10.1016/J.MATPR.2020.11.668

Assunção J, Amaro HM, Malcata FX, Guedes AC (2022) Cyanobacterial pigments: photosynthetic function and biotechnological purposes. In: The pharmacological potential of cyanobacteria. Academic Press, New York, pp 201–256. https://doi.org/10.1016/B978-0-12-821491-6.00008-9

Baran PS, Richter JM (2005) Enantioselective total syntheses of welwitindolinone A and fischerindoles I and G. J Am Chem Soc 127(44):15394–15396. https://doi.org/10.1021/JA056171R

Burja AM, Banaigs B, Abou-Mansour E, Grant Burgess J, Wright PC (2001) Marine cyanobacteria—a prolific source of natural products. Tetrahedron 57(46):9347–9377. https://doi.org/10.1016/S0040-4020(01)00931-0

Chittora D, Meena M, Barupal T, Swapnil P (2020) Cyanobacteria as a source of biofertilizers for sustainable agriculture. Biochem Biophys Rep 22:100737. https://doi.org/10.1016/J.BBREP.2020.100737

do Amaral SC, Santos AV, da Cruz Schneider MP, da Silva JKC, Xavier LP (2020) Determination of volatile organic compounds and antibacterial activity of the Amazonian cyanobacterium *Synechococcus* sp. Strain GFB01. Molecules (Basel, Switzerland) 25(20). https://doi.org/10.3390/MOLECULES25204744

Falkowski PG (2002) The ocean's invisible forest. Sci Am 287(2):54. https://doi.org/10.1038/SCIENTIFICAMERICAN0802-54

Govindasamy R, Gayathiri E, Sankar S, Venkidasamy B, Prakash P, Rekha K, Savaner V, Pari A, Thirumalaivasan N, Thiruvengadam M (2022) Emerging trends of nanotechnology and genetic engineering in cyanobacteria to optimize production for future applications. Life (Basel, Switzerland) 12(12). https://doi.org/10.3390/LIFE12122013

Hassan S, Meenatchi R, Pachillu K, Bansal S, Brindangnanam P, Arockiaraj J, Kiran GS, Selvin J (2022) Identification and characterization of the novel bioactive compounds from microalgae and cyanobacteria for pharmaceutical and nutraceutical applications. J Basic Microbiol 62(9): 999–1029. https://doi.org/10.1002/JOBM.202100477

Hattab MA, Ghaly A (2015) Microalgae oil extraction pre-treatment methods: critical review and comparative analysis. J Fundam Renew Energy Appl 05(04). https://doi.org/10.4172/2090-4541.1000172

Heimann K (2016) Novel approaches to microalgal and cyanobacterial cultivation for bioenergy and biofuel production. Curr Opin Biotechnol 38:183–189. https://doi.org/10.1016/J.COPBIO.2016.02.024

Hoshino R, Ogawa M, Murakami K, Wahyudiono K, H., & Goto, M. (2017) Extraction of lipids from wet *Arthrospira platensis* by liquefied dimethyl ether. Solvent Extr Res Dev Jpn 24(1): 47–60. https://doi.org/10.15261/SERDJ.24.47

Huang IS, Zimba PV (2019) Cyanobacterial bioactive metabolites—a review of their chemistry and biology. Harmful Algae 86:139–209. https://doi.org/10.1016/J.HAL.2019.05.001

Joutsen S, Keiko T (2015) Cyanobacterial bioactive compounds: biosynthesis, evolution, structure and bioactivity. University of Helsinki Open Repository

Kamble S, Gaikar R, Padalia RB, Shinde KD (2013) Extraction and purification of C-phycocyanin from dry *Spirulina* powder and evaluating its antioxidant, anticoagulation and prevention of DNA damage activity. J Appl Pharm Sci. 3(8):149–153

Khalifa SAM, Shedid ES, Saied EM, Jassbi AR, Jamebozorgi FH, Rateb ME, Du M, Abdel-Daim MM, Kai GY, Al-Hammady MAM, Xiao J, Guo Z, El-Seedi HR (2021) Cyanobacteria-from the oceans to the potential biotechnological and biomedical applications. Mar Drugs 19(5). https://doi.org/10.3390/MD19050241

Kothari R, Singh HM, Azam R, Goria K, Bharti A, Singh A, Bajar S, Pathak A, Pandey AK, Tyagi VV (2023) Potential avenue of genetic engineered algal derived bioactive compounds:

influencing parameters, challenges and future prospects. Phytochem Rev 2023:1–34. https://doi.org/10.1007/S11101-023-09859-Y

Kovaleski G, Kholany M, Dias LMS, Correia SFH, Ferreira RAS, Coutinho JAP, Ventura SPM (2022) Extraction and purification of phycobiliproteins from algae and their applications. Front Chem 10:1065355. https://doi.org/10.3389/FCHEM.2022.1065355/BIBTEX

Li Q, Du W, Liu D (2008) Perspectives of microbial oils for biodiesel production. Appl Microbiol Biotechnol 80(5):749–756. https://doi.org/10.1007/S00253-008-1625-9

Mazard S, Penesyan A, Ostrowski M, Paulsen IT, Egan S (2016) Tiny microbes with a big impact: the role of cyanobacteria and their metabolites in shaping our future. Mar Drugs 14(5):97. https://doi.org/10.3390/MD14050097

Meléndez-Martínez AJ, Stinco CM, Mapelli-Brahm P (2019) Skin carotenoids in public health and Nutricosmetics: the emerging roles and applications of the UV radiation-absorbing colourless carotenoids phytoene and phytofluene. Nutrients 11(5). https://doi.org/10.3390/NU11051093

Metcalf JS, Codd GA (2000) Microwave oven and boiling waterbath extraction of hepatotoxins from cyanobacterial cells. FEMS Microbiol Lett 184(2):241–246. https://doi.org/10.1111/J.1574-6968.2000.TB09021.X

Michalak I, Chojnacka K (2014) Algal extracts: technology and advances. Eng Life Sci 14(6): 581–591. https://doi.org/10.1002/ELSC.201400139

Minchev I, Petkova N, Milkova-Tomova I (2020) Ultrasound-assisted extraction of chlorophylls and phycocyanin from *Spirulina platensis*. Biointerface Res Appl Chem 11(2):9296–9304. https://doi.org/10.33263/BRIAC112.92969304

Mohanpuria P, Rana NK, Yadav SK (2008) Biosynthesis of nanoparticles: technological concepts and future applications. J Nanopart Res 10(3):507–517. https://doi.org/10.1007/S11051-007-9275-X/TABLES/1

Morais Junior WG, Gorgich M, Corrêa PS, Martins AA, Mata TM, Caetano NS (2020) Microalgae for biotechnological applications: cultivation, harvesting and biomass processing. Aquaculture 528:735562. https://doi.org/10.1016/J.AQUACULTURE.2020.735562

Onay M, Sonmez C, Oktem HA, Yucel M (2016) Evaluation of various extraction techniques for efficient lipid recovery from thermo-resistant microalgae, *Hindakia*, *Scenedesmus* and *Micractinium* species-comparison of lipid extraction methods from microalgae. Am J Anal Chem 7:141–150. https://doi.org/10.4236/ajac.2016.72012

Pagels F, Pereira RN, Vicente AA, Guedes AC (2021) Extraction of pigments from microalgae and cyanobacteria—a review on current methodologies. Appl Sci 11(11):5187. https://doi.org/10.3390/APP11115187

Pan J, Muppaneni T, Sun Y, Reddy HK, Fu J, Lu X, Deng S (2016) Microwave-assisted extraction of lipids from microalgae using an ionic liquid solvent [BMIM][HSO4]. Fuel 178:49–55. https://doi.org/10.1016/J.FUEL.2016.03.037

Pattanaik B, Lindberg P (2015) Terpenoids and their biosynthesis in cyanobacteria. Life (Basel, Switzerland) 5(1):269–293. https://doi.org/10.3390/LIFE5010269

Pieber S, Schober S, Mittelbach M (2012) Pressurized fluid extraction of polyunsaturated fatty acids from the microalga *Nannochloropsis oculata*. Biomass Bioenergy 47:474–482. https://doi.org/10.1016/J.BIOMBIOE.2012.10.019

Priyanka, Kumar C, Chatterjee A, Wenjing W, Yadav D, Singh PK (2020) Cyanobacteria: potential and role for environmental remediation. In: Abatement of environmental pollutants: trends and strategies. Elsevier, pp 193–202. https://doi.org/10.1016/B978-0-12-818095-2.00010-2

Raja R, Hemaiswarya S, Ganesan V, Carvalho IS (2016) Recent developments in therapeutic applications of cyanobacteria. Crit Rev Microbiol 42(3):394–405. https://doi.org/10.3109/1040841X.2014.957640

Safi C, Ursu AV, Laroche C, Zebib B, Merah O, Pontalier PY, Vaca-Garcia C (2014) Aqueous extraction of proteins from microalgae: effect of different cell disruption methods. Algal Res 3(1):61–65. https://doi.org/10.1016/J.ALGAL.2013.12.004

Santos-Merino M, Singh AK, Ducat DC (2019) New applications of synthetic biology tools for cyanobacterial metabolic engineering. Front Bioeng Biotechnol 7(FEB). https://doi.org/10.3389/FBIOE.2019.00033

Santoyo S, Rodríguez-Meizoso I, Cifuentes A, Jaime L, García-Blairsy Reina G, Señorans FJ, Ibáñez E (2009) Green processes based on the extraction with pressurized fluids to obtain potent antimicrobials from *Haematococcus pluvialis* microalgae. LWT Food Sci Technol 42(7): 1213–1218. https://doi.org/10.1016/J.LWT.2009.01.012

Shi Y, Sheng J, Yang F, Hu Q (2007) Purification and identification of polysaccharide derived from *Chlorella pyrenoidosa*. Food Chem 103(1):101–105. https://doi.org/10.1016/J.FOODCHEM.2006.07.028

Sierra LS, Dixon CK, Wilken LR (2017) Enzymatic cell disruption of the microalgae *Chlamydomonas reinhardtii* for lipid and protein extraction. Algal Res 25:149–159. https://doi.org/10.1016/J.ALGAL.2017.04.004

Singh R, Parihar P, Singh M, Bajguz A, Kumar J, Singh S, Singh VP, Prasad SM (2017) Uncovering potential applications of cyanobacteria and algal metabolites in biology, agriculture and medicine: current status and future prospects. Front Microbiol 8. https://doi.org/10.3389/FMICB.2017.00515

Sitther V, Wyatt LD, Jones C, Yalcin Y (2022) Bioactive compounds and pigments from cyanobacteria: applications in the pharmaceutical industry. In: Expanding horizon of cyanobacterial biology. Elsevier, pp 65–90. https://doi.org/10.1016/B978-0-323-91202-0.00009-9

Sivonen K, Leikoski N, Fewer DP, Jokela J (2010) Cyanobactins-ribosomal cyclic peptides produced by cyanobacteria. Appl Microbiol Biotechnol 86(5):1213–1225. https://doi.org/10.1007/S00253-010-2482-X

Sosa-Hernández JE, Escobedo-Avellaneda Z, Iqbal HMN, Welti-Chanes J (2018) State-of-the-art extraction methodologies for bioactive compounds from algal biome to meet bio-economy challenges and opportunities. Molecules (Basel, Switzerland) 23(11). https://doi.org/10.3390/MOLECULES23112953

Spiden EM, Yap BHJ, Hill DRA, Kentish SE, Scales PJ, Martin GJO (2013) Quantitative evaluation of the ease of rupture of industrially promising microalgae by high pressure homogenization. Bioresour Technol 140:165–171. https://doi.org/10.1016/J.BIORTECH.2013.04.074

Teo CL, Idris A (2014) Enhancing the various solvent extraction method via microwave irradiation for extraction of lipids from marine microalgae in biodiesel production. Bioresour Technol 171: 477–481. https://doi.org/10.1016/J.BIORTECH.2014.08.024

Thajuddin N, Subramanian G (2005) Cyanobacterial biodiversity and potential applications in biotechnology. Curr Sci

Udayan A, Arumugam M, Pandey A (2017) Nutraceuticals from algae and cyanobacteria. In: Algal green chemistry: recent progress in biotechnology. Elsevier, pp 65–89. https://doi.org/10.1016/B978-0-444-63784-0.00004-7

Vasas G, Borbely G, Nanasi P, Nanasi PP (2010) Alkaloids from cyanobacteria with diverse powerful bioactivities. Mini Rev Med Chem 10(10):946–955. https://doi.org/10.2174/138955710792007231

Ventura SPM, e Silva FA, Quental MV, Mondal D, Freire MG, Coutinho JAP (2017) Ionic-liquid-mediated extraction and separation processes for bioactive compounds: past, present, and future trends. Chem Rev 117(10):6984–7052. https://doi.org/10.1021/ACS.CHEMREV.6B00550/ASSET/IMAGES/LARGE/CR-2016-00550T_0030.JPEG

Vincent WF (2009) Cyanobacteria. In: Encyclopedia of inland waters. Wiley, pp 226–232. https://doi.org/10.1016/B978-012370626-3.00127-7

Zbinden MDA, Sturm BSM, Nord RD, Carey WJ, Moore D, Shinogle H, Stagg-Williams SM (2013) Pulsed electric field (PEF) as an intensification pretreatment for greener solvent lipid extraction from microalgae. Biotechnol Bioeng 110(6):1605–1615. https://doi.org/10.1002/BIT.24829

Żymańczyk-Duda E, Samson SO, Brzezińska-Rodak M, Klimek-Ochab M (2022) Versatile applications of cyanobacteria in biotechnology. Microorganisms 10(12):2318. https://doi.org/10.3390/MICROORGANISMS10122318

Chapter 10
Threats, Challenges and Issues of Large-Scale Cyanobacterial Cultivation

Ashutosh Kumar, Bhavya Mishra, and Meenakshi Singh

Abstract Cyanobacterial farming is an emerging trend to harness solar energy to produce multiple bioproducts. Additionally, cyanobacterial biomass production has potential to resolve the global warming problem by mitigating the impact of greenhouse gases. However, the massive cultivation of cyanobacteria is still at the budding stage because of technical issues in large-scale cultivation, optimization of strains, and calibration of sustenance medium and production medium in sterile conditions. Based on the abovementioned factors, this chapter discusses different aspects of cyanobacterial production systems that will cover the issues faced in photoautotrophic, heterotrophic and mixotrophic conditions. Their viability, growth, economic profitability and sustainability in open, closed and photobioreactors, respectively. Further, the challenges faced in the optimization process of versatile strains, which are commercially exploited for bioenergy, functional foods and high-value chemicals, were analysed on multi-parameters. We described various risks involved in downstream processing and adaptation to cultivation media specifications. This chapter will help to understand the glitches that occurred during large-scale production systems and summarize new strategies to support the circular bioeconomy.

Keywords Cyanobacteria · Large-scale production systems · Multi-parameter optimization · Challenges in downstream processing · Adaptation in production media

A. Kumar
Department of Biotechnology, Smt. Chandibai Himathmal Mansukhani College, Mumbai University, Ulhasnagar, Maharashtra, India

B. Mishra
Department of Botany, The Maharaja Sayajirao University of Baroda, Vadodara, Gujarat, India

M. Singh (✉)
Department of Ecology and Biodiversity, Sustaina Greens LLP, Vadodara, Gujarat, India

M. A. Mehmood et al. (eds.), *Pharmaceutical and Nutraceutical Potential of Cyanobacteria*, https://doi.org/10.1007/978-3-031-45523-0_10

10.1 Introduction

Cyanobacteria are photoautotrophic organisms, which are microscopic but visible on crusts or blooms. They can fix carbon dioxide (CO_2), nutrients and water (H_2O) by photosynthesis and the Calvin-Benson cycle (CBC) to produce biomass, thus leading to many applications such as carbon sequestration, soil pollution reduction, wastewater treatment and bioaccumulation and bioremediation of persistent organic pollutants (POPs), heavy metals, herbicides, pesticides and dyes (Pathak et al. 2018; Zahra et al. 2020). Moreover, a few cyanobacterial species such as *Nostoc* and *Oscillatoria* are light-independent, capable of growing in dark and anaerobic conditions, and can fix the nitrogen for which they are also termed diazotrophic cyanobacteria (Panjiar et al. 2017; Suikkanen et al. 2021).

The cyanobacterial contribution is noteworthy for its immense role from atmospheric creation to organic substance production as the source of food for eubacteria (Nicoletti 2016; Rahmatpour et al. 2021). Rapid industrialization and overexploitation of natural resources led to increased use of cyanobacteria in diverse applications because of fossil depletion, energy security and environmental issues. The diverse application of Cyanobacteria in renewable bioenergy production, carbon sequestration potential, and versatile high value chemicals are in great demand. Besides these applications, the possibility of generating biofuel from cyanobacteria rather than traditional crops such as wheat, soybean, peanuts, sunflower, sesame, switchgrass, sugarcane, etc. will be revolutionary since solar energy conversion is 10% high; plus this will also have no effect on food economy which is prevalent with first and second generation of biofuels (Zahra et al. 2020).

In recent times, genetically modified cyanobacteria have gained lots of attention as potential feedstocks for the generation of solar-powered biofuels (Zahra et al. 2020). Additionally, some cyanobacterium species are used as biofertilizers like *Anabaena variabilis, Aulosira fortissima, Nostoc muscorum* and *Tolypothrix tenuis* (Vu et al. 2020). Their role as a product coating in pesticides, nutrients and biofertilizers enables targeted delivery (Pathak et al. 2018). This subsequently prevents the wastage of those products caused by their distribution to undesired locations, saves cost and probably fixes eutrophication (Parambath et al. 2022). Cyanobacteria have the potential in reversing desertification and are also called plant growth promoters (PGRs) since their filtrates can increase the production of plant growth regulators such as auxins, cytokinins and gibberellin (Pathak et al. 2018; El-Naggar et al. 2020). However, fear of cyanotoxins-based soil pollution cannot be neglected, and this issue deserves equal attention to be worked upon.

One mode of large-scale cyanobacterial cultivation is strain and/or species identification; their characterization can help in preparing the best optimization conditions specific to each cyanobacterial strain. Apart from characterization, their statistical study also holds significant importance, and the Taguchi model is one such renowned method employable for identifying the set of influential parameters and optimum conditions responsible for high biomass production (Banerjee et al. 2019). Keeping this in mind, the chapter discloses the state-of-the-art overview of the large-scale cultivation of cyanobacteria by focusing on the issues related to photoautotrophic culture and heterotrophic culture along with the challenges in photobioreactor (PBR) upgradation through construction planning for mixotrophic culture. Furthermore, different examples of strains about commodities particular to bioenergy, functional foods and high-value chemicals; possible strategies such as strain selection, gene manipulation and multi-parameter optimization; and challenges associated with circular bioeconomy have been discussed in brief.

10.2 Different Aspects of Cyanobacterial Production Systems

There are mainly three types of cultivation: phototrophic, heterotrophic and mixotrophic which are described in this chapter, and each type holds its significance (Cruz et al. 2018). Figure 10.1 depicts different production systems such as phototrophic cultivation that utilizes photons or light as the energy source along with CO_2 as an inorganic carbon source to synthesize organic components or chemical energy for growth, metabolic function and sustenance of their life. Heterotrophic cultivation involves growth in absence of light using organic carbon sources through respiratory and fermentative routes, and mixotrophic cultivation can survive in both conditions. Cyanobacteria can utilize inorganic carbon (CO_2) as seen in phototrophic cultivation as well as organic carbon as seen in heterotrophic cultivation, both in the presence and absence of light. Another type of photoheterotrophic cultivation is based on cyanobacterial employment of organic carbon as observed

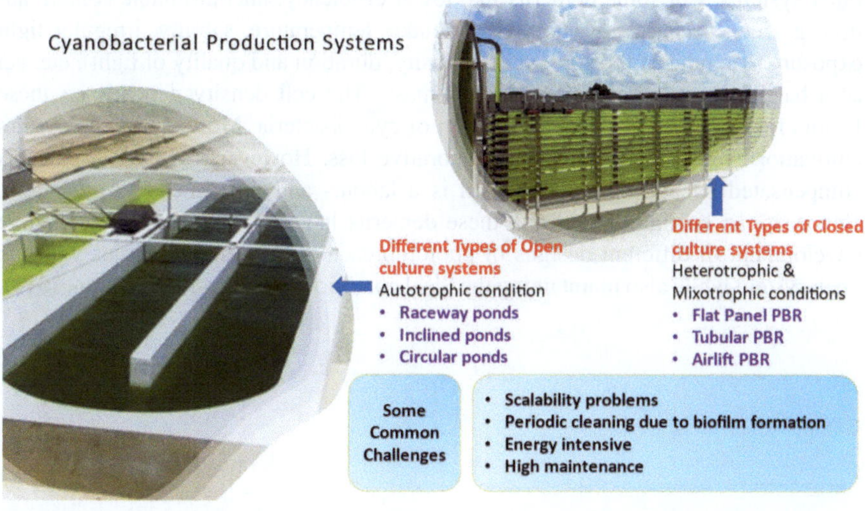

Fig. 10.1 Cyanobacterial cultivation systems

with the heterotrophic type. However, they require light similar to phototrophic cultivation rather than heterotrophic cultivation which use carbon source strictly in dark conditions. Moreover, according to (Cruz et al. 2018), the mixotrophic and photoheterotrophic mode of cultivation is rarely applied since they contaminate easily and require costly photobioreactors for proper distribution of light (Khan et al. 2018).

10.2.1 Phototrophic Culture in Open Systems

The open system based on freely available natural sunlight is the most widely deployed cultivation system. The reasons are their cost-effectiveness and simplicity compared to other systems whose construction, operation and maintenance are far more expensive (Costa and de Morais 2014). There are different classifications of the open system such as circular ponds, raceway ponds (the most widely used), inclined ponds and unmixed ponds depending upon their design, configuration/geometrical pattern and operation (Satpati et al. 2022).

Raceways are well-accepted open systems that are constructed on a closed-loop recirculation channel and are equipped with a rotatory paddle wheel which not only prevents sedimentation but also assists in the proper mixing of phototrophic culture along with nutrient media so that none of the cells remained to receive their energy source light. Although the paddle wheels, pumps, airlifts, jet mixers, mixing boards and gravity flow are responsible for the reduction of energy output and high expenditure, other components such as HDPE, PVC liners, polypropylene and geometric liners used as pond liners further add to the overall cost (Satpati et al. 2022). However, the expenditure on the open system is still less in comparison to other systems. But some demerits like lower efficiency, uncontrollable contamination, grazers, bacterial invasion, CO_2 uptake, temperature, salinity, irregular light exposure (irregular in the context of intensity, duration and quality of light), etc. act as a barrier to achieving maximum biomass. The cell density depends on these factors for the inherent circadian rhythm of cyanobacteria. Moreover, open system cultivation is also entangled with evaporative loss. However, the loss of water is compensated by adding water, which is a labour- and time-intensive procedure. Hence, it can be rightly said that these demerits have necessitated the innovative development of different designs of photobioreactors to overcome issues with the open system while also maintaining the ideal conditions (Costa and de Morais 2014).

10.2.2 Heterotrophic Culture in Closed Photobioreactor (PBR) Systems

Closed system photobioreactors (PBRs) are quite expensive but serve the purpose of their developers by providing 13-fold higher biomass production compared to open system-raceway ponds (Satpati et al. 2022). All credit goes to PBRs offering significant advantages like optimal growth conditions, light (exposure, duration & intensity), mixing, gas exchange, evaporation, high surface-to-volume ratio, and contamination control. Like the open system, closed systems also have been classified into different models such as Flat-Panel PBR, Tubular PBR and Airlift Technology (Fig. 10.1). These are far superior compared to phototrophic in terms of high biomass productivity and are also appreciated due to lower land requirement, low surface area to volume ratio, ability to feed on a diverse variety of carbon sources, easily soluble as well as lower expenditure on downstream processes (Rösch et al. 2019). The efficient utilization of biomass leads to the production of high value end-products like biofuel, fatty acids, proteins etc. Despite its advantages, closed heterotrophic cultivation is less desirable than autotrophic cultivation due to high costs associated with carbon sources, software expense for real-time monitoring, big glass or plastic vessels for uniform light exposure of the cultures for maximum biomass production (Rösch et al. 2019). Apart from that, the removal of oxygen produced during photosynthesis and maintaining sterile conditions add cost and are labour-intensive procedures. However, we should also not ignore the fact that these costs are infinitesimal or small in comparison with high-value compounds.

10.2.3 Mixotrophic Culture in Photobioreactors (PBRs)

As we already discussed, certain cyanobacterial strains can utilize organic carbon as well as inorganic carbon while photosynthesizing concurrently. Such species are not dependent on only one energy source unlike the phototrophic or heterotrophic mode of cultivation. These two kinds of carbon sources play an important role in the continuous supply of carbon which promotes rapid algal growth even during dark conditions. This ability notably forms the basis of a mixotrophic cultivation strategy (MCS) for achieving maximum biomass production (Patel et al. 2021). Indeed, the production is highest in comparison with phototrophic and heterotrophic cultivation. This was evident in (Schwarz et al. 2020) study, in which cyanobacterium *Trichocoleus sociatus* reported a higher cell concentration of 3.77 gL^{-1} (dry weight) grown using 2.53 gL^{-1} of raffinose, while *Nostoc muscorum* reported higher cell cultivation of 2.46 gL^{-1} (dry weight) after 14 days in mixotrophic fed-batch cultivations (fed-batch is cultivation type in which nutrients/feeds/substrates are repeatedly added to increase the duration of culture and produce higher cell densities, and modification of batch cultivation where no additional feed is added). The yield of both cyanobacterial strains cultivated in mixotrophic fed-batch conditions was high compared to phototrophic and heterotrophic cultivation (Schwarz et al. 2020). Moreover, an interesting pattern was seen regarding the role of carbon sources of different types in aiding the yield (Schwarz et al. 2020). *T. sociatus* showed a similar yield no matter whether the carbon source was glucose, galactose, fructose, raffinose or potassium acetate. But *N. muscorum* yielded high cell concentration only when glucose as the carbon source was employed (Schwarz et al. 2020). It requires further research to understand the role of different nutrient compositions on content of value-added compounds. For example, *Arthrospira platensis* grown in mixotrophic cultivation containing acetate and glycerol but lacking nitrogen and phosphorus in media produced fourfold higher biomass compared to autotrophic cultivation; however, phycocyanin and polyhydroxybutyrate (PHB) productivity was negatively affected (Nematollahi et al. 2020). Interestingly, *A. platensis* reported a higher specific growth rate, higher carbohydrate content and maximum biomass yield of 2.98 ± 0.07 gL^{-1} when grown through mixotrophic cultivation (Pereira et al. 2019). This approach is also cost-effective since the growth medium contains whey (dairy industry wastewater by-product) and avoids the need for expensive chemical nutrients. Likewise, a medium consisting of 2.5 mM sucrose reported an increase in biomass productivity by 1.32-fold in *A. platensis* under mixotrophic cultivation (Velioğlu Tosuner and Öztürk Ürek 2020).

10.3 Challenges in Multi-parameters Optimization of Cyanobacterial Biomass

Multi-parameter optimization is an integral part and parcel of cyanobacterial farming, particularly in the wake of immense industrial demand. Consequently, this commercial need has spiked the research work to make use of different scientific approaches in cyanobacterial cultivation to enhance its biomass production, titre and yield of high-quality products while reducing the production and maintenance cost at the same time. This includes the exploration of unknown cyanobacterial strains having higher production capabilities, modifying them via genetic engineering techniques and calibrating the sustenance medium and production medium while maintaining sterile cultivation conditions (Pathak et al. 2018; Vu et al. 2020). The biomass composition and yield differ from species to species. In that context, every industry should prudently select the type of species depending upon the product they produce. For clarification, the industry dealing with bioenergy production will have to look for species involved in producing high lipid but low protein biomass composition as they can be completely utilized for producing biofuels (Pathak et al. 2018). Moreover, the species with some other biomass composition like sacran (a macromolecular polysaccharide) in high amounts can be particularly employed for cosmetics and will remain unutilized by the bioenergy industry. This will impose a negative effect in terms of net profitability. Moreover, not only the biomass composition varies from one species to another species, but their ability to withstand environmental conditions (light quality and duration of its exposure, pH, temperature, salinity, nutrients availability, CO_2, contaminants, grazers, etc.) also varies from species to species (Pathak et al. 2018). However, we have limited knowledge of strains due to the species-restricted study being conducted so far, and hence, there is an utmost emphasis on research dealing with the identification of the best species which exhibit their full biotechnological potential of producing biomass without affecting their compositions irrespective of the type of the weather or geographical location so that they can be exploited commercially throughout the year.

As far as genetic engineering is concerned, we cannot ignore the fact that genetically modified cyanobacterial species either for bioenergy generation or high-value chemical generation can impose a threat to the environment; hence, thorough analysis of any potential concerns should go hand in hand before cultivating genetically engineered cyanobacteria at commercial scale (Vu et al. 2020). Skerker et al. (2009) have briefly discussed the consequences of synthetic biology involving genetic engineering. The novel safer techniques like Clustered Regularly Interspaced Short Palindromic Repeats (CRISPR) for cyanobacterial genome engineering are still in the nascent phase for biomass enhancement. This model works by the generation of carbon by cyanobacteria which are subsequently fed by heterotrophs producing secondary metabolites. In addition, lower expenditure on maintenance and low capital spending are some other plus points of the co-cultured method of cultivation as described in Sect. 10.3.1.

Three essential nutrients including carbon, nitrogen and phosphorus are not so economical. Ammonia could have been an economical solution but imposes an issue by fluctuating the pH levels through releasing the hydronium (H^+) ions; however, agro-industrial wastewater containing loads of nitrogen and phosphorus is an appealing solution since it can cut some expenditure on nutrient media (Kamravamanesh et al. 2019). Besides nutrient media, the quality and quantity of light also determine biomass composition and productivity by drastically affecting the efficiency of cyanobacterial photosynthesis. Another issue is photoinhibition, which is contributed by light exposure but at an intensity higher than the optimum level leading to a decrease in the photosynthetic efficiency along with biomass growth (Grama et al. 2022). Notably, high light intensity is not always directly proportional to high lipid production or other biochemical composition and varies from one species to another species (Nzayisenga et al. 2020). In particular, high light intensity favours increased biodiesel production in *Synechococcus* sp. 6803, but reduced lipid production is due to photoinhibition. Employing high light intensity also does not favour biodiesel production in *Synechococcus* sp. 6803 (Prabha et al. 2022) and poses certain issues like light intensity, light duration and lipid production. However, in GMO cyanobacteria where scientists suppressed its operon CPC to make phycobilisome (PBS) antennae complex to smaller size impart the ability to prevent the over-absorption of sunlight, this means that now they can only harvest incident sunlight in a quantity which is required to them for their metabolic function and survival (Grama et al. 2022). Moreover, the higher light intensity poses another issue of overheating of outdoor PBRs, where water bath immersion and shading are two approaches that are being undertaken to bring the temperature under control.

Additionally, appropriate location selection for algal cultivation, where the local temperature doesn't fluctuate so often during night-time and the climate is moderate as well, i.e., neither too warm nor too cold, has also been suggested to address the issue of overheating. The direction and position of PBRs also matter as per the study which reports that outdoor PBRs located in the east-west direction in a vertical position can harvest 5% higher solar energy compared to PBRs located in the north-south direction in a horizontal position (Prabha et al. 2022). Another experimentation for exploring alternative light such as fluorescent, red, white, yellow LED, red LED, etc. provided optimum condition for large-scale cultivation of any cyanobacterial species (Satpati et al. 2022). In fact, selecting the right LED light can reduce the cost spent on electric power and positively affect the composition of lipids, proteins, pigments or value-added compounds (Prabha et al. 2022). The principle behind reduced cost and increased pigment content lies in the fact that specific LED light inhibits photosynthesis. Hence, for hassle-free photosynthesis, the cyanobacteria start producing other light-harvesting pigments, for example, phycobiliproteins that can now help cyanobacteria to absorb light at a visible range of wavelength; and the light is subsequently transferred to the reaction centre containing chlorophyll a. This is how phycobiliproteins are produced which also decreases cost by increasing the absorption rate of photons or light energy.

Stability is another benefit considering the sunlight exposure which keeps on fluctuating due to climatic and seasonal changes. Stability is an advantage, especially given the fluctuating sunlight exposure caused by climate and seasonal variations. While not cost-effective, this approach can still maximize biomass production, as demonstrated with microalgae. (Satpati et al. 2022). Apart from extreme light exposure, overheating, oxygen build-up, water loss due to evaporation, bacterial invasion, poor CO_2 supply, salinity, cultivation vessel, contamination and uncontrollable mixing rate leading to shear stress-cell destruction are other critical issues presented by open raceway systems. The solution for overcoming the limitations of the open pond is closed PBR systems that provide better regulation of process parameters with minimum or no impact from the outside environment (Gupta et al. 2015; Satpati et al. 2022).

However, the designing, manufacturing, operating, surface sterilizing and cooling of the PBRs are relatively expensive and difficult in comparison with open-raceway ponds resulting in lower biomass yield and higher harvesting cost. Besides, biofouling (a phenomenon caused by the adhesion of unwanted organisms on PBR surfaces which can block sunlight) can trouble the bioprocesses. Another issue with PBRs is using glass which not only is expensive but easily broken as well. Solving these issues is quite challenging but possible to fix by upgrading the design, producing cheaper and robust components and making the system quite automated through technological development. There are a few pieces of evidence already hinting at the progress in PBR improvement. For example, Sitther et al. (2020) reported that the use of gold nanoparticles (AuNPs) in PBRs can reduce cost since suspended AuNPs do not contaminate easily and can be re-used at least five times owing to their recycling ability. The gold nanoparticles-constructed PBR designs (interchangeably called plasmonic mini-photobioreactors) can reduce the cost of cultivation of cyanobacteria, since they have robust ability to scatter, amplify and control the wavelength of incident light while also lowering the damage due to photoinhibition, all of which can increase chlorophyll and carotenoid accumulation and can favour the biomass. The next example of nanotechnology which can prevent the biofouling issue in PBRs is based on using silver nanoparticles (AgNPs) coating which works by killing the organisms on contact while also not affecting the viable algal cells in bulk media. AgNP coatings start disintegrating or decreasing their bacterial activity after a certain duration of time (Hamida et al. 2020). Another problematic challenge faced in pilot-plant known to depict the "time-space-yield" relationship aligned with the prime objective along with other considerations such as whether the product is nature friendly, sustainable and safe (Kamravamanesh et al. 2019). For example, biohydrogen is a pollution-free and sustainable energy resource and more considerable than coal which causes air pollution, global warming and subsequent diseases such as asthma, cancer, chronic obstructive pulmonary disease (COPD), etc. (Gasparotto and da Boit Martinello 2021). This will either help in the selection of particular species for cultivation and in predetermination to avoid certain species which will save time and money or to find alternative approaches until its cultivation gives feasible yields. Surprisingly, a few pilot tests still carry the gaps that are needed to be filled before cultivating at industrial scale. Simply relying on the laboratory scale cultivation data concerning optimum growth parameters (nutrients, pH, temperature, light intensity/quality/duration, etc.) will always show higher yield or productivity but the same won't favour large-scale production since there is a huge difference in terms of parameters, sterility conditions, environmental conditions/factors, workload and productivity. There is always a significant difference in the yield when cultivated in open-outdoor systems. Hence, to see a clear picture and to obtain accurate data, we need to cultivate cyanobacterial cultures in conditions similar to outdoor cultivation systems and study the effects of different parameters on the compositions of biomass for varied industrial applications.

10.4 Cyanobacterial Strains in Bioenergy

10.4.1 Biohydrogen Production

A widespread craze amongst scientists for the development of cyanobacteria-based bioenergy can be witnessed in recent years. Biohydrogen, known for its green, sustainable, and low-cost attributes, has the potential to replace fossil fuels, and cyanobacteria can serve as a source for its production. (Parambath et al. 2022). The fact that cyanobacteria can produce H_2 is revolutionary as H_2 has the highest energy per unit weight (141.65 MJ kg^{-1}) compared to all other fuels as per our current knowledge (Vu et al. 2020). There are three routes by which cyanobacteria can execute hydrogen production. They are (i) bidirectional hydrogenase-mediated hydrogen production, (ii) via introduced hydrogenase from another organism and (iii) nitrogenase-mediated hydrogen production (Sharma and Stal 2013).

There are electrochemical and thermochemical methods such as electrolysis and steam reforming which can be employed to produce hydrogen (Khetkorn et al. 2017). However, both processes tend to consume lots of energy and cause pollution by releasing pollutants such as CO and CO_2, while the latter can also be produced by fossil fuels or natural resources (Khetkorn et al. 2017). Similarly, the members of *Enterobacteriaceae* are also capable of hydrogen production via dark fermentation reactions. However, the problem with this is incomplete decomposition, i.e. only 20% of the stored substrate is decomposed while the rest 80% of storage remains undecomposed (Sharma and Stal 2013). Consequently, the yield is significantly low due to contamination and substrate requirements, all of which do not make hydrogen production via dark fermentation worth the investment of time and money despite its fast reaction rate (Sharma and Stal 2013). But unlike *Enterobacteriaceae*, cyanobacterial cultivation does not require an expensive substrate as it can feed on waste material plus it has a greater ability to convert solar energy to H_2. Water, CO_2 and light are the only cyanobacterial requirements during oxygenic photosynthesis to produce H_2 where H_2O act as the source of electrons and CO_2 as carbon while avoiding the need for rare metals which makes them even more lucrative for bioenergy production (Lupacchini et al. 2021; Tamagnini et al. 2007). The photoinhibition phenomenon is called oxygen intolerance in which oxygen binding with hydrogenase consequently leads to blockage of hydrogenase activity and instantly stops H_2 production (Grama et al. 2022). It should be noticed that reducing agents, i.e. ferredoxin and NADPH (nicotinamide adenine dinucleotide phosphate), work on oxygen reduction during respiration.

The issue of constant release of O_2 interrupting the hydrogenase enzyme can be simply tackled by engineering oxygen-tolerant hydrogenase while also speeding the flow of electrons towards the hydrogenase enzyme (Singh et al. 2016). Another study by Masukawa et al. (2012) on the genes disruption of ***Anabaena*** sp. PCC 7120 resulted in the inactivation of both hydrogenase-producing unpaired mutants ΔHup and ΔHox and paired mutant ΔHupΔHox boosted the hydrogen production from four to seven-fold and yielded up to 50 μmol H2/(mg Chl a*h). A recent study by Pansook et al. (2022) on *Aphanothece halophytica* reported amazing results especially by treating with simazine inhibitors (25 μM) which led to the production of 58.88 ± 0.22 μmol H_2 g^{-1} dry weight h^{-1} and H_2 accumulation at 356.21 ± 6.04 μmol H_2 g^{-1} dry weight in dark condition. This was the highest yield compared to other inhibitors due to the encouragement of dark respiration while inhibiting photosynthetic O_2 evolution.

Another perspective about increased H_2 production is species identification through which species having desired characteristics can be found and cultivated. The diazotrophic cyanobacteria produce H_2 as a by-product while fixing nitrogen into ammonia. According to Martínez-Merino et al. (2013), diazotrophic cyanobacteria showed increased H_2 productivity as compared to non-diazotrophic cyanobacteria (incapable of nitrogen fixation). Therefore, the selection of heterocystous bacteria over non-heterocystous cyanobacteria can be another strategy employable for higher H_2 productivity. The newly discovered strain *Anabaena siamensis* TISTR 8012 isolated from a paddy (rice) field in Thailand could produce H_2 of

0.057 μmolH_2 mg chl a h^{-1} grown in BG11 medium along with 4 μM nickel ion supplementation (Taikhao and Phunpruch 2017). These newly found strains will be promising regarding H_2 production because they contain nitrogenase and bidirectional hydrogenase, but there is need to work on synthetic biology and metabolic engineering techniques for strain improvement.

Production can even increase beyond by targeted genetic engineering of nitrogenase enzyme and metabolic pathways of heterocystous cyanobacteria (Khetkorn et al. 2017). *Nostoc* sp. PCC 7422 was engineered by mutating its *hupL* gene which produced threefold higher H_2 with about 100 μmol mg^{-1} chlorophyll compared to its native or wild-type strain (Singh et al. 2016). Recently, a new technique of hydrogen production based on the fusion of photosystem 1 and hydrogenase in vivo (inside living organisms) rather than in vitro (outside living organisms) was implemented. In vivo has added benefits since it encourages self-repair and saves maintenance costs over in vitro. For this, fusion was made between NiFe-hydrogenase HoxYH and photosystem I subunit PsaD of cyanobacterium *Synechocystis* sp. PCC 6803. The process of fusion was in close vicinity of 4Fe4S cluster F_B which has function of donation of electrons to ferredoxin. The newly developed mutant called *psaD-hoxYH* reported higher hydrogen productivity of 500 μM anaerobically without taking back the hydrogen produced by them (Appel et al. 2020).

10.4.2 Bioethanol Production

The use of cyanobacteria has also been reported in several studies for ethanol/ethyl-alcohol production which exclude the need for yeast culture. To be specific, this is unlike ethanol production by fermenting the crops such as sugarcane and corn which require yeasts; however, this is already in practical implementation for serving the daily bioenergy need of automobiles in Brazil and the USA, respectively (Quintana et al. 2011). This very method is attractive in terms of clean and green energy, usable in diesel engines without the need for technological upgradation and the only solution which is practically employable but unattractive since they require crops as raw material which is utilized as food and hence economically not viable, especially in a situation in which a huge mass of the population is already facing food scarcity (Quintana et al. 2011). Likewise, despite economic and environmental merits, cyanobacterial-based ethanol also presents the issue of low productivity, the reasons for which are discussed further. We know that cyanobacteria obtain their abundant energy via photosynthesis in presence of light and facilitate O_2 respiration costing some energy. However, in dark and anaerobic conditions, the cyanobacteria also synthesize ethanol as a by-product (Deng and Coleman 1999); they prudently avoid respiring or producing oxygen as energy-saving strategies. Here, again comes the importance of genetic engineering techniques firstly employed by Deng and Coleman, in concern to ethanol production for which they utilized the gene isolated from *Zymomonas mobilis* (a gram-negative rod-shaped bacterium) to transform

Synechococcus sp. PCC 7942 and *Synechocystis* sp. PCC 6803, after which they develop the ability to even perform fermentation in oxygenic photosynthesis, i.e. in the presence of light and aerobic conditions. The expression of these genes was initiated by two different promoter regions: (1) rbcLS operon promoter of the cyanobacterium and (2) lac promoter of *Escherichia coli*. The expression assisted in the production of two types of enzymes, alcohol dehydrogenase and pyruvate decarboxylase, and the process was followed by the degradation of sugars into pyruvate and subsequently fermentation (Vu et al. 2020). This finally led to a respective increase in the ethanol productivity of about 54 nmol OD_{730} $unit^{-1}$ L^{-1} day^{-1} and 5.2 mmol OD_{730} $unit^{-1}$ L^{-1} day^{-1} (around 0.23 gL^{-1}) (Quintana et al. 2011; Vu et al. 2020).

In fact, a study also reports that integrating genes such as *Z. mobilis pdc* and *adh* genes via *E. coli* enhances the ability of cyanobacteria to utilize a variety of substrates for ethanol production (Deng and Coleman 1999). In another approach, the researchers first obtained cellulose synthase genes extracted from *Gluconacetobacter xylinus* and administered them into *Synechococcus* sp. PCC 7942 (Quintana et al. 2011). This newly developed transformed gene is now able to synthesize extracellular non-crystalline cellulose and is attractive owing to its enhanced hydrolysis ability which also favours enhancement in ethanol productivity (Quintana et al. 2011). Further improvement was observed in the in silico study in which the ethanol productivity of *Synechocystis* sp. PCC 6803 was about 235% greater compared to productivity reported in the laboratory study.

During ethanol production, it is noticed that ethanol production is greatly affected based on the concentration of the salt medium. This was witnessed when ethanol production was 100-fold higher in NaCl concentration of 1.24 M compared to a lower NaCl concentration of 0.24 M (Quintana et al. 2011). Mussgnug et al. (2007) also mentioned in the study about modulating the antenna size as a strategy that led to enhancement in photosynthetic efficiency up to a certain extent. Still, there are several issues with cyanobacteria-based ethanol which need to be solved. One issue is the high solubility rate of cyanobacterial ethanol in water which requires distillation; consuming high energy plus other toxic and hygroscopic nature of cyanobacterial ethanol makes its distribution through pipelines quite problematic (Sharma and Stal 2013).

10.4.3 Butanol Production

Butanol (C_4H_9OH) is the alcohol characterized by four carbon atoms, sometimes represented as i-BuOH or isobutanol. This represents that they have been produced through some biological methods such as fermentation. Isobutanol being a renewable fuel seems appealing and sustainable owing to its lesser carbon number. It has more potential in replacing gasoline (32.5 MJ L^{-1}) compared to ethanol as they have relatively high energy density and octane number (90% high compared to gasoline with energy density of 29.2 MJ L^{-1}) and low corrosion rate, flammability, hygroscopicity and volatility. It can be employed as solvent directly or as its esters in combustion engines (Mathan Raj et al. 2018). Owing to their lower oxygen content, they can also be blended with gasoline fuel. The other merits of isobutanol are that it doesn't completely dissolve in water, unlike ethanol whose miscibility rate with water is 100%, and doesn't impose stress preventing the pipeline's destruction. Moreover, isobutanol as an advanced biofuel can be produced through genetically engineered microorganisms through ABE processes (Brownstein 2015).

According to Vu et al. (2020), we must accompany thiolase (atoB) as a catalyst for the thiolysis of acetoacetyl, since AtoB assists the metabolic flux from the acetyl-CoA pool towards the 1-butanol biosynthesis; however, we should avoid the condensation of two acetyl-CoA molecules (Singh et al. 2016). Moreover, sensitivity to oxygen and toxicity caused by high amount of butanol are two major issues that can impair cyanobacterial growth. The first issue demands more studies employing synthetic biology to enhance their adaptability to oxygen while others in terms of photobioreactors designing or quite probably finding some alternative medium with some anti-turbidity material. Therefore, to increase the productivity of butanol, engaging synthetic biology tools such as engineering small RNA and promoter, RBS and genome editing seems promising (Singh et al. 2016).

10.4.4 Biodiesel Production

Diesel production using cyanobacteria received lesser attention due to its lipid content, scarcity of information on the biochemical profile, energy content as well as the growth rate of cyanobacterial species even after their identification (Quintana et al. 2011). Hence, the selection of cyanobacteria is itself one prominent issue to find the most potent species characterized by high lipid amount, high productivity and growth rate (doubling time) for large-scale diesel production. Not only this, lipid characteristics such as its chain length and degree of saturation meaning whether the lipid is saturated (having a single bond in its hydrocarbon chain) or unsaturated (having one or more double bonds in its hydrocarbon chain) specifically MUFA (mono-unsaturated fatty acids) not PUFA (poly-unsaturated fatty acids) also hold significant importance in determining the quality of diesel. A recent study by Prabha et al. (2022) on *Synechococcus* sp. 6803 reports increased lipid production when grown in increased intensity of light. Likewise, *Synechococcus nidulans* reported promising biodiesel production of saturated fatty acids up to 50% under combined red and yellow LED light. Interestingly, multiple mutations at several targeted sites reported higher diesel productivity, while single-gene targeted mutagenesis reported relatively lower production of lipids. Additionally, factor like high light intensity correlates to a decrease in PUFA and an increase in MUFA in cyanobacteria, thus favouring high-quality diesel production (Quintana et al. 2011). Other factors like combined N_2 negatively affect the quality of diesel by decreasing the lipid and carbohydrates and increasing the protein content. Hence, the correct form of nitrogen should be identified.

Currently, the toxicity of cyanobacterial cells due to the production of fatty alcohols and high fatty acid concentration is one big issue that can be dealt by regulation of their genes (Singh et al. 2016). Apart from emphasizing the study on the filamentous strain, the other challenge can be understanding the mechanism behind unicellular strain producing a significant amount of diesel compared to other types so that the same can be moulded or mimicked to other species as well using genetic engineering or synthetic biology. Hereafter, the next challenge can be exploring the other factors which can enhance large-scale diesel production.

10.4.5 Bioelectricity Production

Photoelectrochemical devices are the new trending devices that utilize microorganisms to produce bioelectricity (Ucar et al. 2017). Owing to the integration of different kinds of microbes, such devices are also referred to as microbial fuel cells (MFCs). The bacteria employed here are referred to by various terminologies such as electrochemically active bacteria (EAB), electricigens and exoelectrogenic (Parambath et al. 2022). These bacteria act as biocatalysts that oxidize organic contaminants leading to the respective production of electrons along with protons via an external circuit and semipermeable membrane (Clark and Pazdernik 2016). These electrons are then channelized to an electron mediator which subsequently transferred to an electrode contributing to electric power production. Notably, since they utilize different kinds of metallic and toxic contaminants in the form of wastewater, it makes this technology more appealing in the sense that they are improving the environment through wastewater treatment (bioremediation), denitrification and dealing with global warming since the energy production via this technology does not emit carbon-dioxide, as seen with current technology which utilizes fossil fuels (Parambath et al. 2022). Moreover, *Anabaena, Synechococcus* or *Synechocystis* are the species mentioned in the study for their employment in the MFCs (Quintana et al. 2011). However, there are a few concerning issues such as the continuous need of maintaining aeration and oxygen supply which raise the expenditure and spoil the overall profitability (Parambath et al. 2022). However, with rapidly growing technological advancement, we hope that very soon we would be able to witness this emerging technology.

10.5 Cyanobacterial Strains in Functional Foods

Functional foods as the name suggests are foods with added biologically active compounds and nutrients to provide energy and aid good health, well-being and longevity of the person by positively influencing the biological functions including growth and development of the body, fast metabolism, normal digestion, cardiovascular, cognitive and mental health, etc. when consumed regularly in the right amount (Tur and Bibiloni 2016; Donato-Capel et al. 2014). The high demand heading the day-to-day global expansion of the nutraceutical market to worth over 100 billion USD very well illustrates the importance of functional food in daily life (Nwosu and Ubaoji 2020). However, a trend shows the transition from herbs to cyanobacteria since their products are stable in varying conditions such as pH, temperature and water solubility while maintaining high nutrient content (Fig. 10.2). Their nutritional significance reveals through statistical data that 1 kg of *Arthrospira* and 1000 kg of fruits and vegetables provides equal nutrients (Panjiar et al. 2017). The popularity of cyanobacteria-based nutraceutical products despite their safety and immense health benefits in comparison with pharmaceuticals or herbal products is still preferred.

10.5.1 Carbohydrates and Fibres

The term carbohydrates is interchangeably used with another term called saccharide and comprises cellulose, sugars and starch, which make up most of the human diets worldwide (Awuchi and Ikechukwu 2021). The cyanobacterium species *Scytonema bohneri* isolated from sulphur spring water cultivated on BG-11 medium (a nutrient medium lacking nitrogen) contained the highest carbohydrates (28.4%) compared to the average of 15–20% as contained by *Arthrospira* sp. and another six species such as *Calothrix fusca*, *Gloeocapsa livida*, *Lyngbya limnetica*, *Oscillatoria acuminata*, *Oscillatoria calcuttensis* and *Oscillatoria foreaui* (Panjiar et al. 2017; Rajeshwari and Rajashekhar 2011). However, a recent study by Bounnit et al. (2022) reported a

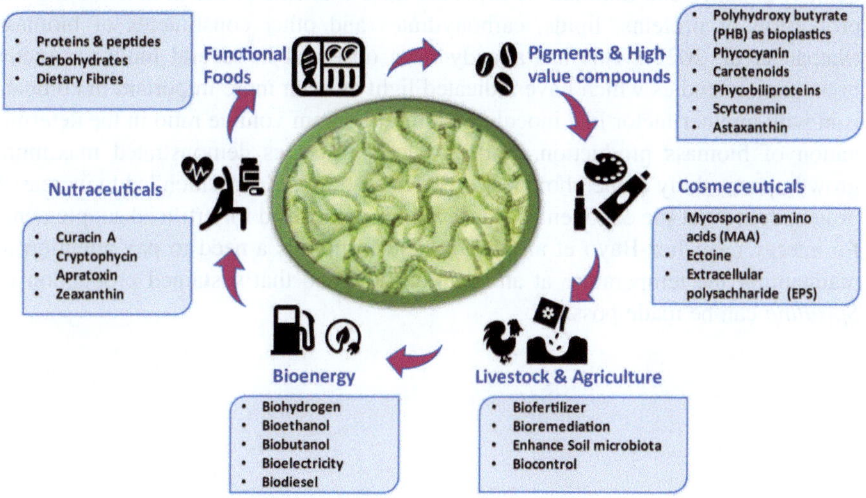

Fig. 10.2 Circular economy approach of cyanobacteria cultivation

comparatively higher carbohydrate content of 29% contained in strains QUCCCM 34 (*Chroococcidiopsis* sp.) and QUCCCM 77 (*Euhalothece* sp.). Probably, the reason behind such ability of high carbohydrate production lies in polysaccharide sheaths' which play a role against desiccation in the harsh natural environment. Furthermore, *Arthrospira* sp. has been well documented for its total dietary fibre content of about 8.45% (Raczyk et al. 2022) (Fig. 10.2).

10.5.2 Proteins and Peptides

Lack of protein diet causes life-threatening deficiency diseases like marasmus and kwashiorkor. It is a matter of concern worldwide, but this is greatly affecting children especially belonging to developing countries. According to UNICEF (2018a), around 3.1 million children die every single year due to undernutrition and hunger. Cyanobacteria can be the hope in this situation as well-balanced food owing to its nutrition, fast growth and affordability. China is already adding cyanobacteria as a constituent with baked barley sprouts for a baby's good health and sound development. *Arthrospira maxima* potentially considered for its significant protein content (65%) compared to other cyanobacterial species (Sánchez-Bayo et al. 2020). Moreover, the superfood *Arthrospira* sp. also ranks on top for its higher protein content (dried Spirulina protein, 51–71%) amongst day-to-day consumed food such as soybean flour (protein, 36%), parmesan cheese (protein, 36%), wheat germ (protein, 27%), peanut (protein, 26%), chicken (protein, 24%), fish (protein, 22%) and egg (protein, 12%) (Zahra et al. 2020).

In concern to parameters affecting large-scale cultivation, light has been of prime importance which not only provides energy for the cell metabolism, but also its quality, quantity and duration of exposure have a drastic effect in determining the proportion of proteins, lipids, carbohydrates and other constituents of biomass (Satpati et al. 2022). This has already been observed in several multi-parameter optimization studies which have indicated light as even more important in comparison with another factor like inoculum/culture medium volume ratio in the determination of biomass production. This is because species demonstrated maximum growth, particularly at the photoperiod of 12: 12 h light/dark which led to increased protein content at the exponential phase avoiding the need for artificial supplements for energy (Sánchez-Bayo et al. 2020). Hence, there is a need to pay attention to maintaining the temperature at an optimum level so that sustained production of *Spirulina* can be made possible.

10.6 Cyanobacterial Strains in High-Value Chemicals

10.6.1 Polyhydroxybutyrate (PHB) as a Sustainable Bioplastic

Polyhydroxybutyrate (PHB) is a carbon polymer synthesized by microorganisms, characterized by its biodegradable nature which belongs to the polyhydroxyalkanoate family. Interestingly, its functions differ in accordance with the organisms which produce it (Koch and Forchhammer 2021). PHB is one of the commercially important value-added compounds amongst many others as discussed below which is synthesized by cyanobacteria using CO_2 and sunlight through photosynthesis. However, issues and challenges are the same; its cultivation cost is so high that makes its commercialization just impractical/unfeasible. Again, we can rely on techniques such as optimizing the cultivation conditions and genetic engineering techniques for PHB productivity enhancement. *A. platensis* MUR126 cultures grown in autotrophic conditions reported increased PHB productivity of 33% on the 14th day due to additional CO_2 supplementation. The same study also suggests the role of organic carbon in increased PHB production, whereas the condition in which media lacked nitrogen and phosphorus had a negative effect on PHB production (Nematollahi et al. 2020; Sreenikethanam and Bajhaiya 2022). Another study reported the increased (R)-3-hydroxybutyrate yield up to 1.84 gL^{-1} just within 10 days by *Synechocystis* species. The approach was simply the optimization of the acetoacetyl-CoA reductase binding site of *Synechocystis* species which resulted in 263 mg L^{-1} day^{-1} as the highest production from CO_2, as per the same report (Kamravamanesh et al. 2019).

Another novel approach called random mutagenesis gave more impressive results. In context with high PHB production, researchers applied the random mutagenesis technique to *Synechocystis* sp. PCC 6714 which uses ultraviolet (UV) rays as a mutagen to develop a mutated strain called MT_a24. As a shred of evidence, this newly developed strain demonstrated increased PHB production of 134 mg L^{-1} day^{-1} constituting 37 ± 4% dry cell weight (DCW) in an optimized condition where phosphorus and nitrogen were under-supplied (Kamravamanesh et al. 2019). The UV mutated strain (MT_a24) has the potential to produce (PHB) even up to1.16 gL^{-1} through media optimization. It is also worth noticing that there is an inverse relationship between glycogen production and PHB synthesis meaning the lesser the glycogen production, the higher the PHB productivity. However, simply blocking the pathways associated with glycogen has been linked to stunted cyanobacteria growth consequently affecting their survival (Kamravamanesh et al. 2019). Hence, the future challenge in terms of PHB content lies in manipulating the glycogen-producing pathways so that there is a high PHB yield during glycogen production too, either through genetic engineering or by looking at some alternative pathway modification.

10.6.2 Cyanobacterial Strains and Their Pigment Potential in the Food Industry

To gather photons or light, cyanobacteria produce a diverse range of pigments such as chlorophyll, carotenoids and phycobiliproteins (the last two are capable of absorbing light over a diverse range of wavelengths which chlorophyll is unable to do, thereby preventing chlorophyll from photooxidation). These natural resources have huge commercial applications in the food industry being an environment-friendly natural colourant. Phycocyanin (blue-coloured phycobiliproteins) are exploited as natural food colouring to impart colour in food such as candies, chewing gum, ice cream, etc. (Vu et al. 2020). Phycocyanin is produced in ample quantity by cyanobacterial species such as *Anabaena*, *Arthrospira* and *Synechococcus* (Vu et al. 2020). Moreover, nitrogen supplementation is an important parameter since phycocyanin pigment is the by-product of nitrogen stored by cyanobacteria (Nur and Buma 2019). The same recommendation of inorganic nitrogen addition is emphasized even for accelerating algal growth as well (Nur and Buma 2019). The study also suggests the use of crude glycerol which apart from enhancing phycocyanin content can also decrease the cost of cultivation being the by-product of the biodiesel industry. In particular, crude oil in the amount of 3.07 gL^{-1} is reported to enhance the phycocyanin content up to 10.8% by *A. platensis* axenic cultures grown in mixotrophic conditions (Corrêa and Teixeira 2021).

Another study reported CO_2 supplementation's role in increasing phycocyanin productivity in mixotrophic conditions. The light source is another parameter that can positively or negatively affect pigment concentration. For example, *A. platensis* reported 40% higher productivity of C-phycocyanin (C-PC) content under white LED light compared to fluorescent lamp light while also saving cost on electricity (Prabha et al. 2022). A study conducted on the non-heterocystous cyanobacterium *N. antarctica* LEGE13457 exhibited a high carotenoid content of 63.9 μg mg^{-1}, while *C. gracile* LEGE 12431 displayed a comparatively lower carotenoid content of 57.8 μg mg^{-1} in the same high light conditions. The reason behind the decreased content in high light is thought to be a strategy adopted by cyanobacteria to prevent photo-oxidative damage due to free radical production (Lopes et al. 2020).

10.6.3 Cyanobacteria Possess Value-Added Compounds for the Cosmetics Industry

The bioactive compounds play a prominent role in protecting the cyanobacteria from the harmful radiation of the sun, and that is why they are also called photoprotective compounds. These cyanobacterial compounds demonstrate great commercial potential in manufacturing several cosmeceuticals and skin care products. Mycosporine-like amino acids (MAAs) and their derivatives (tetrahydropyridines, shinorine, etc.) are bioactive compounds that can be extracted from filamentous cyanobacterium

Chlorogloeopsis fritschii PCC 6912 which can be used to produce sunscreen as natural UV blockers owing to its adsorption maxima in the UV range (Vu et al. 2020; Llewellyn et al. 2020). A recent study has revealed that ultraviolet light (UV) induces the production of MAAs and other sunscreens-like compounds in cyanobacterium as physiological adjustment to prevent themselves from ultraviolet-damaging effects (Nowruzi et al. 2020). While it remains unconfirmed whether far-red light (FR) directly induces the production of MAAs, the study has determined that, at a minimum, far-red light (FR) promotes the synthesis of MAAs (Llewellyn et al. 2020). Such value-added compounds can block harmful UV rays of the sun protecting the skin, as they have antioxidant activity and can prevent inflammation (Fig. 10.2).

The other cyanobacterial pigment ectoine exhibits depigmentation, anti-melanogenic membrane protection, osmoprotectant, UV ray protection and skin-whitening properties all of which just illustrate how valuable they are for the cosmetics industry (Sitther et al. 2020; Hseu et al. 2020). Zahra et al. (2020) have particularly illustrated that two cyanobacterial strains *Pseudomonas fluorescens* PGM37 and *Aphanothece sacrum* contain extracellular polysaccharide (EPS) and sacran. Both compounds hold amazing skin moisturizing abilities and massive potential for the future cosmetic market. A study has identified sacran as having a better ability to absorb water compared to a costly and largely employed ingredient hyaluronic acid (Vu et al. 2020). They even possess the ability to replace hyaluronic acid since they will be more economical, safer and effective to be used in a moisturizer which also would not cause any side effects being a natural source.

Another bioactive compound lauric acid is primarily derived from virgin coconut oil and palm kernel oil and acts as the burden on global food supply and the economy; however, cyanobacterial-derived lauric acid has potential to fix such issues (Anzaku et al. 2017). Its industrial demand is due to anti-fungal activity and broad-spectrum antimicrobial activity against fungi like *Candida* spp. and certain types of bacteria with lipid coating. Phycocyanin produced by cyanobacterial species is another pigment of commercial cosmeceuticals, which are important constituent of cosmetics products like facial masks, eyeliner, eyeshadows and lipsticks.

10.6.4 Cyanobacterial Value-Added Compounds for the Pharmacological Industry

In general, cyanobacteria remarkably garnered anti-allergic, antibacterial, antiviral, anti-inflammatory and immune-suppressant properties. Hence, revolutionize the pharmaceutical market by replacing synthetic medicines which have adverse side effects to not only the human body but aquatic body systems as well on consumption. Regardless of their toxicity, cyanotoxins hold utmost promise in their use as biopesticides, but here we will just focus on their role in cancer therapies. Curacin A extracted from filamentous cyanobacterium *Lyngbya majuscula* is currently in phase I clinical trials and has application in the treatment of breast cancer, while cryptophycin derived from *Nostoc* species seems promising in the therapy of solid tumour or multi-resistant tumours concerning breast, colon, lungs, ovarian, pancreas, prostate and brain (Zahra et al. 2020). Similarly, apratoxin derived from the marine cyanobacterium *Lyngbya majuscula* and borophycin derived from *Nostoc spongiaeforme/Nostoc linckia* also hold potential in cancer drug development (Qamar et al. 2021) (Fig. 10.2). Phycocyanin from *A. platensis* is worth mentioning since it can be used in anti-cancer drugs as they can inhibit the proliferation of cancer cells in vitro and in vivo and secrete fibrinolytic enzymes which can treat thrombosis disease (a condition characterized by clot formation inside the blood vessels if cultivated under mixotrophic condition along with corn steep liquor (Braune et al. 2021; De Barros et al. 2020). Growth inhibition of human chronic myelogenous leukaemia blast crisis K562 cells has already been reported when the phycocyanin pigment was administered in a dose-dependent manner (Liu et al. 2000).

10.6.5 Zeaxanthin Biosynthesis by Cyanobacteria

Zeaxanthin extracted from *Synechocystis* sp. and *M. aeruginosa* exhibits antioxidant and anti-inflammatory characteristics. It has also been documented to reduce the progression of polyps (noncancerous tumours or neoplasms) in males and other types of cancer in both males and females. (Okuyama et al. 2014). Additionally, they also work in treating macular degeneration. There are also some studies of cyanobacteria producing 24-ethyl cholesterol in very small amounts which have applications for hyperlipidaemia/hyperlipoproteinemia (the condition of blood containing lipids such as cholesterol and triglycerides in high amounts). However, data is very limited and seems quite doubtful and thus needs more research work (Leblond et al. 2011). So, this is how cyanobacterial compounds can contribute to the prevention as well as treatment of diseases being listed. In the future, we expect to see some more discoveries regarding novel compound isolations and their pharmacological applications.

10.6.6 Cyanobacterial Pigment Potential in Research and Development

Cyanobacterial pigments which are involved in harvesting light have also tremendous applications in research areas since they possess the ability to fluoresce/glow at a certain wavelength and can bind to antibodies (Vu et al. 2020). Allophycocyanin, phycocyanin and phycoerythrin can be employed in flow cytometry and other detection techniques such as immunofluorescence technique as a chemical tag.

Phycobilin-producing species has already been discussed in food industry section (Zahra et al. 2020; Vu et al. 2020).

10.7 Risk Involved in Circular Bioeconomy Frameworks

A circular bioeconomy is sustained by natural resources. It is a new economic model that promotes the utilization of renewable natural capital and focuses on waste minimization to replace the large range of non-renewable products. The purpose of 3R's (reduce, recycle, reuse) in the bioeconomy is to develop a circular economy that prevents excessive and unwanted waste (Dahiya et al. 2022). Microalgae have a lot of potential as a sustainable feedstock in a biorefinery and undoubtedly gained global interest in circular bioeconomy. Fundamental ideas like sustainability, cascading usage and waste hierarchy can be included to create microalgal biorefineries. For example, the cyanobacterial cultivation process involves the reuse of wastewater after harvesting biomass; further leftover biomass after metabolite extraction can be reused as biofertilizer or other products applicable in fodder or bioenergy industry.

Cyanobacteria biorefinery products have capacity to synthesize a wide spectrum of bioactive chemicals (Achparaki et al. 2012). For example, polyhydroxyalkanoates (PHA) synthesized from cyanobacteria can assist in resolving the existing environmental issues with the pursuit of sustainable bioplastic production while trying to cut production costs, reuse waste, reduce CO_2, promote bioremediation and make better use of cyanobacteria metabolites in many industries (Gomes Gradíssimo et al., 2020)

The majority of cyanobacteria-based biorefineries are focused on single products, failing to satisfy the goal of efficient biomass utilization. Multiple product recovery cyanobacterial biorefineries, on the other hand, can efficiently valorize biomass with little to no waste formation. However, cyanobacterial biorefineries come with several obstacles and challenges. Most of them are linked to production processes and downstream strategies, both of which are difficult to handle commercially.

10.7.1 Constraints in Downstream Processing of Cyanobacteria

The high expense of the downstream process is the most significant obstacle. When comparing the economics of different biorefinery stages, it has been observed that downstream processing costs are at least 50% higher than cyanobacterial production costs (Mitra and Mishra 2019). By using low-cost harvesting, cell disruption and extracting procedures, as well as state-of-the-art technology for multi-product recovery and sequential extraction of multiple value-added products, the cost of biomass production and downstream can be reduced (Prabha et al. 2022).

The fundamental restriction in the downstream processing of cyanobacteria and microalgae is high water content and high N and P content (Parmar et al. 2011). Biomass extraction is the process of concentrating cyanobacterial bulk cultures into thick cakes to extract useful products from them. Filtration, coagulation/flocculation, sedimentation and flotation can all be used to collect cyanobacteria (Satpati et al. 2022). Cell harvesting accounts for 20–30% of total biomass production costs; therefore, large-scale production necessitates cost-effective biomass recovery for affordable end products (Yashavanth et al. 2021). The harvesting method may also influence the properties of the end-product.

Downstream processing should be as efficient and low energy consuming as possible. As a result, the design of downstream processes for cyanobacterial proteins should now focus on the functioning of the produced constituents rather than the end protein quantity in the extract. Heat treatment is used in several processes, such as pasteurization. Heat treatment of biomass can cause adverse protein denaturation, which can reduce some functionality. A heating step can also raise the viscosity of the solution, which can cause chaos in downstream processes like pumping (Grossmann et al. 2020). This entails choosing the right species and growing conditions, as well as designing a cost-effective downstream procedure. According to recent developments, the manufacturing of polyhydroxybutyrate (PHB), a bioplastic derived from cyanobacteria, is now possible (Karan et al. 2019). However, no economic research has been performed to determine the phototrophic PHB production costs. According to reports, heterotrophic organisms may produce PHB for between 2 and 5 € kg^{-1}. The cost of producing PHB in cyanobacteria could be more expensive than that of heterotrophic microorganisms (>5 € kg^{-1}) due to the significantly lower time space yield and biomass productivity in cyanobacteria as well as difficulties with the downstream processing (Kamravamanesh et al. 2018). Using open pond raceways with wastewater-born cyanobacteria instead of freshwater strains is another option to expand the facility or reactor (for cyanobacteria-producing PHB) while simultaneously lowering production costs. However, it should be noted that the downstream processing effort will directly rise as the volume increases (Drosg et al. 2015).

According to the order of recovery of high to low market value products from cyanobacteria, numerous products have been sequentially extracted. Sequential extraction for the recovery of lipids, total fatty acids and phycocyanin has been studied in detail in the cyanobacterium *A. platensis*. It has been discovered that sequential product recovery, with phycocyanin extraction as the initial stage, followed by lipid and polysaccharide extraction, will produce a significant yield (Prabha et al. 2022). A thorough study has been conducted on the successful extraction of lipids and amino acids from *Lyngbya* species. They came to the conclusion that UV exposure would be a good way to get the *Lyngbya* to start producing UV protection and high levels of lipids with a strong saturation index. Compared to UVA, UVB causes a higher generation of MAAs. In a sustainable biorefinery technique, the same feedstock might be used to recover both lipid and MAAs. Following the extraction of the mycosporine amino acids (MAA), they isolated lipids from the leftover biomass (Chandra et al. 2019). This method implies that UV exposure and recovery of UV protectant with biodiesel are a more sustainable choice for improved fuel productivity, and the sequential extraction methodology presented can definitely be taken into consideration for use in microalgae biorefineries (Chandra et al. 2019). However, to assess the projected cost of the MAA recovery, additional research on *Lyngbya* sp. reports 17.40 mg of dried methanolic residue/mL of potentially recovered MAAs. The findings from these investigations might enable the commercialization of a larger industrial output in new market products (Zahra et al. 2020; Fuentes-Tristan et al. 2019). Many phycologists are taking on the unique challenges of producing high-quality sustainable biomass feedstock from cyanobacteria; in this context, artificial ecosystems are being created by mixing various vigorous cyanobacteria species and raising them in a common habitat or culture media with the same concentrations of various ingredients (Patel et al. 2018).

10.7.2 Adaptation in Production Media for High Yields

Scaling up cultures to the extremely large levels necessary for commercial production is a difficult task. The key operational requirements for large-scale commercial microalgal culture comprises production of the inoculum effectively and reliably in the shortest amount of time possible; effective monitoring; the detection of any potential issues; the development of operating protocols to maintain reliability, high productivity and product quality for prolonged periods; and strategies for preventing culture collapse if equipment failure interrupts regular operations. To begin, think about how to improve the process of producing enough inoculum while minimizing the time and cost involved. It is also critical to control large-scale cultures to avoid severe contamination or collapse to reduce the requirement for re-inoculation from stock cultures (Borowitzka and Vonshak 2017). Long-term, stable, high-productivity large-scale cultures, which are usually maintained under prevailing outdoor conditions of variable irradiance, temperature and rainfall, present additional challenges, the majority of which are not encountered in the constant environment experienced by small-scale laboratory cultures.

Biosynthesis of polyunsaturated fatty acids (PUFA) from cyanobacteria is a more cost-effective option. Organic residues and wastes created by a variety of industries are significant carbon load-containing materials that can be employed as growth media for microorganisms that produce PUFAs. Recent research on the genus *Arthrospira* has found that growth was evaluated in batch cultures for 14 days under the optimum conditions for each species, along with the monitoring of pH, biomass composition and nutrient uptake. The values of pH achieved by *A. maxima* were greater than 8, which corresponds to the ideal values of species of the genus *Arthrospira* which can thrive in alkaline environments. The pH of the *C. vulgaris* culture increased from its initial value of 7.5 to 9.0 throughout the course of the first 48 hours. This is related to how quickly CO_2 is used, which causes the pH to increase towards alkaline values. Compared to other microalgae, *N. gaditana* can grow at high pH levels of 8 to 10 (Sánchez-Bayo et al. 2020). The study demonstrates that by lowering the amount of $NaNO_3$ in the *A. platensis* medium, higher quantities of biomass can be produced. The findings show that *A. platensis* can be produced at lower costs and with less effluent generation without impacting cultivation efficiency (de Castro et al. 2015).

The *Cyanobacterium* sp. SCSIO-45682 was isolated from an open pond of a marine green alga (*Picochlorum* sp. SCSIO-45015). The effects of initial sodium bicarbonate ($NaHCO_3$) concentrations on *Cyanobacterium* sp. SCSIO-45682 growth and biochemical composition were examined. The study revealed that *Cyanobacterium* sp. SCSIO-45682 adapted well to 16.8 gL^{-1} $NaHCO_3$. Furthermore, under higher $NaHCO_3$ concentrations, biomass, polysaccharide, chlorophyll-a and phycocyanin yields were increased. At 8.4 gL^{-1} $NaHCO_3$, the greatest final biomass concentration of 2.5 gL^{-1} was found (Chen et al. 2021). Recent research has shown that an excess of phosphorus in media might lead to a decrease in protein content. The large-scale cultivation of *Synechocystis* sp. PCC6803 was done in semi-continuous open raceway ponds in municipal wastewater and reached a growth rate of 0.21 gL^{-1} day^{-1}. It is estimated to generate 94.5 tones dry biomass ha^{-1} year. By employing municipal wastewater as a growth medium, production costs are reduced by between 32 and 35%. With the aid of this method, the price of lipid extraction is also decreased, and yields of up to 90% are obtained (Prabha et al. 2022). Some cyanobacteria, which tend to accumulate amino acid polymers (such as cyanophycin) in phosphorus-depleted circumstances as a stress response, have shown this reduction. Even if the ultimate cyanophycin content in the cell dry mass may be higher in *Synechocystis* sp. cultures, the productivity drops in phosphorus-limited culture conditions, which may lead to a decreased overall amount of biomass since cell division is inhibited (Grossmann et al. 2020). Identification of cyanobacterial species that can withstand changing environmental conditions and provide reliable, commercially acceptable volumes of biomass all year long for varied uses is necessary. In order to increase the biomass production and financial success of cyanobacterial farming, efforts must also be made to increase the fitness of already-known species under a variety of environmental conditions (Pathak et al. 2018). Table 10.1 shows the literature about growth, biomass yield and valuable compounds extracted from cyanobacteria in various culture media and their associated risk factors. However, the selection of robust strains and establishment of their synthetic consortia for the maximum production of high-quality biomass and commodity chemicals necessitate the engineering of the cyanobacteria community in an adapted habitat (Patel et al. 2018).

Several products from a cyanobacterial biorefinery can be extracted and further purified to achieve maximum biomass valorization. However, from the standpoint of economic viability, it is possible to choose and extract the most suitable viable co-products from the biorefinery. It is essential to assess the forms of cyanobacteria under consideration, any potential value-added products that might be retrieved and the system's techno-economic viability. The economic feasibility studies are limited to a few cyanobacterial species, and the new species receives less attention, which is one of the biggest limitations. To advance innovative strains for the future, research

Table 10.1 Improved productivity of cyanobacterium strains by adaptations in media compositions and risks involved to generate high-value compounds

Cyanobacterial strains	Biomass production (mg L^{-1})	Specific media composition	Biomass yield (mg L^{-1})	Risk factors	Valuable compound	References
Cyanobacterium sp. SCSIO-45682	0.04	$NAHCO_3$	122.14 ± 0.07	Contamination and maintenance of the alkaline system	Polysaccharide	Chen et al. (2021)
Synechococcus elongatus UTEX 2973	4.11	Cobg-11	295.5	Maintenance of pH, NaCl conc. and alkaline environment	Sucrose	Zhang et al. (2020)
Synechococcus sp. PCC 11901	1.54	Availability of cobalamin in the growth medium	18.3	Biological contamination increases the salinity of the growth medium	Free fatty acids	Włodarczyk et al. (2020)
Synechocystis PCC 6803	–	$BG11^{+}$growth media	–	–	Carbohydrates	Patel et al. (2018)
A. platensis	153.3	Sodium bicarbonate	3270	–		de Castro et al. (2015)
Lyngbya sp.	10–1500	Methanol extraction	17.40	Exposure to UV-B is necessary	Mycosporine amino acids	Prabha et al. ((2022)
Anabaena doliolum		Molybdenum deficient media	5200	Molybdenum quantity	Carotenoid	(Rathore et al., 1993)

on their techno-economic viability is required. The development of novel strategies that can optimize the culture conditions for the mass generation of novel cyanobacterial strains should be the focus of further research (Prabha et al. 2022).

10.8 Conclusion and Prospects

Cyanobacteria could be potential feedstocks to overcome the problems which are created due to the faster depletion of fossil fuels and crude oils. Cyanobacteria can harness solar energy and can utilize wastewater streams and potential greenhouse gases for the generation of biofuel and high-value compounds. The ongoing research on circular economy is critical in utilizing wastes in the cyanobacterial value chain. However, the large-scale production of these cyanobacteria is plagued with several limitations on the technical, physiological and process parameter fronts. Therefore, to ensure transition from lab to market, autotrophic, heterotrophic and mixotrophic cyanobacterial biomass production strategies are discussed. Discussion on the selection of production systems takes into account various factors such as strain characteristics, the production site's natural environment, and how strains respond to different process variables like light, darkness, medium composition, pH, and agitation. This analysis covers their viability, growth, economic potential, and the optimization of multiple parameters to align with a circular economy approach. It is projected that the risks involved in genetic manipulation of cyanobacteria, downstream processing and adaptation to production media specifications need to be focused on, in the future. The impact and role of these parameters' associated risk factors were evaluated for the implementation of circular bioeconomy in near future.

References

Achparaki M, Thessalonikeos E, Tsoukali H, Mastrogianni O, Zaggelidou E, Chatzinikolaou F, Vasilliades N, Raikos N, Isabirye M, Raju DVN, Kitutu M, Yemeline V, Deckers J, Poesen J (2012) We are IntechOpen, the world' s leading publisher of open access books built by scientists, for scientists, vol 13. IntechOpen

Anzaku AA, Akyala JI, Juliet A, Obianuju EC (2017) Antibacterial activity of Lauric acid on some selected clinical isolates. Ann Clin Lab Res 5. https://doi.org/10.21767/2386-5180.1000170

Appel J, Hueren V, Boehm M, Gutekunst K (2020) Cyanobacterial in vivo solar hydrogen production using a photosystem I–hydrogenase (PsaD-HoxYH) fusion complex. Nat Energy 5:458–467. https://doi.org/10.1038/s41560-020-0609-6

Awuchi C, Ikechukwu A (2021) Biochemistry and nutrition of carbohydrates. Global J Res Agric Life Sci. 1:4–12

Banerjee S, Singh H, Das D, Atta A (2019) Process optimization for enhanced biodiesel production by *Neochloris oleoabundans* UTEX 1185 with concomitant CO_2 sequestration. Ind Eng Chem Res 58:15760–15771. https://doi.org/10.1021/acs.iecr.8b05730

Borowitzka MA, Vonshak A (2017) Scaling up microalgal cultures to commercial scale. Eur J Phycol 52:407–418. https://doi.org/10.1080/09670262.2017.1365177

Bounnit T, Saadaoui I, Ghasal G, al Rasheed R, Dalgamouni T, Jabri H, al Leroy E, Legrand J (2022) Assessment of novel halo- and thermotolerant desert cyanobacteria for phycobiliprotein production. Process Biochem 118:425–437. https://doi.org/10.1016/j.procbio.2022.04.017

Braune S, Krüger-Genge A, Kammerer S, Jung F, Küpper J-H (2021) Phycocyanin from Arthrospira platensis as potential anti-cancer drug: review of in vitro and in vivo studies. Life 11:91. https://doi.org/10.3390/life11020091

Brownstein AM (2015) Isobutanol. In: Renewable motor fuels. Elsevier, New York, pp 47–56. https://doi.org/10.1016/B978-0-12-800970-3.00005-4

Chandra R, Das P, Vishal G, Nagra S (2019) Factors affecting the induction of UV protectant and lipid productivity in *Lyngbya* for sequential biorefinery product recovery. Bioresour Technol 278:303–310. https://doi.org/10.1016/j.biortech.2019.01.084

Chen Z, Li T, Yang B, Jin X, Wu H, Wu J, Lu Y, Xiang W (2021) Isolation of a novel strain of *Cyanobacterium* sp. with good adaptation to extreme alkalinity and high polysaccharide yield. J Oceanol Limnol 39:1131–1142. https://doi.org/10.1007/s00343-020-0113-7

Clark DP, Pazdernik NJ (2016) Environmental biotechnology. In: Biotechnology. Elsevier, New York, pp 393–418. https://doi.org/10.1016/B978-0-12-385015-7.00012-0

Corrêa PS, Teixeira CMLL (2021) Polyhydroxyalkanoates and pigments coproduction by *Arthrospira* (Spirulina) platensis cultivated in crude glycerol. J Appl Phycol 33:1487–1500. https://doi.org/10.1007/s10811-021-02373-8

Costa JAV, de Morais MG (2014) An open pond system for microalgal cultivation. In: Biofuels from algae. Elsevier, New York, pp 1–22. https://doi.org/10.1016/B978-0-444-59558-4.00001-2

Cruz YR, Aranda DAG, Seidl PR, Diaz GC, Carliz RG, Fortes MM, da Ponte DAMP, de Paula RCV (2018) Cultivation systems of microalgae for the production of biofuels. In: Biofuels - state of development. InTech. https://doi.org/10.5772/intechopen.74957

Dahiya D, Sharma H, Rai AK, Nigam PS (2022) Application of biological systems and processes employing microbes and algae to reduce, recycle, reuse (3Rs) for the sustainability of circular bioeconomy. AIMS Microbiol. https://doi.org/10.3934/microbiol.2022008

De Barros PDS, Costa e Silva PEC, Nascimento TP, Costa RMPB, Bezerra RP, Porto ALF (2020) Fibrinolytic enzyme from *Arthrospira platensis* cultivated in medium culture supplemented with corn steep liquor. Int J Biol Macromol 164:3446–3453. https://doi.org/10.1016/j.ijbiomac.2020.08.217

de Castro GFP, Rizzo RF, Passos TS, dos Santos BNC, Dias D, Domingues J, de Lima Araújo KG (2015) Biomass production by *Arthrospira platensis* under different culture conditions. Food Sci Technol (Brazil) 35:18–24. https://doi.org/10.1590/1678-457X.6421

Deng M-D, Coleman JR (1999) Ethanol synthesis by genetic engineering in cyanobacteria. Appl Environ Microbiol 65:523–528. https://doi.org/10.1128/AEM.65.2.523-528.1999

Donato-Capel L, Garcia-Rodenas CL, Pouteau E, Lehmann U, Srichuwong S, Erkner A, Kolodziejczyk E, Hughes E, Wooster TJ, Sagalowicz L (2014) Technological means to modulate food digestion and physiological response. In: Food structures, digestion and health. Elsevier, New York, pp 389–422. https://doi.org/10.1016/B978-0-12-404610-8.00014-1

Drosg B, Fritz I, Gattermayr F, Silvestrini L (2015) Photo-autotrophic production of poly (hydroxyalkanoates) in cyanobacteria. Chem Biochem Eng Q. https://doi.org/10.15255/CABEQ.2014.2254

El-Naggar NE-A, Hussein MH, Shaaban-Dessuuki SA, Dalal SR (2020) Production, extraction and characterization of *Chlorella vulgaris* soluble polysaccharides and their applications in AgNPs biosynthesis and biostimulation of plant growth. Sci Rep 10:3011. https://doi.org/10.1038/s41598-020-59945-w

Fuentes-Tristan S, Parra-Saldivar R, Iqbal HMN, Carrillo-Nieves D (2019) Bioinspired biomolecules: mycosporine-like amino acids and scytonemin from Lyngbya sp. with UV-protection potentialities. J Photochem Photobiol B Biol 201:111684. https://doi.org/10.1016/j.jphotobiol.2019.111684

Gasparotto J, da Boit Martinello K (2021) Coal as an energy source and its impacts on human health. Energy Geosci 2:113–120. https://doi.org/10.1016/j.engeos.2020.07.003

Gomes Gradíssimo D, Pereira Xavier L, Valadares Santos A (2020) Cyanobacterial polyhydroxyalkanoates: a sustainable alternative in circular economy. Molecules 25(18):4331. https://doi.org/10.3390/molecules25184331

Grama SB, Liu Z, Li J (2022) Emerging trends in genetic engineering of microalgae for commercial applications. Mar Drugs 20:285. https://doi.org/10.3390/md20050285

Grossmann L, Hinrichs J, Weiss J (2020) Cultivation and downstream processing of microalgae and cyanobacteria to generate protein-based techno-functional food ingredients. Crit Rev Food Sci Nutr. https://doi.org/10.1080/10408398.2019.1672137

Gupta PL, Lee S-M, Choi H-J (2015) A mini review: photobioreactors for large scale algal cultivation. World J Microbiol Biotechnol 31:1409–1417. https://doi.org/10.1007/s11274-015-1892-4

Hamida RS, Ali MA, Redhwan AMO, Bin-Meferij MM (2020) Cyanobacteria – a promising platform in green nanotechnology: a review on nanoparticles fabrication and their prospective applications. Int J Nanomedicine 15:6033–6066. https://doi.org/10.2147/ijn.s256134

Hseu Y-C, Chen X-Z, Vudhya Gowrisankar Y, Yen H-R, Chuang J-Y, Yang H-L (2020) The skin-whitening effects of Ectoine via the suppression of α-MSH-stimulated melanogenesis and the activation of antioxidant Nrf2 pathways in UVA-irradiated keratinocytes. Antioxidants 9:63. https://doi.org/10.3390/antiox9010063

Kamravamanesh D, Lackner M, Herwig C (2018) Bioprocess engineering aspects of sustainable polyhydroxyalkanoate production in cyanobacteria. Bioengineering. https://doi.org/10.3390/BIOENGINEERING5040111

Kamravamanesh D, Kiesenhofer D, Fluch S, Lackner M, Herwig C (2019) Scale-up challenges and requirement of technology-transfer for cyanobacterial poly (3-hydroxybutyrate) production in industrial scale. Int J Biobased Plast 1:60–71. https://doi.org/10.1080/24759651.2019.1688604

Karan H, Funk C, Grabert M, Oey M, Hankamer B (2019) Green bioplastics as part of a circular bioeconomy. Trends Plant Sci. https://doi.org/10.1016/j.tplants.2018.11.010

Khan MI, Shin JH, Kim JD (2018) The promising future of microalgae: current status, challenges, and optimization of a sustainable and renewable industry for biofuels, feed, and other products. Microb Cell Factories 17:36. https://doi.org/10.1186/s12934-018-0879-x

Khetkorn W, Rastogi RP, Incharoensakdi A, Lindblad P, Madamwar D, Pandey A, Larroche C (2017) Microalgal hydrogen production – a review. Bioresour Technol 243:1194–1206. https://doi.org/10.1016/j.biortech.2017.07.085

Koch M, Forchhammer K (2021) Polyhydroxybutyrate: a useful product of chlorotic cyanobacteria. Microb Physiol 31:67–77. https://doi.org/10.1159/000515617

Leblond JD, Timofte HI, Roche SA, Porter NM (2011) Sterols of glaucocystophytes. Phycol Res 59:129–134. https://doi.org/10.1111/j.1440-1835.2011.00610.x

Liu Y, Xu L, Cheng N, Lin L, Zhang C (2000) Inhibitory effect of phycocyanin from *Spirulina platensis* on the growth of human leukemia K562 cells. J Appl Phycol 12:125–130. https://doi.org/10.1023/A:1008132210772

Llewellyn CA, Greig C, Silkina A, Kultschar B, Hitchings MD, Farnham G (2020) Mycosporine-like amino acid and aromatic amino acid transcriptome response to UV and far-red light in the cyanobacterium *Chlorogloeopsis fritschii* PCC 6912. Sci Rep 10:20638. https://doi.org/10.1038/s41598-020-77402-6

Lopes G, Clarinha D, Vasconcelos V (2020) Carotenoids from cyanobacteria: a biotechnological approach for the topical treatment of psoriasis. Microorganisms 8:302. https://doi.org/10.3390/microorganisms8020302

Lupacchini S, Appel J, Stauder R, Bolay P, Klähn S, Lettau E, Adrian L, Lauterbach L, Bühler B, Schmid A, Toepel J (2021) Rewiring cyanobacterial photosynthesis by the implementation of an oxygen-tolerant hydrogenase. Metab Eng 68:199–209. https://doi.org/10.1016/j.ymben.2021.10.006

Martínez-Merino V, Gil MJ, Cornejo A (2013) Biological hydrogen production. In: Renewable hydrogen technologies. Elsevier, New York, pp 171–199. https://doi.org/10.1016/B978-0-444-56352-1.00008-8

Masukawa H, Kitashima M, Inoue K, Sakurai H, Hausinger RP (2012) Genetic engineering of cyanobacteria to enhance biohydrogen production from sunlight and water. Ambio 41:169–173. https://doi.org/10.1007/s13280-012-0275-4

Mathan Raj V, Ganapathy Subramanian LR, Manikandaraja G (2018) Experimental study of effect of isobutanol in performance, combustion and emission characteristics of CI engine fuelled with cotton seed oil blended diesel. Alex Eng J 57:1369–1378. https://doi.org/10.1016/j.aej.2017.06.007

Mitra M, Mishra S (2019) Multiproduct biorefinery from Arthrospira spp. towards zero waste: current status and future trends. Bioresour Technol 291:121928. https://doi.org/10.1016/j.biortech.2019.121928

Mussgnug JH, Thomas-Hall S, Rupprecht J, Foo A, Klassen V, McDowall A, Schenk PM, Kruse O, Hankamer B (2007) Engineering photosynthetic light capture: impacts on improved solar energy to biomass conversion. Plant Biotechnol J 5(6):802–814. https://doi.org/10.1111/j.1467-7652.2007.00285.x

Nematollahi MA, Laird DW, Hughes LJ, Raeisossadati M, Moheimani NR (2020) Effect of organic carbon source and nutrient depletion on the simultaneous production of a high value bioplastic and a specialty pigment by *Arthrospira platensis*. Algal Res 47:101844. https://doi.org/10.1016/j.algal.2020.101844

Nicoletti M (2016) Microalgae nutraceuticals. Foods 5:54. https://doi.org/10.3390/foods5030054

Nowruzi B, Sarvari G, Blanco S (2020) The cosmetic application of cyanobacterial secondary metabolites. Algal Res 49:101959. https://doi.org/10.1016/j.algal.2020.101959

Nur MMA, Buma AGJ (2019) Opportunities and challenges of microalgal cultivation on wastewater, with special focus on palm oil mill effluent and the production of high value compounds. Waste Biomass Valori 10:2079–2097. https://doi.org/10.1007/s12649-018-0256-3

Nwosu OK, Ubaoji KI (2020) Nutraceuticals: history, classification and market demand, in: functional foods and nutraceuticals. Springer, Cham, pp 13–22. https://doi.org/10.1007/978-3-030-42319-3_2

Nzayisenga JC, Farge X, Groll SL, Sellstedt A (2020) Effects of light intensity on growth and lipid production in microalgae grown in wastewater. Biotechnol Biofuels 13(1):4. https://doi.org/10.1186/s13068-019-1646-x

Okuyama Y, Ozasa K, Oki K, Nishino H, Fujimoto S, Watanabe Y (2014) Inverse associations between serum concentrations of zeaxanthin and other carotenoids and colorectal neoplasm in Japanese. Int J Clin Oncol 19(1):87–97. https://doi.org/10.1007/s10147-013-0520-2

Panjiar N, Mishra S, Yadav AN, Verma P (2017) Functional foods from cyanobacteria. In: Microbial functional foods and nutraceuticals. Wiley, Chichester, UK, pp 21–37. https://doi.org/10.1002/9781119048961.ch2

Pansook S, Incharoensakdi A, Phunpruch S (2022) Simazine enhances dark fermentative H2 production by unicellular halotolerant *cyanobacterium Aphanothece halophytica*. Front Bioeng Biotechnol 10. https://doi.org/10.3389/fbioe.2022.904101

Parambath SD, Bhombal Y, Kumar A, Singh M, Chavali M, Chandrasekhar K (2022) The multifaceted microalgal approach to wastewater treatment to generate energy and essential chemicals. In: Micro-algae: next-generation feedstock for biorefineries, cultivation and refining processes. Springer, pp 139–176. https://doi.org/10.1007/978-981-19-0793-7_6

Parmar A, Singh NK, Pandey A, Gnansounou E, Madamwar D (2011) Cyanobacteria and microalgae: a positive prospect for biofuels. Bioresour Technol. https://doi.org/10.1016/j.biortech.2011.08.030

Patel VK, Sundaram S, Patel AK, Kalra A (2018) Characterization of seven species of cyanobacteria for high-quality biomass production. Arab J Sci Eng 43:109–121. https://doi.org/10.1007/s13369-017-2666-0

Patel AK, Singhania RR, Sim SJ, Di Dong C (2021) Recent advancements in mixotrophic bioprocessing for production of high value microalgal products. Bioresour Technol 320: 124421. https://doi.org/10.1016/j.biortech.2020.124421

Pathak J, Rajneesh, Maurya PK, Singh SP, Häder D-P, Sinha RP (2018) Cyanobacterial farming for environment friendly sustainable agriculture practices: innovations and perspectives. Front Environ Sci 6. https://doi.org/10.3389/fenvs.2018.00007

Pereira MIB, Chagas BME, Sassi R, Medeiros GF, Aguiar EM, Borba LHF, Silva EPE, Neto JCA, Rangel AHN (2019) Mixotrophic cultivation of *Spirulina platensis* in dairy wastewater: effects on the production of biomass, biochemical composition and antioxidant capacity. PLoS One 14: e0224294. https://doi.org/10.1371/journal.pone.0224294

Prabha S, Vijay AK, Paul RR, George B (2022) Cyanobacterial biorefinery: towards economic feasibility through the maximum valorization of biomass. Sci Total Environ 814:152795. https://doi.org/10.1016/j.scitotenv.2021.152795

Qamar H, Hussain K, Soni A, Khan A, Hussain T, Chénais B (2021) Cyanobacteria as natural therapeutics and pharmaceutical potential: role in antitumor activity and as nanovectors. Molecules 26:247. https://doi.org/10.3390/molecules26010247

Quintana N, van der Kooy F, van de Rhee MD, Voshol GP, Verpoorte R (2011) Renewable energy from cyanobacteria: energy production optimization by metabolic pathway engineering. Appl Microbiol Biotechnol 91:471–490. https://doi.org/10.1007/s00253-011-3394-0

Raczyk M, Polanowska K, Kruszewski B, Grygier A, Michałowska D (2022) Effect of spirulina (*Arthrospira platensis*) supplementation on physical and chemical properties of semolina (*Triticum durum*) based fresh pasta. Molecules 27:355. https://doi.org/10.3390/molecules27020355

Rahmatpour N, Hauser DA, Nelson JM, Chen PY, Villarreal AJC, Ho M-Y, Li F-W (2021) A novel thylakoid-less isolate fills a billion-year gap in the evolution of cyanobacteria. Curr Biol 31: 2857–2867.e4. https://doi.org/10.1016/j.cub.2021.04.042

Rajeshwari KR, Rajashekhar M (2011) Biochemical composition of seven species of cyanobacteria isolated from different aquatic habitats of Western Ghats, Southern India. Braz Arch Biol Technol 54(5):849–857. https://doi.org/10.1590/S1516-89132011000500001

Rathore DS, Kumar A, Kumar HD (1993) Lipid content and fatty acid composition in N2-fixing cyanobacterium Anabaena doliolum as affected by molybdenum deficiency. World J Microbiol Biotechnol 9(5):508–510. https://doi.org/10.1007/BF00386284

Rösch C, Roßmann M, Weickert S (2019) Microalgae for integrated food and fuel production. GCB Bioenergy 11:326–334. https://doi.org/10.1111/gcbb.12579

Sánchez-Bayo A, Morales V, Rodríguez R, Vicente G, Bautista LF (2020) Cultivation of microalgae and cyanobacteria: effect of operating conditions on growth and biomass composition. Molecules 25:2834. https://doi.org/10.3390/molecules25122834

Satpati GG, Kumar A, Parambath SD, Bagchi T, Singh M, Chavali M, Chandrasekhar K (2022) Recent development in microalgal cultivation and biomass harvesting methods. In: Algal biorefineries and the circular bioeconomy. CRC Press, Boca Raton, FL, pp 61–91. https://doi.org/10.1201/9781003195429-2

Schwarz A, Walther J, Geib D, Witthohn M, Strieth D, Ulber R, Muffler K (2020) Influence of heterotrophic and mixotrophic cultivation on growth behaviour of terrestrial cyanobacteria. Algal Res 52:102125. https://doi.org/10.1016/j.algal.2020.102125

Sharma NK, Stal LJ (2013) The economics of cyanobacteria-based biofuel production: challenges and opportunities. In: Cyanobacteria. Wiley, Chichester, UK, pp 167–180. https://doi.org/10.1002/9781118402238.ch10

Singh V, Chaudhary DK, Mani I, Dhar PK (2016) Recent advances and challenges of the use of cyanobacteria towards the production of biofuels. Renew Sust Energ Rev 60:1–10. https://doi.org/10.1016/j.rser.2016.01.099

Sitther V, Tabatabai B, Fathabad SG, Gichuki S, Chen H, Arumanayagam ACS (2020) Cyanobacteria as a biofuel source: advances and applications. In: Advances in cyanobacterial biology. Elsevier, New York, pp 269–289. https://doi.org/10.1016/B978-0-12-819311-2.00018-8

Skerker JM, Lucks JB, Arkin AP (2009) Evolution, ecology and the engineered organism: lessons for synthetic biology. Genome Biol 10:114. https://doi.org/10.1186/gb-2009-10-11-114

Sreenikethanam A, Bajhaiya A (2022) Algae based bio-plastics: future of green economy. In: Biorefineries - selected processes. IntechOpen. https://doi.org/10.5772/intechopen.100981

Suikkanen S, Uusitalo L, Lehtinen S, Lehtiniemi M, Kauppila P, Mäkinen K, Kuosa H (2021) Diazotrophic cyanobacteria in planktonic food webs. Food Webs 28:e00202. https://doi.org/10.1016/j.fooweb.2021.e00202

Taikhao S, Phunpruch S (2017) Effect of metal cofactors of key enzymes on biohydrogen production by nitrogen-fixing cyanobacterium *Anabaena siamensis* TISIR 8012. Energy Procedia 138:360–365. https://doi.org/10.1016/j.egypro.2017.10.166

Tamagnini P, Leitão E, Oliveira P, Ferreira D, Pinto F, Harris DJ, Heidorn T, Lindblad P (2007) Cyanobacterial hydrogenases: diversity, regulation and applications. FEMS Microbiol Rev 31: 692–720. https://doi.org/10.1111/j.1574-6976.2007.00085.x

Tur JA, Bibiloni MM (2016) Functional foods. In: Encyclopedia of food and health. Elsevier, pp 157–161. https://doi.org/10.1016/B978-0-12-384947-2.00340-8

Ucar D, Zhang Y, Angelidaki I (2017) An overview of electron acceptors in microbial fuel cells. Front Microbiol 8. https://doi.org/10.3389/fmicb.2017.00643

UNICEF (2018) (n.d.). Retrieved October 11, 2023, from https://www.unicefusa.org/press/unicef-too-manychildren-dying-malnutrition

Velioğlu Tosuner Z, Öztürk Ürek R (2020) Evaluation of sucrose as carbon source in mixotrophic culture of *Arthrospira platensis* Gomont 1892. Aquat Res 1–12. https://doi.org/10.3153/AR20001

Vu HP, Nguyen LN, Zdarta J, Nga TTV, Nghiem LD (2020) Blue-green algae in surface water: problems and opportunities. Curr Pollut Rep 6:105–122. https://doi.org/10.1007/s40726-020-00140-w

Włodarczyk A, Selão TT, Norling B, Nixon PJ (2020) Newly discovered *Synechococcus* sp. PCC 11901 is a robust cyanobacterial strain for high biomass production. Commun Biol 3. https://doi.org/10.1038/s42003-020-0910-8

Yashavanth PR, Das M, Maiti SK (2021) Recent progress and challenges in cyanobacterial autotrophic production of polyhydroxybutyrate (PHB), a bioplastic. J Environ Chem Eng 9 (4):105379. https://doi.org/10.1016/j.jece.2021.105379

Zahra Z, Choo DH, Lee H, Parveen A (2020) Cyanobacteria: review of current potentials and applications. Environments 7:13. https://doi.org/10.3390/environments7020013

Zhang L, Chen L, Diao J, Song X, Shi M, Zhang W (2020) Construction and analysis of an artificial consortium based on the fast-growing cyanobacterium *Synechococcus elongatus* UTEX 2973 to produce the platform chemical 3-hydroxypropionic acid from CO2. Biotechnol Biofuels 13. https://doi.org/10.1186/s13068-020-01720-0

Chapter 11
Cyanobacterial Exopolysaccharides: Extraction, Processing, and Applications

Faryal Yousaf, Sayyad Ali Raza Bukhari, Hafiz Abdullah Shakir, Muhammad Khan, Marcelo Franco, and Muhammad Irfan

Abstract Exopolysaccharide (EPS) extraction from cyanobacteria is an effective biotechnological approach with imperious applications in different industries. Cyanobacteria being photoautotrophic organisms show diverse cellular arrangements under stressful environmental conditions like exposure to desiccation, ultraviolent radiation, and high temperature or pressure. These cyanobacteria, because of their environment-friendly nature and fast growth, are considered as a renewable source to produce exopolysaccharides with applications in pharmaceutical, food, and cosmetics production, treatment of wastewater, heavy metal removal, and many more. Most of the research is conducted on the extraction of exopolysaccharides, like the use of conventional alcoholic precipitation, tangential filtration, and certain other alternative methods in addition to the bioprocessing of exopolysaccharides from cyanobacteria. Currently, scientists are working on the manipulation of cyanobacterial exopolysaccharides by developing cultures and processes, so that better production of exopolysaccharides can be attained. This chapter highlights the structural and physicochemical characterization of the EPS. Furthermore, it highlights the protocols for the enhancement, processing, and extraction of EPS along with the potential applications of cyano-EPS in different fields of bio-industry.

Keywords Exopolysaccharides (EPS) · Cyanobacteria · Extraction · Purification · Applications

F. Yousaf · S. A. R. Bukhari · M. Irfan (✉)
Department of Biotechnology, University of Sargodha, Sargodha, Pakistan
e-mail: irfan.ashraf@uos.edu.pk

H. A. Shakir · M. Khan
Institute of Zoology, University of the Punjab New Campus, Lahore, Pakistan

M. Franco
Department of Exact Science, State University of Santa Cruz, Ilhéus, Brazil

M. A. Mehmood et al. (eds.), *Pharmaceutical and Nutraceutical Potential of Cyanobacteria*, https://doi.org/10.1007/978-3-031-45523-0_11

11.1 Introduction

Production of extracellular polysaccharides particularly from microorganisms has been expanded over the last few years because of their applications in various arenas like pharmaceutical, food, and removal of heavy metals (Abdo et al. 2012; Shaieb et al. 2014; Baldev et al. 2015). To produce diverse chemical products, it has been obligatory to discover native as well as innovative strains of microbes with the proficiency of synthesizing the naturally active components (Tiwari et al. 2020). Among these microbes, cyanobacteria have been considered as an effective source of different polymers (Khattar et al. 2010; Bhatnagar et al. 2014). Cyanobacteria are gram negative, oxygen-evolving phototrophic prokaryotes that are capable of fixing nitrogen (Kumar et al. 2018). Nearby the cells of cyanobacteria, a distinctive mucilaginous casing is present which is mainly composed of carbohydrates called exopolysaccharides (EPS). These are known as exopolysaccharides due to their extracellular secretory nature. The formation of exopolysaccharide in different microbes including cyanobacteria has been found to be very beneficial to protect the cells specifically from stress as well as other extreme conditions. Because of these polysaccharides, cyanobacteria prevent pressure generated due to dehydration in hyper saline or desert places. Such exopolysaccharides can exit in two forms either as released exopolysaccharides (RPS) or as capsular exopolysaccharides (CPS). Capsular exopolysaccharides being mucilaginous cause encapsulation of cyanobacterial cells, while released exopolysaccharides are being released into the extracellular surroundings (Freitas et al. 2011). Composition of these cyanobacteria EPS involved monosacharaides that have been particularly categorized into inorganic constituents, pentose, and acidic sugars (Challouf et al. 2011). Up to 13 or even more monosaccharides are being described as cyanobacterial exopolysaccharides (Rossi and De Philippis 2015). Extraordinary quantities of neutral sugars such as arabinose, mannose, xylose, and glucose have been perceived in these EPS. The EPS from cyanobacteria own enhanced level of sulfate as well as uronic acid groups that provide adhesive nature and negative charge to these exopolysaccharides (Singh et al. 2019).

The processing of cyanobacterial exopolysaccharide is comprised of three phases. The first phase involves collection or selection of cyanobacteria host. Second is optimizing the parameters required for exopolysaccharide production, and last is the bioprocessing of EPS by downstream processing (Cruz et al. 2020). Extraction of exopolysaccharide has been found to be very useful as their features might be pretentious upon isolation or purification processes. For the extraction of exopolysaccharides from cyanobacteria, different physical methods and chemical processes have been used. Physical ways include heating, centrifugation, sonication, and thaw procedures, while chemical procedures involve the use of several chemical mediators like EDTA, formaldehyde, and sodium hydroxide (Mishra et al. 2011). It was found that manufacturing of exopolysaccharide (EPS) from different cyanobacteria has been an essential source for several purposes like elimination of solid particles as well as heavy metals from water bodies, for improving the water holding capacity and stabilization of soil and many more (Ozturk and Aslim 2008). For the manufacturing of commercially valuable and advantageous components of medicine, these cyanobacteria are available as cost-effective sources because of their ability to grow easily and in large amounts (Singh et al. 2019). Additionally, they can be used as food additives or as thickeners (Kumar et al. 2018). Different studies regarding cyanobacterial EPS from various sources have been done, like *Phormidium persicinum* from a marine environment having the potential to act as a bioflocculant (Baldev et al. 2015). Another EPS production from *Phormidium* sp. and *Oscillatoria* sp. was found to act as an effective antibiofilm agent by showing multi-drug resistance (MDR) against *Pseudomonas aeruginosa* (ArunKumar et al. 2018). Cyanobacterial EPS from dry lands like *Nostoc flagelliforme* showed significant protection or tolerance towards acid deposition and dryness stress (Gao et al. 2019). This book chapter focuses on general aspects of cyanobacterial EPS, including their composition, characterization, extraction processes, and factors influencing the growth of EPS under different cultivation conditions. Moreover, the role of these EPS in many different industries is also highlighted.

11.2 Cyanobacterial Exopolysaccharides (EPS)

Polysaccharides being found in all living beings are mixture of polymeric carbohydrates along with multifaceted sugars. From past few years, these polysaccharides have established much consideration, having the potential to perform various biological functions and withstanding stress conditions (de Jesus Raposo et al. 2015; Kraan 2012; Misurcova et al. 2015). These EPS have been categorized into three subcategories: slime, sheaths, and capsule (Rossi and Philippis 2016). Among these, capsules or sheaths are found to be highly compressed or compact enabling cyanobacteria protection from biotic and abiotic pressure generated in the surrounding environment. Slime on the other hand has less compact nature as compared to capsules or sheaths causing slithering motion in cyanobacteria (Sciuto and Moro 2015).

On the other hand, continuous oxygenation, the production cost primarily for carbon substrate, and demand for extreme energy minimize the general uses of this technology (Kamravamanesh et al. 2018). Cyanobacterial exopolysaccharides comprised non-saccharide constituents that are associated with their physicochemical attributes. This includes succinyl, pyruvyl and acetyl groups containing organic components and inorganic components like phosphate or sulfate (Kehr and Dittmann 2015; Rossi et al. 2018).

11.3 Chemical, Physicochemical, and Rheological Properties of Exopolysaccharides

For the analysis of chemical composition of exopolysaccharides, different spectrometric and chromatography techniques have been used. Cyanobacteria EPS comprised different key sugars like galactose, glucose orthomethyl, xylose, rhamnose, mannose, fucose, arabinose, and some acidic remains of galacturonic and glucuronic acids (Adhikary 1998; Bertocchi et al. 1990). Among these sugar molecules, glucose comprised the major ratio. Besides sugar molecules, sulfate group is also found in exopolysaccharide. These sulphate groups are found mainly in archaea and eukaryotic exopolysaccharides while occurring rarely in eubacteria. The anionic feature of exopolysaccharide in cyanobacteria is due to the presence of acidic substituents like galacturonic acid along with glucuronic acid and sulfate groups. This negative charge (anionic) provides an adhesive nature to the polysaccharide that results in attraction towards cationic metal ions (Kumar et al. 2018) which in turn is a valuable feature for bioremediation (Pereira et al. 2011). Availability of certain groups likes deoxy sugars (rhamnose and fucose), ester associated acetyl groups, and some other peptide elements results in hydrophobic character to some exopolysaccharides. These hydrophobic EPS have emulsion trait essential for industrial usage (Shepherd et al. 1995). The composition of cyanobacterial EPS may alter with culture conditions, light intensity, temperature and varying concentration of sulfur (S), potassium (K), and other different metallic ions. Some of the chemical compositions of cyanobacteria EPS have been shown in Table 11.1.

Cellular morphology is also being influenced by certain alterations in physiological parameters (Kumar et al. 2018). Otero and Vincenzini (2003) found that the availability of nitrate inside the media causes the capsular strains of *Nostoc* (PCC 7936 and PCC 8113) to inhibit the release of slime and become uncapsulated. Such alterations affected the rheological traits and viscosity of the culture medium. It was also observed that upon exposure to non-nitrate media, these uncapsulated strains of cyanobacteria returned to capsular slimy form (Kumar et al. 2018). These exopolysaccharides are classified into two types based on their composition. These are heteropolysaccharides and homopolysaccharides. Homopolysaccharides as the term is indicating involved monosaccharides of single kind, mostly sucrose produced by means of enzyme called sucrase (Pereira et al. 2009), while heterosaccharides are composed of two or more different kinds of sugars, produced by numerous sorts of glycosyltransferases enzyme (Kumar et al. 2018).

Table 11.1 Chemical composition of exopolysaccharides derived from diverse cyanobacteria

Cyanobacteria strains	Composition	Polysaccharide	References
Aphanothece sacrum	Glucose, rhamnose, fucose, galactose, mannose, xylose, alpha- glucose, α-galactose	Capsular polysaccharide	Mota et al. (2020)
Nostoc sphaeroides Küt	Glucose, galactose mannose, xylose	Capsular polysaccharide	Liu et al. (2018)
Cyanothece PCC 0010	Mannose, glucose, uronic acids, galactose, xylose, rhamnose, fucose, arabinose	Released polysaccharide	Ngatu et al. (2012)
Arthrospira platensis	Arabinose, fucose, uronic acid, xylose, rhamnose, glucose, galactose, mannose, amino acid	Exopolysaccharide	Trabelsi et al. (2013)
Nostoc verrucosum	Xylose, mannose (+) glucose (+), amino acid, sulfate, uronic acid	Exopolysaccharide	Sakamoto et al. (2011)
Arthrospira platensis MMG-9	Fructose, fucose, ribose, xylose, galactose, glucose (+), mannose, rhamnose, uronic acid	Exopolysaccharides	Ahmed et al. (2014)
Nostoc carneum	Xylose (+), glucose	Exopolysaccharides	Hussein et al. (2015)
Chlamydomonas reinhardtii	Galactose, rhamnose, arabinose, glucose, ribose, xylose, alpha- galactose	Exopolysaccharide	Bafana (2013)
Dunaliella tertiolecta UTEX LB 999 (Ch)	Glucose	Exopolysaccharide	Goo et al. (2013)
Microcoleus vaginatus (Cy)	Mannose, arabinose, galactose, glucose, glucuronic acid, xylose, fucose, rhamnose	Released polysaccharides	Ge et al. (2014)
Nostoc commune (Cy)	Mannose, glucuronic, xylose, glucose, acid, α-galactose, and some unidentified uronic acid	Exopolysaccharide	Sakamoto et al. (2011)
Porphyridium purpureum CCAP 1380/1A (Rh)	Galactose, glucose, fucose, xylose, α-galactose	Exopolysaccharides	Roussel et al. (2015)
Rhodella violacea LMGEIP 001 (Rh)	Galactose, glucose, arabinose, rhamnose, xylose, α-glucose	Exopolysaccharides	Villay et al. (2013)
Nostoc verrucosum (cy)	Xylose, mannose, glucose, alpha glucose	Exopolysaccharides	Sakamoto et al. (2011)
Porphyridium marinum CCAP 1380/10 (Rh)	Fucose, xylose, galactose, glucose, α-glucose	Exopolysaccharides	Roussel et al. (2015)
Chlamydomonas reinhardtii	Arabinose, rhamnose, ribose, xylose, galactose, glucose, uronic acid	Exopolysaccharide	Bafana (2013)
Dunaliella tertiolecta	Galactose, xylose, glucose, ribose, uronic acid	Exopolysaccharide	Goo et al. (2013)

11.4 Pathways Involved in the Biosynthesis of Cyanobacteria Exopolysaccharides

EPS production is strongly regulated by cellular processes involving a variety of genes. There are two ways for the identification of these genes. One is indirect way known as mutation-associated method, while the other is direct way known as probe-based method or PCR method. Mutation-associated methods involved the use of mutants. These mutants can be involved either spontaneously or as inducing agents like UV light, chemical, or transposons, while PCR involved the use of DNA sequences for the identification of genes. Probe-based method is found to be very powerful for the rapid identification of genes (Griffin et al. 1996). Majorly cyanobacterial EPS is synthesized by Wzx-Wzy-dependent pathway, ABC transporter-dependent pathway (Wzm/Wzt), and synthase-dependent pathway.

11.4.1 Wzx-Wzy-Dependent Pathway

In this mechanism, assemblies of discrete units that are present at the inner membrane are being associated with undecaprenyl diphosphate or C_{55} by means of glycosyltransferases. Translocation of these units alongside the cytoplasmic membrane occurred by a protein called Wzx (O-antigen [O-Ag] flippase) (Islam and Lam 2014). Additionally, another protein Wzy (O-Ag polymerase) is involved in polymerization of repeating units before they released the cell externally (Singh et al. 2019). Heteropolymers known as highly variable sugar patterns are also being formed in this pathway (Schmid et al. 2015).

11.4.2 ABC Transporter-Dependent Pathway

This pathway is also known as Wzm-Wzt pathway. This is mostly involved where capsular polysaccharide is required. It results in homopolymers due to the presence of GT-surrounding operons. For the exportation of repeated units, ABC transporters are involved. This transporter covers both families of periplasmic proteins known as outer membrane polysaccharide export (OPX) and polysaccharide copolymerase protein (PCP) (Whitney and Howell 2013). Some of the ABC transporters involved in cyanobacteria have been discussed in Table 11.2.

Table 11.2 ABC transporters involved in the uptake of nutrients, transport, and secretion of cyanobacterial EPS

Bacteria	ABC transporters	Predicted functions	References
Synechocystis sp. PCC 6803	EpsT,	Export of EPS	Wang et al. (2017)
Synechococcus elongates	EpsM, EpsP	Secretion of EPS	Yoo and Kim (2003)
Synechocystis sp. PCC 6803	EpsP, EpsT	Secretion of EPS	He et al. (2015)
Anabaena sp. PCC 7120	ALL4388-ALL4390, HepA	Transport of polysaccharides	Shvarev and Maldener (2018)
Anabaena sp.	RND, MFS	Antibiotic or multi-drug resistance	Shvarev et al. (2019)
Anabaena sp. PCC 7120	fhuBCD, futABC, fecBCD	Uptake of iron	Rudolf et al. (2015), Stevanovic et al. (2012), Nicolaisen et al. (2008)
Anabaena sp. PCC 7120	GlsC, GlsP	Uptake of esculin (sucrose analog)	Nieves-Morión and Flores (2018)
Anabaena sp. PCC 7120	AzuABC	Uptake of zinc	Napolitano et al. (2012)
Rhizobium leguminosarum bv. *viciae* 3841	RL2975, RL2976, RL2977	Tolerance to desiccation and formation of biofilm	Vanderlinde et al. (2010)
Synechocystis sp. PCC 6803	Syp1	Secretion of EPS	Jin and Kim (2005)
Nostoc sp. PCC 7120	EptA	Production and secretion of EPS	Oda et al. (2009)
Anabaena sp. PCC 7120,	EpsD, EpsE, EpsM	Production and secretion of EPS	Van Alstyne and Sadowsky (2000)
Anabaena variabilis ATCC 29413-FD, *Anabaena* sp. PCC 7120)	devBCAO	Formation of laminated layer of heterocysts	Fiedler et al. (2001)
Anabaena sp. PCC 7120	NrtABCD	Uptake of nitrate and nitrite	Cai and Wolk (1997), Frias et al. (1997)

11.4.3 Synthase-Dependent Pathway

This pathway is of utmost importance because this involves a single synthase protein that is sufficient for polymerization and translocation of repeating units. In this process, homopolymers like alginates, hyaluronic acid (HA), or cellulose are generated (Chong et al. 2005; Rehm 2010). In general, biosynthetic genes for exopolysaccharides are being assembled and constant between several species but in case of cyanobacteria, genes for EPS biosynthesis have been distributed throughout the chromosome and assembled in small repeating units (Kehr and Dittmann

2015). Until now, very few studies have been made on the exportation and biosynthesis of EPS from cyanobacteria (Simkovsky et al. 2012; Fisher et al. 2013; Singh et al. 2016).

11.5 Approaches to Enhance the Production of EPS

Exopolysaccharides from cyanobacteria are vital for biofilm generation and development of stress tolerance against environmental stresses such as oxidative stress, temperature variation, radiation (UV), dehydration, etc. (Hu et al. 2003; Adhikary 1998; Potts 2004). Different excretory pigments are known to have UV absorbent properties. For instance, in cyanobacterial sheaths one of the UV absorbents called scytonemin has been found (Keshari and Adhikary 2013). For the protection of capsular exopolysaccharides from the harm of UV emission, another UV absorbent known as mycosporine has been reported in the coverings of different species of cyanobacteria like *Nostoc commune* and *Tolypothrix byssoidea* (Roy et al. 1998). In the presence of adhesive EPS, nitrogen fixing at higher oxygen level was done by *Nostoc cordubensis*, which in the absence of such mucilage was not possible. These EPS also protect cyanobacteria from dehydration. Because of these interior and exterior polysaccharides, bacteria develop resistance toward water loss and make them adaptable to survive in water-deficient conditions. Another important role of cyanobacteria EPS is their ability to protect the cells from direct exposure to heavy metals, thus defending cells from toxic effect of metals (Kumar et al. 2018).

There is a need to increase the production of cyanobacterial EPS because of their applications in various fields. For the enriched production of such polysaccharides, several approaches have been employed. Some of these approaches are depicted in Fig. 11.1.

11.5.1 Starvation of Nitrogen, Sulfate, and Phosphorus

Inadequacy of nutrients has been considered as a standard approach for the buildup of directed exopolysaccharides, though there is drop of growth in such situations (González-Fernández and Ballesteros 2012). It was found that in the presence of combined nitrogen, there was increased production of cyanobacteria EPS. On the other hand, enhanced production of polysaccharide was also observed in numerous cyanobacteria upon nitrogen shortage (Singh et al. 2019). Like in nitrogen-deficient BG11 culture media (without $NaNO_3$), increased production of EPS was observed in *Parachlorella* sp. BX1.5 (Sasaki et al. 2020). Similarly, in the presence of 2 mM KNO_3, high EPS production was observed in *Botryococcus braunii* UC 58, while decrease in the quantity of EPS was found when 8 mM KNO_3 was supplied in a culture medium (Lupi et al. 1994). Phosphorus and sulfate are two of the vital nutrients that play a significant role in the synthesis of cyanobacteria EPS. Just

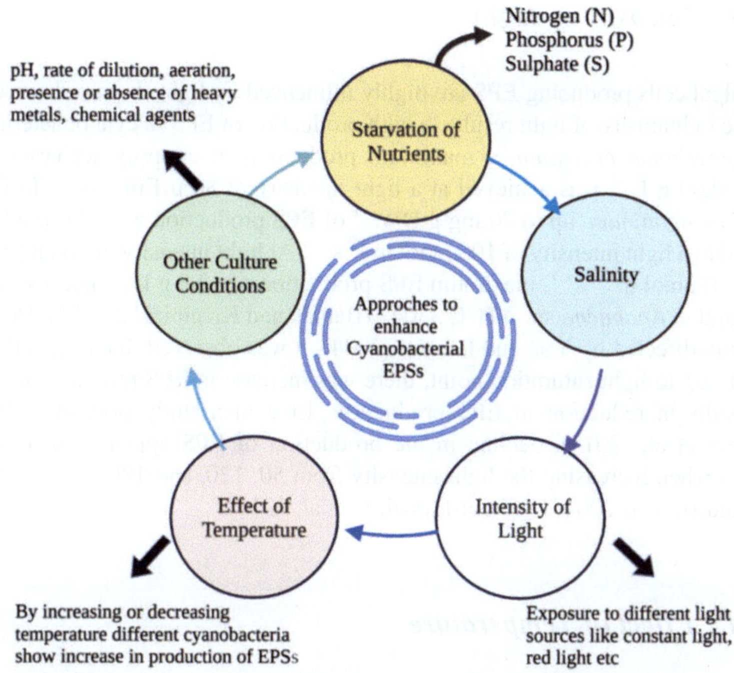

Fig. 11.1 Approaches for the enhanced production of exopolysaccharides from cyanobacteria

like shortage of nitrogen, starvation or decreased level of phosphorus and sulfate contributed to enhance manufacturing of EPS (Delattre et al. 2016).

11.5.2 Salinity

Salinity is associated with the amount of salt dissolved and is considered an effective source to enhance the production of exopolysaccharides in cyanobacteria (Su et al. 2007; Nicolaus et al. 1999). The salt-tolerant cyanobacteria process different structural changes like low respiration rate and enhance production of carbohydrate sources like sucrose that will provide osmotic solute protection to the cyanobacterial cells in harsh environments. Under 0.5 molar salt stresses, a large amount of EPS production was observed in different cyanobacteria such as *Cyanothece* sp. 113, *Anabaena* sp. ATCC, and *Spirulina* sp. (Pereira et al. 2009). The highest EPS production up to 1205.27 ± 39.08 fg C cell^{-1} was found in *Synechococcus* strain CCAP1405 under high salinity condition (Bemal and Anil 2018).

11.5.3 Intensity of Light

Microalgal cells producing EPS are highly influenced by light. It was observed that increase in intensity of light results in high production of EPS in cyanobacteria. Like in *Porphyridium cruentum*, a maximum production of exopolysaccharide up to 0.95 ± 0.09 g L^{-1} was achieved at a light intensity of 80 μ Em^{-2} s^{-1}. In *Cyanobacterium aponinum*, up to 70 mg g DW^{-1} of EPS production was observed, when exposed to a light intensity of 100 μmol m^{-2} s^{-1}. At light intensity of 460 μ Em^{-2} s^{-1} and 120 μmol m^{-2} s^{-1}, maximum EPS production of 12.5 g L^{-1} and 0.454 g g^{-1} was found in *Anabaena* sp. ATCC 33047 (Babiak and Krzemińska 2021). However, in a study directed by You and Barnett (2004), it was observed that at specific light intensity up to light saturation point, there was increase in EPS production, beyond this results in reduction in EPS production. Like in a study done by Clément-Larosière et al. (2014), decline in the production of EPS appeared in *Chlorella vulgaris* when increasing the light intensity from 50, 120, and 180 μmol $m^{-2}s^{-1}$ C (You and Barnett 2004; Clément-Larosière et al. 2014).

11.5.4 Effect of Temperature

Variation in temperature influences the cyanobacterial EPS production. However, the impact of temperature on extracellular polysaccharide production is not yet completely understood (Delattre et al. 2016). Moreno et al. (1998) reported that *Anabaena* sp. cyanobacterium upon interaction with high temperature up to 40–45 ° C resulted in major increase in the level of exopolysaccharides. Furthermore, in different microalgal species like *Graesiella* sp., *Dictyosphaerium chlorelloides*, and *Botryococcus braunii* UC 58, increase in temperature results in better production of EPS. Like between 25 and 30 °C, EPS production of 4500–5500 mg L^{-1} was observed in *Botryococcus braunii* UC 58. Similarly, at temperatures of 25.7 °C and 40 °C, EPS production of 1075 mgL^{-1} and 11.7 mgL^{-1} per day was determined in *Dictyosphaerium chlorelloides* and *Graesiella* sp., respectively (Moreno et al. 2000). On the other hand, a decrease in the production of these cyanobacteria EPS was observed by Nicolaus et al. (1999) in the cyanobacterium *Spirulina* sp. upon exposure to high temperature (Nicolaus et al. 1999). Another study done by Otero and Vincenzini (2004) showed that high temperature neither increases nor decreases the EPS production in cyanobacterium strain *Nostoc* sp. PCC 7936 (Otero and Vincenzini 2004). Impact of temperature and abovementioned factors on the cyanobacterial EPS is shown in Table 11.3.

Table 11.3 Impact of different factors on the production of EPS by different cyanobacteria

Organisms	Factors	Effect on the production of cyanobacterial EPS	References
Cyanobacterium aponinum	Presence of light	Positive	Gris et al. (2017)
Chlorella vulgaris	Presence of light	Positive	Clément-Larosière et al. (2014)
Porphyridium cruentum	Presence of light	Positive	Liqin et al. (2008)
Aphanocapsa halophyta MN11	Presence of light, nitrogen, NaCl	Positive	Sudo et al. (1995), Matsunaga et al. (1996)
Anabaena sp. ATCC 33047	Presence of combined nitrogen, NaCl	Negative	Moreno et al. (1998)
Spirulina sp.	Presence of combined nitrogen, increase in temperature	Negative	Nicolaus et al. (1999)
Cyanothece sp. 113	Presence of light, combined nitrogen, NaCl, high temperature	Positive	Su et al. (2007)
Phormidium sp.	Presence of light, combined nitrogen, continuous air	Positive	Nicolaus et al. (1999)
Anabaena sp. ATCC 33047	Presence of light	Positive	Moreno et al. (1998)
Nostoc sp. PCC 7936	Presence of high temperature	No change	Otero and Vincenzini (2003, 2004)
Nostoc sp. PCC 7936	Presence of combined nitrogen	Negative	Otero and Vincenzini (2003, 2004)

11.5.5 *Effect of Other Culture Conditions*

Other parameters like variation in pH, rate of dilution, aeration, presence or absence of heavy metals, chemical agents like EDTA, and others have been associated with cyanobacteria exopolysaccharide production (Delattre et al. 2016). Studies regarding the genes and certain pathways of metabolism and biosynthesis need to be recognized (Singh et al. 2019). Aeration was also found to impact the production of EPS. Few studies have observed that increase in continual source of aeration results in enhanced production of EPS (Su et al. 2007; Nicolaus et al. 1999; Moreno et al. 1998). Aeration indirectly results in an increase in turbulence of culture, causing better stirring of culture cells as well as good penetration of light (Pereira et al. 2009). Very limited reports have been given related to the influence of metals on the production of EPS. In one of the strains of cyanobacterium *Synechocystis* sp., increased content of EPS was observed when 35 ppm of metals such as cadmium, chromium, or both was applied (Ozturk et al. 2014). A study conducted by Mona and

Kaushik (2015) on the impact of cadmium and cobalt on EPS production in *Nostoc linckia* revealed fivefold more EPS making while fourfold when exposed with chromium; the maximum synthesis of EPS was being observed when applied with 40 mgL^{-1} cobalt and chromium metals (Mona and Kaushik 2015).

11.6 Methods for the Extraction of Cyanobacterial Exopolysaccharides

For the extraction of exopolysaccharides, numerous methods have been available, among which extraction of cyanobacterial EPS at pilot scale along with purification procedures was given by Li et al. (2011). Besides this, many other conventional and refined procedures are also designed and tested (Singh et al. 2019). Most frequently used tests for the estimation of total sugars are orcinol/phenol-sulfuric acid. For the identification of polysaccharide composition based on vibrational functional group, Fourier transform infrared spectroscopy (FTIR) has been used (Mathlouthi and Koenig 1987).

Extraction of cyanobacterial exopolysaccharides has been found crucial for examining their chemical compositions, commercialization, quality determination, purification, and product generation. From industrial point of view, pureness of exopolysaccharides as well as efficacy of the procedure is very significant. Although there is no such universal procedure for the extraction of exopolysaccharide, yet already available extraction procedures are being modified by scientists to improve the extraction process. The choice or selection of extraction protocol is very crucial to avoid EPS contamination with other components as well as for the identification of cyanobacterial EPS composition (Singh et al. 2019). The general process of cyanobacterial EPS extraction has been depicted in Fig. 11.2.

In the case of bound exopolysaccharides, the use of EDTA, NaOH, treatment with resins (ionic based), sonication or washing of cell using distilled water, etc. have been used for the isolation of cyanobacterial EPS (Pierre et al. 2014). Different factors like concentration of biomass, pH, temperature (30 to 95 °C), and period for treatment (1–4 h) largely impact the extraction of bound exopolysaccharide depending upon different cyanobacterial species. De Philippis et al. (2000) suggested removing cyanobacterial cells from media through centrifugation at 14,000 g and 10 °C for 20 min followed by treatment of medium with 2-propanol to attain crude exopolysaccharides. Dispersion of released polysaccharides (RPS) to DW (distilled water) as well as lyophilization was done for the refining of crude polymers. Then the extraction of different constituents like uronic acid and hexose-amine and carbohydrates from the sample containing released exopolysaccharides was being conducted by carbazole, phenol, H_2SO_4, and Ehrlich's reagent. With the help of automatic elemental analyzer, composition of elements within the RPSs was evaluated calorimetrically. Modification of this method has been reported by Parikh and Madamwar (2006) who suggested performing centrifugation at 15,000 g for

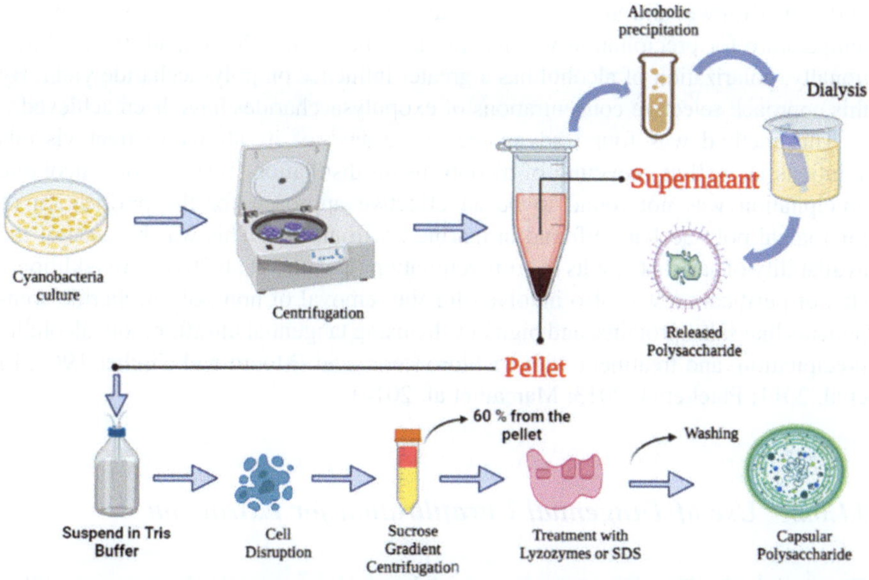

Fig. 11.2 Steps for the extraction of cyanobacteria EPS (modified from Kumar et al. 2018)

40 min at 15 °C for cell separation from medium. For concentration of supernatant, up to one fourth of volume magnetic stirring was done for 10 h at 60 °C. After that precipitation of the concentrated EPS was carried out by the addition of cold acetone to an equal volume to that of supernatant and placed them overnight at 4 °C. Milli-Q water was used for redissolving the exopolysaccharide being precipitated out. The process of precipitation and redissolving was carried out repeatedly for the extraction of crude exopolysaccharides which were then refined by dialysis at 4 °C for about 20 h (Kumar et al. 2018).

11.6.1 Extraction Using Alcoholic Precipitation

The first experimental strategy for the isolation of *Porphyridium* sp.-encapsulated polysaccharide was done by Ramus (1972). After the polysaccharide has been solubilized in culture medium, the biomass was de-pigmented by ethanol and acetone. The cells of *Porphyridium* sp. containing exopolysaccharide gel dissolved in hot water. For the extraction of exopolysaccharide, precipitation was done by treating with cetyl-pyridinium chloride and re-precipitated by ethanol. Centrifugation or microfiltration is used for the cyanobacterial removal and for the extraction of solubilized EPS (Li et al. 2011; Ye et al. 2005; Zhang et al. 2008). After centrifugation the supernatant or filtrate obtained was being precipitated out by means of different alcohols like isopropanol, methanol, or ethanol (in 1 volume of filtrate or supernatant, use 2–3 volumes of alcohol) (Moore and Tischer 1964; Gloaguen et al.

2004; De Brouwer and Stal 2004; Liu et al. 2015; Patel et al. 2013). Besides this the temperature for precipitation was usually kept between – 20 °C and 20 °C. Additionally, polarization of alcohol has a greater influence on polysaccharide yield. By this approach selective concentrations of exopolysaccharides have been achieved.

This method was found advantageous because of its ability to treat viscous solutions as well as recycling of alcohol using distillation. Besides this, alcoholic precipitation was not found to be an effective approach for the purification of microalgal polysaccharide found in marine environments. This was because of the availability of salt that results in co-precipitation (Delattre et al. 2016). An additional step of purification was also involved for the removal of non-polysaccharide components like salts, proteins, and pigments by using tangential ultrafiltration, alcoholic precipitation, and treatment with trichloroacetic acid (Moore and Tischer 1964; Li et al. 2011; Patel et al. 2013; Marcati et al. 2014).

11.6.2 Use of Tangential Ultrafiltration for Extraction

Tangential ultrafiltration procedures had been given as a preference over other conventional approaches for extraction of exopolysaccharides using precipitated ethanol. Such techniques were being found favorable in both upstream and downstream practices to recover various natural macromolecules at low temperature and pressure. In Tangential Flow Filtration (TFF) also known as cross-flow filtration, the stream of feed is passed in parallel direction over the membrane. After that the retained material was recirculated back to the feed tank. The process of extraction and refinement on pilot-scale for biologically active cyanobacterial exopolysaccharides like *Spirulina platensis*, *Chlorella pyrenoidosa*, *Nostoc sphaeroides, Haematococcus pluvialis, Nostoc commune* and *Chaetoceros muelleri* has been described (Li et al. 2011). This process of EPS extraction from microalgae relies on microfiltration using polypropylene membrane and for cyanobacteria relies on TFF by using polyethersulfone membrane with molecular weight cut-off (MWCO) of 5000 Da that result in 20–40 times concentrated exopolysaccharides. Further extension to the two-step membranous method by using polyethersulfone membrane with MWCO of 300 or 10 Kilo Daltons for the separation of *Porphyridium* sp. polysaccharide has been reported (Marcati et al. 2014).

Ultrafiltration was found to be an effective strategy for the extraction and refinement of exopolysaccharides from microalgal culture (Li et al. 2011; Patel et al. 2013). Certain drawbacks associated with tangential ultrafiltration were also observed, like the use of large quantity of water for the isolation, purification as well as concentration of desired EPS and membrane fouling (Li et al. 2011). This results in low production of EPS as compared to EPS production by alcoholic precipitation. Furthermore, any problem associated with membrane has been directly linked with the viscosity of exopolysaccharides, which affects the flow rate which ultimately causes increase in operational prices (Delattre et al. 2016). Recently, a two-phase culture process is done by Chentir et al. (2017) for the extraction of exopolysaccharide from *Arthrospira* sp. This process involves the use of Whatman filter paper at the initial stage for the filtration of culture and for the separation of cells from culture. After that, supernatant was 20 times concentrated by using Tangential ultrafiltration Millipore system (TFF). This TFF uses the Millipore membranes with 5 kDa MWCO. For desalting, optimum washing is done to get eight times concentrated ultra-pure water (v/5v). After that, deproteination was done by maintaining pH to 10 and suspending 20 percent w/v of exopolysaccharides in water. For deproteination 1% alkaline protease was added, and the process then proceeds for about 24 h at 50 °C. Before centrifugation, the resulting fraction has been cooled down to 4 °C for about 30 min. After centrifugation, precipitation was carried out by overnight treatment of supernatant with absolute ethanol (96% v/v) at 4 °C. The obtained pellet was re-dissolved in DW and freeze dried. By using this approach, considerable exopolysaccharides along with other constituents like 5.14 ± 0.32% of LSP (low soluble proteins), 2.42 ± 0.12% sulfate, and 67.3 ± 1.1% carbohydrate contents were obtained (Bhunia et al. 2018).

11.6.3 Alternative Methods for EPS Extraction

Alternatives such as MAE (microwave-assisted extraction) and UAE (ultrasound-assisted extraction) have been proposed to enhance the extraction of EPS (Herrero et al. 2006; Plaza et al. 2010; Rodriguez-Jasso et al. 2011; Budarin et al. 2012; Kadam et al. 2013; Kurd and Samavati 2015). In ultrasound-assisted extraction approach, ultrasound waves are used to disperse the solvent of extraction. As a result of mechanical force, physical cavitation, and thermal effects, the cell wall is being disrupted causing increase in the permeability of membrane and favorable extraction of biomolecules (Toma et al. 2001; Kadam et al. 2013). Because of its better use, UAE approach has been used for the extraction of polysaccharides (PS) from *Spirulina platensis* (Kurd and Samavati 2015) specifically in food industry (Chemat and Khan 2011; 668; Kadam et al. 2015; Vilkhu et al. 2008). It is an effective strategy for the isolation of cell-bound exopolysaccharides (Delattre et al. 2016). Microwave-assisted extraction on the other hand has been known as a non-ionizable EM (electromagnetic) radiation method. This method is found effective to extract microalgae metabolites specifically intracellular metabolites like phycobiliproteins from *Porphyridium purpureum* (Juin et al. 2015) and carotenoid extraction from *Dunaliella tertiolecta* (Pasquet et al. 2011).

These alternative strategies have certain benefits over conventional processes of extraction. These are (i) less time of extraction requires, (ii) reduced amount of solvent being consumed, and (iii) larger yield of extracted biomolecules (Kadam et al. 2013; Rodriguez-Jasso et al. 2011). Besides this there are certain limitations as it cannot be used for co-extracting intracellular constituents like lipids, pigments, proteins, and certain carotenoids which are dependent on some extra separation methods for the purification of EPS.

11.7 Strategies for the Processing of EPS from Cyanobacteria

For industrial purposes, cyanobacteria polymers have been found as a potential sustainable foundation. For the bioprocessing of cyanobacteria exopolysaccharides, three steps including choice of cyanobacterial strain, optimization of EPS producing parameters, and downstream processing are important to consider. These factors have been briefly discussed below.

11.7.1 Selection of Strain

The typical approach for the selection of a specific EPS-producing cyanobacterial strain starts by observing the EPS content and growth rate. It was found that during screening of different cyanobacterial strains, because of their diverse nature, each strain responds differently under specific cultivation conditions. Screening based on cultivation conditions could be a useful approach for the comparison of multiple cyanobacterial strains in a predefined atmosphere (Cruz et al. 2020).

Most experiments have been done on Oscillatoriales and Nostocales orders by the isolation of 40 strains of Nostoc from Pasteur culture residing in diverse environments. Experimentation has shown that the synthesis of released polysaccharides from these strains was not found to be dependent on outermost morphology investments nor on the habitat. Furthermore, screening was performed on the 166 algal strains found in marine, in which only16 strains were found to be Cyanobacteria. Some of these strains have been found for the first time to have potential of producing exopolysaccharides. After selecting the culture media from culture collector, *Synechococcus* sp. strain was found to be a potent producer of released exopolysaccharides (RPS) up to 0.12 gL^{-1} when screening was being done at 150 μE light intensity and temperature of 20 °C (Gaignard et al. 2019). Besides the presence of nutrients in culture media for the growth of cyanobacterial cells, certain chemical formulations like BG11 and $BG11_0$ can also be used. $BG11_0$ is being utilized by strains which need to fix nitrogen like by Nostocales order. Regardless of some existing nutrients like that often used, desired availability of nutrients has been provided by the culture media for the efficient growth of cyanobacteria. It has been found that in the presence of diastrophic situation, limitation of growth and redirection to nitrogen fixation were preferred over production of polysaccharides specifically capsular exopolysaccharides (Cruz et al. 2020).

In some studies, resources from seawater and some other habitats have also stimulated production of EPS. Rheological property of culture medium has been found to be affected by the production of EPS. Light usually described in the form of light quality and intensity of light (μE) and as light and dark cycle (L/Dh) acts as source of energy for the growth of cyanobacteria. Different variety of cyanobacteria have been characterized based on their temperature adaptability such as mesophiles (30–40 °C), thermophiles (> 40 °C), and psychrophiles (<15 °C). It is essential that homogeneity with respect to specific nutrients and pH of the cyanobacteria cultures should be maintained. For mixing the cultures, aeration has been introduced in some studies as well. For controlling pH, air mixture containing carbon dioxide has been introduced (Cruz et al. 2020).

11.7.2 Optimization of Parameters Involved in Production

Although selection of strains for EPS production could be better enough for the development of whole bioprocess, there is always a need to upgrade or to improve the performance. At this point, it is important to analyze the product formation by the cells in a determined manner. At this stage selection of strategy matters a lot for higher production of EPS. To enhance the production of exopolysaccharides, light, nitrogen source, and salinity stress conditions were found to be efficient inducers (Cruz et al. 2020). Culture medium has major impact on the bioprocessing of exopolysaccharides. The presence and absence of different sources like nitrogen, carbon (organic or inorganic), micronutrients, etc. affect the production of exopolysaccharides. In the presence of nitrogen, 1.8 gL^{-1} production of released polysaccharides was observed in *Cyanothece* sp. CCY 0110 (Mota et al. 2013). In another study the absence of nitrogen is related to enhanced formation of exopolysaccharides by *Anabaena* sp. BTA990 and *Nostoc* sp. BTA97 (Tiwari et al. 2020). Most used N sources are sodium nitrates (Yu et al. 2010; Lama et al. 1996) or potassium nitrates (Khattar et al. 2010; Jindal et al. 2011). In heterocyst-producing species, enhanced formation of EPS has been reported in the presence of inorganic nitrogen (Borah et al. 2018; Han et al. 2017a, b; De Philippis and Vincenzini 1998).

For the efficient production of EPS, pH is to be maintained which is usually done by the presence of carbon (inorganic C source mostly CO_2 in concentration of 0.05–10% v/v) along with combined air. The production of exopolysaccharides is also influenced by the presence of sodium bicarbonate for *A. platensis* cultivation (Vergnes et al. 2019). Organic sources of carbon have also influenced exopolysaccharide production as 290 mgL^{-1} of EPS production was observed in *A. platensis* (Trabelsi et al. 2013). Tremendous effect on the production of exopolysaccharides has been observed in the presence of micronutrients like magnesium sulfate, calcium chloride, and potassium phosphate. Likely induction of micronutrient in cyanobacterium *Cyanothece epiphytica* results in no damage to the structure of cyanobacterial cells even when exposed with ozone at concentration of 0.06 mgL^{-1} (Cruz et al. 2020).

For the bioprocessing of EPS, control of process elements has a major impact on its viability and reproducibility. pH of about 6 to 12 needs to be maintained for better cell growth and production of EPS from cyanobacteria. It also affects the uptake of nutrient solubility, morphology of cell membrane, and activity of enzymes (Tiwari et al. 2020; Khangembam et al. 2016). Shear force/stress was also found to impact growth of exopolysaccharides by means of stimulating orientation and nutrients (Xiao et al. 2016; Fadlallah et al. 2016). Additionally, temperature was also found to have positive impact on the production of exopolysaccharides. Similarly, constant illumination by light and dark cycles favors an increase in the production of exopolysaccharides in Anabaena genus (Cruz et al. 2020). Other factors that can contribute to bioprocess optimization for the efficient production of biomass and EPS include modes of cultivation either fed batch, continuous, or others along with design of experiment like PBR (Cruz et al. 2020). Table 11.4 highlights some of the optimization conditions for cyanobacteria exopolysaccharides. The impact of flue gas on the EPS production of *A. platensis* was monitored by batch cultivation of cyanobacteria in 30 m^2 of photobioreactor (PBR). Environmental parameters such as temperature of 30 to 34 °C and extra illumination were being supplied at night for about 10 h by means of LED during 15 days of experiment. It resulted in the 2.36 gL^{-1} of capsular polysaccharide production with immunogenic property (Chen et al. 2012).

11.8 Downstream Processing

The method for the extraction is the same as described above. To improve the purification of polysaccharides, two approaches have been used. One was the use of Milli-Q water for solubilizing dry polysaccharides, while a second approach was carried out by dialyzing them against water (Cruz et al. 2020). The overall outline showing the phenomenon of exopolysaccharide production from cyanobacteria has been shown in Fig. 11.3.

11.9 Applications of Cyanobacterial Exopolysaccharides in Bioindustry

Cyanobacteria exopolysaccharides have diverse roles in biotechnology applications and bio-industry. These exopolysaccharides from cyanobacteria are of great importance because many of the cyanobacterial species specifically from extreme environments constitute UV-absorbing pigments (like scytonemin) in their sheaths, thus providing protection from harmful effects of UV irradiations (Keshari and Adhikary 2013; Adhikary and Sahu 1998; Potts 2004). Exopolysaccharides also protect the cyanobacteria from stress conditions (Kumar et al. 2018). Exopolysaccharides play role in cellular attachment to sediments as well as in benthic cyanobacteria wherein they provide defense to nitrogenase against toxic effect of oxygen (De Philippis and Vincenzini 1998). These EPS were also found to provide stability to the membranous vesicles. Exopolysaccharides from terrestrial cyanobacterium *N. commune*

Table 11.4 Optimized conditions required for the bioprocessing of cyanobacteria exopolysaccharide

Cyanobacteria	Culture media	Process parameters	Polysaccharide target	References
Anabaena sp. PCC 7120	BG11 medium	Light intensity (100 μmol photons m^{-2} s^{-1}), temperature (30 °C), pH (8.5)	Exopolysaccharides	Roy et al. (2023)
Gloeothece verrucosa	$NaNO_3$ (nitrogen source), continuous light	Light intensity (10 μmol photons m^{-2} s^{-1}), temperature (25 °C)	Exopolysaccharides	Van Camp et al. (2022)
Synechococcus elongates BDU 10144	Diammonium phosphate (nitrogen source), $MgSO_4$ (micronutrient), and potassium	Light intensity (50 μmol photons m^{-2} s^{-1}), temperature (25 °C), pH (10)	Exopolysaccharides	Chandra and Mallick (2022)
Arthrospira platensis F&M-C256	Glucose, rhamnose, galactose, galactose, xylose, continuous light	Light intensity (700 μmol photons m^{-2} s^{-1} w), temperature (35 °C), pH (9.2–9.4)	Released polysaccharide	Zanolla et al. (2022)
Chroococcus sp. FPU101	Galactose, rhamnose, fucose, xylose, mannose, glucose, galacturonic acid, and glucuronic acid (carbon source), NaCl (salinity)	Light intensity (between 10 and 80 μmol m^{-2} s^{-1}), temperature (between 20 and 30 °C)	Exopolysaccharides	Yoshikawa et al. (2021)
Arthrospira platensis	$NaNO_3$, (nitrogen source), K_2HPO_4 (potassium source)	Temperature (32 °C), pH (11.5)	Exopolysaccharide	Li et al. (2022)
Phormidium sp. ETS05	BG11 medium	Light intensity (20 μmol photons m^{-2} s^{-1}), temperature (30 ± 1 °C)	Exopolysaccharide	Zampieri et al. (2020)
Arthrospira platensis	Sodium bicarbonate (carbon source)	Light intensity (280 or 350 μmol m^{-2} s^{-1}), temperature (30 or 35 °C), pH (9.5)	Exopolysaccharide	Vergnes et al. (2019)

(continued)

Table 11.4 (continued)

Cyanobacteria	Culture media	Process parameters	Polysaccharide target	References
Nostoc flagelliforme	BG11 medium, NaCl (salinity)	Light intensity (60 μmol photons m^{-2} s^{-1}), temperature (25 °C)	Capsular polysaccharide	Shen et al. (2019)
Cyanothece epiphytica AUS-JR/DB/NT-021	BG11 medium, absence of nitrogen source, sodium chloride (salinity), magnesium sulfate (micro-nutrients)	Light intensity (34 μmol photons m^{-2} s^{-1}), temperature (28–36 °C)	Capsular as well as released polysaccharide	Borah et al. (2018)
Cyanobacterium aponinum	BG11 medium, 5% CO_2, $NaHCO_3$, NaCl (salinity)	Light intensity (35–40 μmol photons m^{-2} s^{-1}), temperature (29 ± 1 °C), pH (8)	Released polysaccharide	Gris et al. (2017)
Anabaena sp. BTA997	BG11 medium	Initial pH (8.5)	Released polysaccharide	Khangembam et al. (2016)

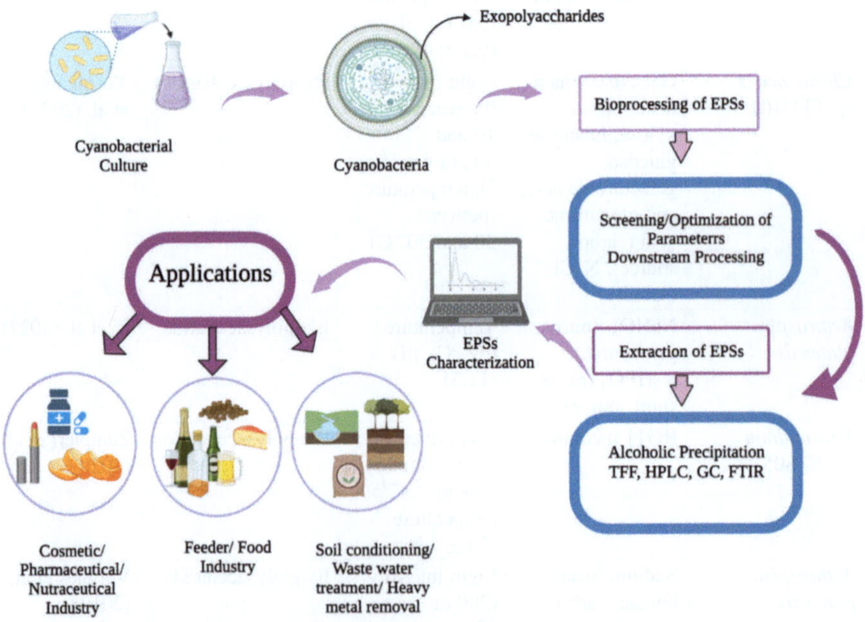

Fig. 11.3 Schematic outline to produce cyanobacterial EPS (modified from Singh et al. 2019)

were found to protect them against desiccation and provide tolerance to freezing. Various studies have been performed on the defensive nature of EPS from Cyanothece, *M. vaginatus,* Anabaena strains Synechocystis, and *Aphanocapsa halophyta.* EPS have important defensive as well as structural characteristics of biofilm (Singh et al. 2019). A study was conducted on the effect of alleviated salt level on cyanobacterial exopolysaccharides. It was found that increased concentration of salt promotes exopolysaccharide production (Mishra and Jha 2009; Ozturk and Aslim 2010; Singh et al. 2016; Verma et al. 2018).

In agriculture and food industries, these cyanobacterial exopolysaccharides have tremendous impact. These complex exopolysaccharides are able to form glycosidic bonds with other groups such as sulfate and acyl or amino acids which are beneficial as gelling or thickening agents (Delattre et al. 2009; Kraan 2012). Cyanobacterial exopolysaccharides also act as stabilizing agent, water binding agent, gelling agent, as well as emulsifying agent which make them useful in food industry (Singh et al. 2019). EPS have anti-coagulating, antibacterial, anti-cancerous, antitumor, anti-oxidative, and anti-inflammatory property as well as provide protection against deleterious effect of radiations (Kraan 2012; Misurcova et al. 2015; de Jesus Raposo et al. 2015). *Spirulina platensis* constitutes sulfur-containing polysaccharides called calcium spinilan that was found effective in inhibition of tumor expansion (metastasis) (Mishima et al. 1998).

For commercialization purposes especially in nutraceutical industries, as nutritious supplements as well as healthy food items, *Chlorella* spp. and *Arthrospira* sp. play vital role as they help to increase the immune response, enhance antitumor activity, and promote growth of animals (Choi et al. 2013; Kang et al. 2013). In vitro effects of immunomodulation by exopolysaccharides have been reported for *Cyanobacterium aponinum* (Gudmundsdottir et al. 2015). In pharmaceutical industry, some of these cyanobacterial EPS like pullulan and dextran have gained remarkable importance (Singh et al. 2019).

In the case of heterotrophic microbes, exopolysaccharides act as a source of carbon for them (Thomas et al. 2008; Miralles et al. 2012; Chen et al. 2014). Exopolysaccharides also constitute flocculating (adhesion) property. Many cyanobacterial EPS from *Anabaenopsis circularis* PCC 6720, *Phormidium persicinum*, *Anabaena* sp. N1444, *Microcoleus* sp., *Calothrix desertica* PCC 7102, *Phormidium* sp. strain J-1, *Anabaena* sp. PC-1, and *Lyngbya* sp. are found to have flocculating features. They are valuable for treatment for wastewater and removal of heavy metals. Exopolysaccharide from *Anabaena spiroides* helped in lessening the concentration of heavy metals by binding to these heavy metals like copper(II), mercury(II), and lead(II). The immobilizing property of EPS due to the presence of negative charges on them aids in the removal of heavy metal ions carrying positive charge (Singh et al. 2019). Some of the biological activities of exopolysaccharides from cyanobacteria have been depicted in Table 11.5.

Table 11.5 Biological activity and characterization of exopolysaccharides from cyanobacteria

Cyanobacteria	Molecular weight (kDa)	Polysaccharides	Activity	References
Nostoc linckia	1310	Exopolysaccharide	Antioxidant activity	Uhliariková et al. (2022)
Aphanothece halophytica	2.0×10^6	Released polysaccharide	Anticancer, adjuvant and antiviral activity	Freitas et al. (2021)
Chroococcus sp. FPU101	5.9×10^3	Exopolysaccharide	Not identified	Yoshikawa et al. (2021)
Phormidium versicolor NCC466 (CFv-PS)	63.79	Heteropolysaccharide	Hepatoprotective, antioxidant, and cytoprotective properties	Belhaj et al. (2018)
Nostoc sphaeroides	137.55	Water-soluble heteropolysaccharide	For medicinal consumption, such as archoptoma, antipyretic, scald, and nyctalopia	Liu et al. (2018)
Cyanothece sp. CCY 0110	12–14	Released polysaccharide	Act as bio-sorbent for heavy metal bioremediation like copper, cadmium, lead	Mota et al. (2016)
Cyanobacterium aponinum	1.06×10^6	Exopolysaccharide	Immunomodulatory	Gudmundsdottir et al. (2015)
Anabaena anomala	864	Exopolysaccharide	Antibacterial activity	Bhatnagar et al. (2014)
Anabaena oryzae	539	Exopolysaccharide	Antioxidant and iron-chelating properties	Bhatnagar et al. (2014)
Cyanothece sp.	4.5×10^4	Water-soluble polysaccharide	Gelation property, removal of heavy metals	Ohki et al. (2014)
Tolypothrix tenuis	1953	Exopolysaccharide	Blood clotting mediators for managing wound	Bhatnagar et al. (2014)
Anabaena sp.	3679	Exopolysaccharide	Better thrombogenicity, slightly hemolytic, short blood coagulation time, non-cytotoxic and antibacterial activity	Bhatnagar et al. (2014)
Aphanothece sacrum	>1000	Sulfated polysaccharide	Gelation behavior, ability to bind with metallic ions present in soil	Okajima-Kaneko et al. (2007)
Aphanothece stagnina	3.14×10^4	Exopolysaccharide	Trivalent metal-mediated gelation	Le Nguyen et al. (2012)

Nostoc flagelliforme	Not identified	Exopolysaccharide	Antiviral	Kanekiyo et al. (2007), Ahmadi et al. (2015)
Arthrospira platensis	Not identified	Exopolysaccharide	Antioxidant, skin healing, anticoagulant, antibacterial, antiviral properties	Majdoub et al. (2009), Ahmed et al. (2014), Challouf et al. (2011), Reichert et al. (2017)

11.10 Conclusion and Prospects

Cyanobacteria are becoming potentially efficient hosts because of their ability to produce exopolysaccharides. Better understandings of proper methods of EPS production as well as their physical, chemical, and rheological properties offer potential applications of these EPS in bioindustry, although very few studies regarding their functional and structural diversity are known. Besides their extraction, processing, and applications that have been discussed, there are certain limitations associated with EPS production: (I) the extraction of EPS from cyanobacterial cells can be a challenging task as they are tightly associated with cell walls, making the extraction process difficult and time-consuming; (II) the cost of producing EPS can be high due to the need for large-scale cultivation and extraction processes, making it challenging for EPS to compete with other biopolymers in the market; (III) there is a lack of standardization in EPS extraction and processing techniques, which can impact the quality and consistency of EPS for commercial applications; and (IV) the quantity of EPS extracted from different cyanobacterial strains can vary widely, which can impact its performance in various applications. Therefore, there is a need to develop highly reliable and accurate processes for the production, extraction, and processing of cyanobacterial EPS. Research efforts have been ongoing to address these challenges for better production of EPS. As the demand for sustainable and biocompatible materials continues to rise, EPS holds a promising future in bioindustry.

References

Abdo S, Hetta M, Samhan F, El-Din R, Ali G (2012) Phytochemical and antibacterial study of five freshwater algal species. Asian J Plant Sci 11(3):109–116

Adhikary S (1998) Polysaccharides from mucilaginous envelope layers of cyanobacteria and their ecological significance. J Sci Ind Res 57(8):454–466

Adhikary SP, Sahu JK (1998) UV protecting pigment of the terrestrial cyanobacterium *Tolypothrix byssoidea*. J Plant Physiol 153(5–6):770–773

Ahmadi A, Zorofchian Moghadamtousi S, Abubakar S, Zandi K (2015) Antiviral potential of algae polysaccharides isolated from marine sources: a review. Biomed Res Int 2015:825203

Ahmed M, Moerdijk-Poortvliet TC, Wijnholds A, Stal LJ, Hasnain S (2014) Isolation, characterization and localization of extracellular polymeric substances from the cyanobacterium *Arthrospira platensis strain MMG-9*. Eur J Phycol 49(2):143–150

ArunKumar M, Divya S, LewisOscar F, Thajuddin N, Nithya C (2018) Cyanobacterial exopolysaccharides: a potent antibiofilm agent against *Pseudomonas aeruginosa*. IJRAR 5(4)

Babiak W, Krzemińska I (2021) Extracellular polymeric substances (EPS) as microalgal bioproducts: a review of factors affecting EPS synthesis and application in flocculation processes. Energies 14(13):4007

Bafana A (2013) Characterization and optimization of production of exopolysaccharide from *Chlamydomonas reinhardtii*. Carbohydr Polym 95(2):746–752

Baldev E, MubarakAli D, Shriraman R, Pandiaraj D, Alharbi N, Thajuddin N (2015) Extraction and partial characterization of exopolysaccharides from marine cyanobacteria and their flocculation property. Res J Environ Sci 9(1):28

Belhaj D, Athmouni K, Ahmed MB, Aoiadni N, El Feki A, Zhou JL, Ayadi H (2018) Polysaccharides from *Phormidium versicolor (NCC466)* protecting HepG2 human hepatocellular carcinoma cells and rat liver tissues from cadmium toxicity: evidence from in vitro and in vivo tests. Int J Biol Macromol 113:813–820

Bemal S, Anil AC (2018) Effects of salinity on cellular growth and exopolysaccharide production of freshwater Synechococcus strain CCAP1405. J Plankton Res 40(1):46–58

Bertocchi C, Navarini L, Cesàro A, Anastasio M (1990) Polysaccharides from cyanobacteria. Carbohydr Polym 12(2):127–153

Bhatnagar M, Parwani L, Sharma V, Ganguly J, Bhatnagar A (2014) Exopolymers from Tolypothrix tenuis and three *Anabaena sp*.(Cyanobacteriaceae) as novel blood clotting agents for wound management. Carbohydr Polym 99:692–699

Bhunia B, Uday USP, Oinam G, Mondal A, Bandyopadhyay TK, Tiwari ON (2018) Characterization, genetic regulation and production of cyanobacterial exopolysaccharides and its applicability for heavy metal removal. Carbohydr Polym 179:228–243

Borah D, Nainamalai S, Gopalakrishnan S, Rout J, Alharbi NS, Alharbi SA, Nooruddin T (2018) Biolubricant potential of exopolysaccharides from the cyanobacterium *Cyanothece epiphytica*. Appl Microbiol Biotechnol 102(8):3635–3647

Budarin V, Ross A, Biller P, Riley R, Clark J, Jones J, Gilmour D, Zimmerman W (2012) Microalgae biorefinery concept based on hydrothermal microwave pyrolysis. Green Chem 14(12):3251–3254

Cai Y, Wolk CP (1997) Nitrogen deprivation of anabaena sp. strain PCC 7120 elicits rapid activation of a gene cluster that is essential for uptake and utilization of nitrate. J Bacteriol 179(1):258–266

Challouf R, Trabelsi L, Ben Dhieb R, El Abed O, Yahia A, Ghozzi K, Ben Ammar J, Omran H, Ben Ouada H (2011) Evaluation of cytotoxicity and biological activities in extracellular polysaccharides released by cyanobacterium *Arthrospira platensis*. Braz Arch Biol Technol 54:831–838

Chandra N, Mallick N (2022) Co-production of bioethanol and commercially important exopolysaccharides from the marine cyanobacterium Synechococcus elongatus BDU 10144 in a novel low-cost seawater-fertilizer-based medium. Int J Energy Res 46(10):13487–13510

Chemat F, Khan MK (2011) Applications of ultrasound in food technology: processing, preservation and extraction. Ultrason Sonochem 18(4):813–835

Chen H-W, Yang T-S, Chen M-J, Chang Y-C, Lin C-Y, Eugene I, Wang C, Ho C-L, Huang K-M, Yu C-C (2012) Application of power plant flue gas in a photobioreactor to grow *Spirulina* algae, and a bioactivity analysis of the algal water-soluble polysaccharides. Bioresour Technol 120: 256–263

Chen L, Rossi F, Deng S, Liu Y, Wang G, Adessi A, De Philippis R (2014) Macromolecular and chemical features of the excreted extracellular polysaccharides in induced biological soil crusts of different ages. Soil Biol Biochem 78:1–9

Chentir I, Hamdi M, Doumandji A, HadjSadok A, Ouada HB, Nasri M, Jridi M (2017) Enhancement of extracellular polymeric substances (EPS) production in Spirulina (*Arthrospira* sp.) by two-step cultivation process and partial characterization of their polysaccharidic moiety. Int J Biol Macromol 105:1412–1420

Choi WY, Kang DH, Lee HY (2013) Enhancement of immune activation activities of *Spirulina maxima* grown in deep-sea water. Int J Mol Sci 14(6):12205–12221

Chong BF, Blank LM, Mclaughlin R, Nielsen LK (2005) Microbial hyaluronic acid production. Appl Microbiol Biotechnol 66(4):341–351

Clément-Larosière B, Lopes F, Gonçalves A, Taidi B, Benedetti M, Minier M, Pareau D (2014) Carbon dioxide biofixation by *Chlorella vulgaris* at different CO2 concentrations and light intensities. Eng Life Sci 14(5):509–519

Cruz D, Vasconcelos V, Pierre G, Michaud P, Delattre C (2020) Exopolysaccharides from cyanobacteria: strategies for bioprocess development. Appl Sci 10(11):3763

De Brouwer J, Stal L (2004) Does warm-water extraction of benthic diatoms yield extracellular polymeric substances or does it extract intracellular chrysolaminaran? Eur J Phycol 39(2): 129–131

de Jesus Raposo MF, De Morais AMB, De Morais RMSC (2015) Marine polysaccharides from algae with potential biomedical applications. Mar Drugs 13(5):2967–3028

De Philippis R, Vincenzini M (1998) Exocellular polysaccharides from cyanobacteria and their possible applications. FEMS Microbiol Rev 22(3):151–175

De Philippis R, Faraloni C, Margheri MC, Sili C, Herdman M, Vincenzini M (2000) Morphological and biochemical characterization of the exocellular investments of polysaccharide-producing *Nostoc* strains from the Pasteur culture collection. World J Microbiol Biotechnol 16(7):655–661

Delattre C, Laroche C, Michaud P (2009) Bacterial and fungal polysaccharides produced by fermentation – an overview. In: Advances in fermentation technology. Asiatech Publishers, New Delhi, pp 484–521

Delattre C, Pierre G, Laroche C, Michaud P (2016) Production, extraction and characterization of microalgal and cyanobacterial exopolysaccharides. Biotechnol Adv 34(7):1159–1179. https://doi.org/10.1016/j.biotechadv.2016.08.001

Fadlallah H, Jarrahi M, Herbert E, Ferrari R, Mejean A, Peerhossaini H (2016) Effects of shear stress on the growth rate of micro-organisms in agitated reactors. In: Fluids Engineering Division Summer Meeting

Fiedler G, Muro-Pastor AM, Flores E, Maldener I (2001) NtcA-dependent expression of the devBCA operon, encoding a heterocyst-specific ATP-binding cassette transporter in Anabaena spp. J Bacteriol 183(12):3795–3799

Fisher ML, Allen R, Luo Y, Curtiss R III (2013) Export of extracellular polysaccharides modulates adherence of the cyanobacterium *Synechocystis*. PLoS One 8(9):e74514

Freitas F, Alves VD, Reis MA (2011) Advances in bacterial exopolysaccharides: from production to biotechnological applications. Trends Biotechnol 29(8):388–398

Freitas F, Torres CA, Araújo D, Farinha I, Pereira JR, Concórdio-Reis P, Reis MA (2021) Advanced microbial polysaccharides. In: Biopolymers for biomedical and biotechnological applications. Wiley, pp 19–62

Frias JE, Flores E, Herrero A (1997) Nitrate assimilation gene cluster from the heterocyst-forming cyanobacterium Anabaena sp. strain PCC 7120. J Bacteriol 179(2):477–486

Gaignard C, Laroche C, Pierre G, Dubessay P, Delattre C, Gardarin C, Gourvil P, Probert I, Dubuffet A, Michaud P (2019) Screening of marine microalgae: investigation of new exopolysaccharide producers. Algal Res 44:101711

Gao X, Liu L-T, Liu B (2019) Dryland cyanobacterial exopolysaccharides show protection against acid deposition damage. Environ Sci Pollut Res 26(23):24300–24304

Ge H, Zhang J, Zhou X, Xia L, Hu C (2014) Effects of light intensity on components and topographical structures of extracellular polymeric substances from *Microcoleus vaginatus (Cyanophyceae)*. Phycologia 53(2):167–173

Gloaguen V, Ruiz G, Morvan H, Mouradi-Givernaud A, Maes E, Krausz P, Strecker G (2004) The extracellular polysaccharide of *Porphyridium sp.*: an NMR study of lithium-resistant oligosaccharidic fragments. Carbohydr Res 339(1):97–103

González-Fernández C, Ballesteros M (2012) Linking microalgae and cyanobacteria culture conditions and key-enzymes for carbohydrate accumulation. Biotechnol Adv 30(6):1655–1661

Goo BG, Baek G, Choi DJ, Park YI, Synytsya A, Bleha R, Seong DH, Lee C-G, Park JK (2013) Characterization of a renewable extracellular polysaccharide from defatted microalgae Dunaliella tertiolecta. Bioresour Technol 129:343–350

Griffin A, Edwards K, Gasson M, Morris V (1996) Identification of structural genes involved in bacterial exopolysaccharide production. Biotechnol Genet Eng Rev 13(1):1–18

Gris B, Sforza E, Morosinotto T, Bertucco A, La Rocca N (2017) Influence of light and temperature on growth and high-value molecules productivity from Cyanobacterium *aponinum*. J Appl Phycol 29(4):1781–1790

Gudmundsdottir AB, Omarsdottir S, Brynjolfsdottir A, Paulsen BS, Olafsdottir ES, Freysdottir J (2015) Exopolysaccharides from cyanobacterium *aponinum* from the blue lagoon in Iceland increase IL-10 secretion by human dendritic cells and their ability to reduce the IL-17+ RORγt+/ IL-10+ FoxP3+ ratio in CD4+ T cells. Immunol Lett 163(2):157–162

Han PP, Yao SY, Guo RJ, Shen SG, Yan RR, Tan ZL, Jia SR (2017a) The relationship between monosaccharide composition of extracellular polysaccharide and activities of related enzymes in *Nostoc flagelliforme* under different culture conditions. Carbohydr Polym 174:111–119. https://doi.org/10.1016/j.carbpol.2017.05.093

Han P-P, Yao S-Y, Guo R-J, Yan R-R, Wu Y-K, Shen S-G, Jia S-R (2017b) Influence of culture conditions on extracellular polysaccharide production and the activities of enzymes involved in the polysaccharide synthesis of *Nostoc flagelliforme*. RSC Adv 7(71):45075–45084

He J, Zhang X, Tan H, Zhang J (2015) Characterization of an ABC transporter involved in exopolysaccharides secretion in the cyanobacterium Synechocystis sp. PCC 6803. PLoS One 10(5):e0125644. https://doi.org/10.1371/journal.pone.0125644

Herrero M, Jaime L, Martín-Álvarez PJ, Cifuentes A, Ibáñez E (2006) Optimization of the extraction of antioxidants from Dunaliella salina microalga by pressurized liquids. J Agric Food Chem 54(15):5597–5603

Hu C, Liu Y, Paulsen BS, Petersen D, Klaveness D (2003) Extracellular carbohydrate polymers from five desert soil algae with different cohesion in the stabilization of fine sand grain. Carbohydr Polym 54(1):33–42

Hussein MH, Abou-ElWafa GS, Shaaban-Dessuuki SA, Hassan NI (2015) Characterization and antioxidant activity of exopolysaccharide secreted by *Nostoc carneum*. Int J Pharmacol 11(5): 432–439

Islam ST, Lam JS (2014) Synthesis of bacterial polysaccharides via the Wzx/Wzy-dependent pathway. Can J Microbiol 60(11):697–716

Jin Y, Kim SS (2005) Identification and characterization of a gene responsible for exopolysaccharide production in the cyanobacterium Synechocystis sp. PCC 6803. Microbiology 151(4):1175–1182

Jindal N, Singh D, Khattar J (2011) Kinetics and physico-chemical characterization of exopolysaccharides produced by the cyanobacterium *Oscillatoria formosa*. World J Microbiol Biotechnol 27(9):2139–2146

Juin C, Chérouvrier J-R, Thiéry V, Gagez A-L, Bérard J-B, Joguet N, Kaas R, Cadoret J-P, Picot L (2015) Microwave-assisted extraction of phycobiliproteins from *Porphyridium purpureum*. Appl Biochem Biotechnol 175(1):1–15

Kadam SU, Tiwari BK, O'Donnell CP (2013) Application of novel extraction technologies for bioactives from marine algae. J Agric Food Chem 61(20):4667–4675

Kadam SU, Álvarez C, Tiwari BK, O'Donnell CP (2015) Extraction of biomolecules from seaweeds. In: Seaweed sustainability. Elsevier, pp 243–269

Kamravamanesh D, Lackner M, Herwig C (2018) Bioprocess engineering aspects of sustainable polyhydroxyalkanoate production in cyanobacteria. Bioengineering 5(4):111

Kanekiyo K, Hayashi K, Takenaka H, Lee J-B, Hayashi T (2007) Anti-herpes simplex virus target of an acidic polysaccharide, nostoflan, from the edible blue-green alga *Nostoc flagelliforme*. Biol Pharm Bull 30(8):1573–1575

Kang H, Salim H, Akter N, Kim D, Kim J, Bang H, Kim M, Na J, Hwangbo J, Choi H (2013) Effect of various forms of dietary chlorella supplementation on growth performance, immune characteristics, and intestinal microflora population of broiler chickens. J Appl Poult Res 22(1): 100–108

Kehr JC, Dittmann E (2015) Biosynthesis and function of extracellular glycans in cyanobacteria. Life 5(1):164–180

Keshari N, Adhikary SP (2013) Characterization of cyanobacteria isolated from biofilms on stone monuments at Santiniketan, India. Biofouling 29(5):525–536

Khangembam R, Tiwari ON, Kalita M (2016) Production of exopolysaccharides by the cyanobacterium *Anabaena sp. BTA992* and application as bioflocculants. J Appl Biol Biotechnol 4:8–11

Khattar JIS, Singh DP, Jindal N, Kaur N, Singh Y, Rahi P, Gulati A (2010) Isolation and characterization of exopolysaccharides produced by the cyanobacterium *Limnothrix redekei PUPCCC 116*. Appl Biochem Biotechnol 162(5):1327–1338

Kraan S (2012) Algal polysaccharides, novel applications and outlook. In: Carbohydrates-comprehensive studies on glycobiology and glycotechnology. IntechOpen

Kumar D, Kaštánek P, Adhikary SP (2018) Exopolysaccharides from cyanobacteria and microalgae and their commercial application. Curr Sci 115(2):234–241

Kurd F, Samavati V (2015) Water soluble polysaccharides from *Spirulina platensis*: extraction and in vitro anti-cancer activity. Int J Biol Macromol 74:498–506. https://doi.org/10.1016/j.ijbiomac.2015.01.005

Lama L, Nicolaus B, Calandrelli V, Manca MC, Romano I, Gambacorta A (1996) Effect of growth conditions on endo-and exopolymer biosynthesis in *Anabaena cylindrica 10 C*. Phytochemistry 42(3):655–659

Le Nguyen QT, Okajima M, Mitsumata T, Kan K, Tran HT, Kaneko T (2012) Trivalent metal-mediated gelation of novel supergiant sulfated polysaccharides extracted from *Aphanothece stagnina*. Colloid Polym Sci 290(2):163–172

Li H, Li Z, Xiong S, Zhang H, Li N, Zhou S et al (2011) Pilot-scale isolation of bioactive extracellular polymeric substances from cell-free media of mass microalgal cultures using tangential-flow ultrafiltration. Process Biochem 46(5):1104–1109

Li Z, Liu Y, Zhou T, Cao L, Cai Y, Wang Y et al (2022) Effects of culture conditions on the performance of Arthrospira platensis and its production of exopolysaccharides. Foods 11(14): 2020

Liqin S, Wang C, Lei S (2008, May) Effects of light regime on extracellular polysaccharide production by Porphyridium cruentum cultured in flat plate photobioreactors. In: 2008 2nd international conference on bioinformatics and biomedical engineering. IEEE, pp 1488–1491

Liu B, Sun Z, Ma X, Yang B, Jiang Y, Wei D, Chen F (2015) Mutation breeding of extracellular polysaccharide-producing microalga *Crypthecodinium cohnii* by a novel mutagenesis with atmospheric and room temperature plasma. Int J Mol Sci 16(4):8201–8212

Liu Y, Su P, Xu J, Chen S, Zhang J, Zhou S et al (2018) Structural characterization of a bioactive water-soluble heteropolysaccharide from *Nostoc sphaeroids kütz*. Carbohydr Polym 200:552–559

Lupi FM, Fernandes HML, Tomé MM, Sá-Correia I, Novais JM (1994) Influence of nitrogen source and photoperiod on exopolysaccharide synthesis by the microalga Botryococcus braunii UC 58. Enzym Microb Technol 16(7):546–550

Majdoub H, Mansour MB, Chaubet F, Roudesli MS, Maaroufi RM (2009) Anticoagulant activity of a sulfated polysaccharide from the green alga *Arthrospira platensis*. Biochim Biophys Acta 1790(10):1377–1381

Marcati A, Ursu AV, Laroche C, Soanen N, Marchal L, Jubeau S et al (2014) Extraction and fractionation of polysaccharides and B-phycoerythrin from the microalga *Porphyridium cruentum* by membrane technology. Algal Res 5:258–263

Mathlouthi M, Koenig JL (1987) Vibrational spectra of carbohydrates. Adv Carbohydr Chem Biochem 44:7–89

Matsunaga T, Sudo H, Takemasa H, Wachi Y, Nakamura N (1996) Sulfated extracellular polysaccharide production by the halophilic cyanobacterium Aphanocapsa halophytia immobilized on light-diffusing optical fibers. Appl Microbiol Biotechnol 45:24–27

Miralles I, Domingo F, Cantón Y, Trasar-Cepeda C, Leirós MC, Gil-Sotres F (2012) Hydrolase enzyme activities in a successional gradient of biological soil crusts in arid and semi-arid zones. Soil Biol Biochem 53:124–132

Mishima T, Murata J, Toyoshima M, Fujii H, Nakajima M, Hayashi T et al (1998) Inhibition of tumor invasion and metastasis by calcium spirulan (Ca-SP), a novel sulfated polysaccharide derived from a blue-green alga, Spirulina platensis. Clin Exp Metastasis 16(6):541–550

Mishra A, Jha B (2009) Isolation and characterization of extracellular polymeric substances from micro-algae *Dunaliella salina* under salt stress. Bioresour Technol 100(13):3382–3386

Mishra A, Kavita K, Jha B (2011) Characterization of extracellular polymeric substances produced by micro-algae *Dunaliella salina*. Carbohydr Polym 83(2):852–857

Misurcova L, Orsavova J, Ambrozova JV (2015) Algal polysaccharides and health. In: Ramawat KG, Merillon JM (eds) Polysaccharides bioactivity and biotechnology. Springer, Cham, Switzerland, pp 110–144

Mona S, Kaushik A (2015) Chromium and cobalt sequestration using exopolysaccharides produced by freshwater cyanobacterium Nostoc linckia. Ecol Eng 82:121–125

Moore BG, Tischer RG (1964) Extracellular polysaccharides of algae: effects on life-support systems. Science 145(3632):586–587

Moreno J, Vargas MA, Olivares H, Rivas J, Guerrero MG (1998) Exopolysaccharide production by the cyanobacterium *Anabaena sp.* ATCC 33047 in batch and continuous culture. J Biotechnol 60(3):175–182

Moreno J, Vargas MA, Madiedo JM, Muñoz J, Rivas J, Guerrero MG (2000) Chemical and rheological properties of an extracellular polysaccharide produced by the cyanobacterium *Anabaena sp.* ATCC 33047. Biotechnol Bioeng 67(3):283–290

Mota R, Guimarães R, Büttel Z, Rossi F, Colica G, Silva CJ et al (2013) Production and characterization of extracellular carbohydrate polymer from *Cyanothece sp. CCY 0110*. Carbohydr Polym 92(2):1408–1415

Mota R, Rossi F, Andrenelli L, Pereira SB, De Philippis R, Tamagnini P (2016) Released polysaccharides (RPS) from *Cyanothece sp.* CCY 0110 as biosorbent for heavy metals bioremediation: interactions between metals and RPS binding sites. Appl Microbiol Biotechnol 100(17): 7765–7775

Mota R, Vidal R, Pandeirada C, Flores C, Adessi A, De Philippis R et al (2020) Cyanoflan: a cyanobacterial sulfated carbohydrate polymer with emulsifying properties. Carbohydr Polym 229:115525

Napolitano M, Rubio MÁ, Santamaria-Gomez J, Olmedo-Verd E, Robinson NJ, Luque I (2012) Characterization of the response to zinc deficiency in the cyanobacterium Anabaena sp. strain PCC 7120. J Bacteriol 194(10):2426–2436

Ngatu NR, Okajima MK, Yokogawa M, Hirota R, Eitoku M, Muzembo BA et al (2012) Anti-inflammatory effects of sacran, a novel polysaccharide from *Aphanothece sacrum*, on 2,4,6-trinitrochlorobenzene–induced allergic dermatitis in vivo. Ann Allergy Asthma Immunol 108(2):117–122

Nicolaisen K, Moslavac S, Samborski A, Valdebenito M, Hantke K, Maldener I et al (2008) Alr0397 is an outer membrane transporter for the siderophore schizokinen in Anabaena sp. strain PCC 7120. J Bacteriol 190(22):7500–7507

Nicolaus B, Panico A, Lama L, Romano I, Manca MC, De Giulio A, Gambacorta A (1999) Chemical composition and production of exopolysaccharides from representative members of heterocystous and non-heterocystous cyanobacteria. Phytochemistry 52(4):639–647

Nieves-Morión M, Flores E (2018) Multiple ABC glucoside transporters mediate sugar-stimulated growth in the heterocyst-forming cyanobacterium Anabaena sp. strain PCC 7120. Environ Microbiol Rep 10(1):40–48

Oda Y et al (2009) Identification and characterization of an ABC transporter gene involved in exopolysaccharide production in the cyanobacterium Nostoc sp. PCC 7120. Microbiology 155(7):2079–2087

Ohki K, Le NQT, Yoshikawa S, Kanesaki Y, Okajima M, Kaneko T, Thi TH (2014) Exopolysaccharide production by a unicellular freshwater cyanobacterium *Cyanothece sp.* isolated from a rice field in Vietnam. J Appl Phycol 26(1):265–272

Okajima-Kaneko M, Ono M, Kabata K, Kaneko T (2007) Extraction of novel sulfated polysaccharides from *Aphanothece sacrum* (Sur.) Okada, and its spectroscopic characterization. Pure Appl Chem 79(11):2039–2046

Otero A, Vincenzini M (2003) Extracellular polysaccharide synthesis by *Nostoc* strains as affected by N source and light intensity. J Biotechnol 102(2):143–152

Otero A, Vincenzini M (2004) Nostoc (cyanophyceae) goes nude: extracellular polysaccharides serve as a sink for reducing power under unbalanced C/N metabolism. J Phycol 40(1):74–81

Ozturk S, Aslim B (2008) Relationship between chromium (VI) resistance and extracellular polymeric substances (EPS) concentration by some cyanobacterial isolates. Environ Sci Pollut Res 15:478–480

Ozturk S, Aslim B (2010) Modification of exopolysaccharide composition and production by three cyanobacterial isolates under salt stress. Environ Sci Pollut Res 17(3):595–602

Ozturk S, Aslim B, Suludere Z, Tan S (2014) Metal removal of cyanobacterial exopolysaccharides by uronic acid content and monosaccharide composition. Carbohydr Polym 101:265–271

Parikh A, Madamwar D (2006) Partial characterization of extracellular polysaccharides from cyanobacteria. Bioresour Technol 97(15):1822–1827

Pasquet V, Chérouvrier JR, Farhat F, Thiéry V, Piot JM, Bérard JB et al (2011) Study on the microalgal pigments extraction process: performance of microwave assisted extraction. Process Biochem 46(1):59–67

Patel AK, Laroche C, Marcati A, Ursu AV, Jubeau S, Marchal L et al (2013) Separation and fractionation of exopolysaccharides from *Porphyridium cruentum*. Bioresour Technol 145:345–350

Pereira S, Zille A, Micheletti E, Moradas-Ferreira P, De Philippis R, Tamagnini P (2009) Complexity of cyanobacterial exopolysaccharides: composition, structures, inducing factors and putative genes involved in their biosynthesis and assembly. FEMS Microbiol Rev 33(5): 917–941

Pereira S, Micheletti E, Zille A, Santos A, Moradas-Ferreira P, Tamagnini P, De Philippis R (2011) Using extracellular polymeric substances (EPS)-producing cyanobacteria for the bioremediation of heavy metals: do cations compete for the EPS functional groups and also accumulate inside the cell? Microbiology 157(2):451–458

Pierre G, Delattre C, Laroche C, Michaud P (2014) Galactans and its 1644 applications. In: Ramawat KG, Mérillon J-M (eds) Polysaccharides: bioactivity and 1645 biotechnology. Springer, Cham, Switzerland, pp 753–794

Plaza M, Amigo-Benavent M, Del Castillo MD, Ibáñez E, Herrero M (2010) Facts about the formation of new antioxidants in natural samples after subcritical water extraction. Food Res Int 43(10):2341–2348

Potts M (2004) Nudist colonies: a revealing glimpse of cyanobacterial extracellular polysaccharide. J Phycol 40(1):1–3

Ramus J (1972) The production of extracellular polysaccharide by the unicellular red alga Porphyridium aerugineum. J Phycol 8(1):97–111

Rehm BH (2010) Bacterial polymers: biosynthesis, modifications and applications. Nat Rev Microbiol 8(8):578–592

Reichert M, Bergmann S, Hwang J, Buchholz R, Lindenberger C (2017) Antiviral activity of exopolysaccharides from *Arthrospira platensis* against koi herpesvirus. J Fish Dis 40(10): 1441–1450

Rodriguez-Jasso RM, Mussatto SI, Pastrana L, Aguilar CN, Teixeira JA (2011) Microwave-assisted extraction of sulfated polysaccharides (fucoidan) from brown seaweed. Carbohydr Polym 86(3): 1137–1144

Rossi F, De Philippis R (2015) Role of cyanobacterial exopolysaccharides in phototrophic biofilms and in complex microbial mats. Life 5(2):1218–1238

Rossi F, Philippis RD (2016) Exocellular polysaccharides in microalgae and cyanobacteria: chemical features, role and enzymes and genes involved in their biosynthesis. In: The physiology of microalgae. Springer, Cham, pp 565–590

Rossi F, Mugnai G, De Philippis R (2018) Complex role of the polymeric matrix in biological soil crusts. Plant Soil 429(1):19–34

Roussel M, Villay A, Delbac F, Michaud P, Laroche C, Roriz D et al (2015) Antimicrosporidian activity of sulphated polysaccharides from algae and their potential to control honeybee nosemosis. Carbohydr Polym 133:213–220

Roy A, Tripathy P, Adhikary SP (1998) UV protecting pigment of epilithic cyanobacteria occurring on the temples of various regions of India. In: Subramanian G, Kaushik BD, Venkataraman GS (eds) Proceedings of international conference of cyanobacterial biotechnology. Oxford & IBH, New Delhi, pp 439–447

Roy K, Bannerjee S, Hazra T, Das D, Pandit S, Lahiri D, Kavisri M (2023) Exopolysaccharide production by *Anabaena sp.* PCC 7120: physicochemical parameter optimization and two-stage cultivation strategy to maximize the product yield. Biomass Convers Biorefin 2023:1–12

Rudolf M, Kranzler C, Lis H, Margulis K, Stevanovic M, Keren N, Schleiff E (2015) Multiple modes of iron uptake by the filamentous, siderophore-producing cyanobacterium, Anabaena sp. PCC 7120. Mol Microbiol 97(3):577–588

Sakamoto T, Kumihashi K, Kunita S, Masaura T, Inoue-Sakamoto K, Yamaguchi M (2011) The extracellular-matrix-retaining cyanobacterium *Nostoc verrucosum* accumulates trehalose, but is sensitive to desiccation. FEMS Microbiol Ecol 77(2):385–394

Sasaki M, Takagi A, Ota S, Kawano S, Sasaki D, Asayama M (2020) Coproduction of lipids and extracellular polysaccharides from the novel green alga Parachlorella sp. BX1. 5 depending on cultivation conditions. Biotechnol Rep 25:e00392

Schmid J, Sieber V, Rehm B (2015) Bacterial exopolysaccharides: biosynthesis pathways and engineering strategies. Front Microbiol 6:496

Sciuto K, Moro I (2015) Cyanobacteria: the bright and dark sides of a charming group. Biodivers Conserv 24(4):711–738

Shaieb FA, Issa AA, Meragaa A (2014) Antimicrobial activity of crude extracts of cyanobacteria *Nostoc commune* and *Spirulina platensis*. Arch Biomed Sci 2(2):34–41

Shen SG, Lin YH, Zhao DX, Wu YK, Yan RR, Zhao HB et al (2019) Comparisons of functional properties of polysaccharides from *Nostoc flagelliforme* under three culture conditions. Polymers 11(2):263

Shepherd R, Rockey J, Sutherland IW, Roller S (1995) Novel bioemulsifiers from microorganisms for use in foods. J Biotechnol 40(3):207–217

Shvarev D, Maldener I (2018) ATP-binding cassette transporters of the multicellular cyanobacterium Anabaena sp. PCC 7120: a wide variety for a complex lifestyle. FEMS Microbiol Lett 365(4):fny012

Shvarev D, Nishi CN, Maldener I (2019) Two DevBCA-like ABC transporters are involved in the multidrug resistance of the cyanobacterium Anabaena sp. PCC 7120. FEBS Lett 593(14): 1818–1826

Simkovsky R, Daniels EF, Tang K, Huynh SC, Golden SS, Brahamsha B (2012) Impairment of O-antigen production confers resistance to grazing in a model amoeba–cyanobacterium predator–prey system. Proc Natl Acad Sci 109(41):16678–16683

Singh S, Verma E, Tiwari B, Mishra AK (2016) Exopolysaccharide production in *anabaena sp. PCC 7120* under different CaCl2 regimes. Physiol Mol Biol Plants 22(4):557–566

Singh S, Kant C, Yadav RK, Reddy YP, Abraham G (2019) Cyanobacterial exopolysaccharides: composition, biosynthesis, and biotechnological applications. In: Cyanobacteria. Academic Press, pp 347–358

Stevanovic M, Hahn A, Nicolaisen K, Mirus O, Schleiff E (2012) The components of the putative iron transport system in the cyanobacterium Anabaena sp. PCC 7120. Environ Microbiol 14(7): 1655–1670

Su C, Chi Z, Lu W (2007) Optimization of medium and cultivation conditions for enhanced exopolysaccharide yield by marine *Cyanothece sp. 113*. Chin J Oceanol Limnol 25(4):411–417

Sudo H, Burgess JG, Takemasa H, Nakamura N, Matsunaga T (1995) Sulfated exopolysaccharide production by the halophilic cyanobacterium Aphanocapsa halophytia. Curr Microbiol 30:219–222

Thomas AD, Hoon SR, Linton PE (2008) Carbon dioxide fluxes from cyanobacteria crusted soils in the Kalahari. Appl Soil Ecol 39(3):254–263

Tiwari ON, Sasmal S, Kataria AK, Devi I (2020) Application of microbial extracellular carbohydrate polymeric substances in food and allied industries. 3 Biotech 10:1–17

Toma M, Vinatoru M, Paniwnyk L, Mason TJ (2001) Investigation of the effects of ultrasound on vegetal tissues during solvent extraction. Ultrason Sonochem 8(2):137–142

Trabelsi L, Ben Ouada H, Zili F, Mazhoud N, Ammar J (2013) Evaluation of *Arthrospira platensis* extracellular polymeric substances production in photoautotrophic, heterotrophic and mixotrophic conditions. Folia Microbiol 58(1):39–45

Uhliariková I, Matulová M, Košťálová Z, Lukavský J, Capek P (2022) Lactylated acidic exopolysaccharide produced by the cyanobacterium Nostoc cf. linckia. Carbohydr Polym 276:118801

Van Alstyne KL, Sadowsky MJ (2000) ABC transporters involved in the production and secretion of exopolysaccharides by anabaena sp. PCC 7120. Microbiology 146(4):879–887

Van Camp C, Fraikin C, Claverie E, Onderwater R, Wattiez R (2022) Capsular polysaccharides and exopolysaccharides from *Gloeothece verrucosa* under various nitrogen regimes and their potential plant defence stimulation activity. Algal Res 64:102680

Vanderlinde EM, Harrison JJ, Muszyński A, Carlson RW, Turner RJ, Yost CK (2010) Identification of a novel ABC transporter required for desiccation tolerance, and biofilm formation in Rhizobium leguminosarum bv. viciae 3841. FEMS Microbiol Ecol 71(3):327–340

Vergnes JB, Gernigon V, Guiraud P, Formosa-Dague C (2019) Bicarbonate concentration induces production of exopolysaccharides by *Arthrospira platensis* that mediate bioflocculation and enhance flotation harvesting efficiency. ACS Sustain Chem Eng 7(16):13796–13804

Verma E, Chakraborty S, Tiwari B, Singh S, Mishra AK (2018) Alleviation of NaCl toxicity in the cyanobacterium *Synechococcus sp. PCC 7942* by exogenous calcium supplementation. J Appl Phycol 30(3):1465–1482

Vilkhu K, Mawson R, Simons L, Bates D (2008) Applications and opportunities for ultrasound assisted extraction in the food industry—a review. Innovative Food Sci Emerg Technol 9(2): 161–169

Villay A, Laroche C, Roriz D, El Alaoui H, Delbac F, Michaud P (2013) Optimisation of culture parameters for exopolysaccharides production by the microalga *Rhodella violacea*. Bioresour Technol 146:732–735

Wang Y, Zhang Y, Liu Y, Sun J (2017) Biosynthesis and export of exopolysaccharides produced by cyanobacteria. Appl Microbiol Biotechnol 101(17):6725–6735

Whitney JC, Howell PL (2013) Synthase-dependent exopolysaccharide secretion in gram-negative bacteria. Trends Microbiol 21(2):63–72

Xiao Y, Li Z, Li C, Zhang Z, Guo J (2016) Effect of small-scale turbulence on the physiology and morphology of two bloom-forming cyanobacteria. PLoS One 11(12):e0168925

Ye Y, Le Clech P, Chen V, Fane AG (2005) Evolution of fouling during crossflow filtration of model EPS solutions. J Membr Sci 264(1–2):190–199

Yoo JS, Kim SS (2003) Optimal production of exopolysaccharides from the cyanobacterium Synechococcus elongatus PCC 7942. Process Biochem 39(1):89–96

Yoshikawa S, Kanesaki Y, Uemura A, Yamada K, Okajima M, Kaneko T, Ohki K (2021) Physiological and genomic analysis of newly-isolated polysaccharide synthesizing cyanobacterium *Chroococcus sp.* FPU101 and chemical analysis of the exopolysaccharide. J Gen Appl Microbiol 67(5):207–213

You T, Barnett SM (2004) Effect of light quality on production of extracellular polysaccharides and growth rate of *Porphyridium cruentum*. Biochem Eng J 19(3):251–258

Yu H, Jia S, Dai Y (2010) Accumulation of exopolysaccharides in liquid suspension culture of *Nostoc flagelliforme* cells. Appl Biochem Biotechnol 160(2):552–560

Zampieri RM, Adessi A, Caldara F, Codato A, Furlan M, Rampazzo C et al (2020) Anti-inflammatory activity of exopolysaccharides from *Phormidium sp.* ETS05, the most abundant cyanobacterium of the therapeutic euganean thermal muds, using the zebrafish model. Biomol Ther 10(4):582

Zanolla V, Biondi N, Niccolai A, Abiusi F, Adessi A, Rodolfi L, Tredici MR (2022) Protein, phycocyanin, and polysaccharide production by *Arthrospira platensis* grown with LED light in annular photobioreactors. J Appl Phycol 34(3):1189–1199
Zhang G, Ji S, Gao X, Liu Z (2008) Adsorptive fouling of extracellular polymeric substances with polymeric ultrafiltration membranes. J Membr Sci 309(1–2):28–35

Chapter 12
Innovations in the Cyanobacteria-Based Biorefineries for Biopharmaceutical Industries

Ayesha Shahid, Fahad Khan, and Muhammad Farooq

Abstract The biopharmaceutical industry witnesses exceptional growth globally since their emergence. It revitalizes our ability to combat antimicrobial resistance by providing alternative treatment solutions to present-day lethal diseases. Currently, it is enduring sweeping changes due to resource limitations, climate change, and financial/regulatory constraints which compelled the researchers to focus on process-to-product innovation. Cyanobacteria are now envisioned as potential renewable raw materials for the biopharmaceutical industry due to their health-promoting metabolic profile. Currently, cyanobacteria have been part of pharmaceutical value chains as whole biomass or as single metabolite constituents. This chapter provides insight into the value proposition of cyanobacterial biorefinery for the biopharmaceutical industry by discussing the constraints and innovations in the processing route. Strain development through genetic engineering, selection of suitable cultivation modes, and sustainable processing of biomass through green chemicals are considered viable to leverage the maximum potential of cyanobacteria. Innovative technologies like industry 4.0-powered continuous photobioreactors and acoustic extraction of metabolites enable the implementation of a centralized cyanobacterial value chain for the biopharmaceutical industry. Encapsulation, immobilization, and transporter engineering pave the way for efficient product recovery in integrated mode. The discussed information provides the basis for the effective inclusion of cyanobacterial biorefinery in the biopharmaceutical industry.

Keywords Bioactive compounds · Cyanopharm · Industry 4.0 · Innovative value chains · Green chemicals · Industrial overview

---

A. Shahid (✉)
National Center for Genome Editing, Center for Advanced Studies, University of Agriculture Faisalabad, Faisalabad, Pakistan

F. Khan
Tasmanian Institute of Agriculture, The University of Tasmania, Hobart, Australia

M. Farooq
Department of Bioinformatics and Biotechnology, Government College University Faisalabad, Faisalabad, Pakistan

M. A. Mehmood et al. (eds.), *Pharmaceutical and Nutraceutical Potential of Cyanobacteria*, https://doi.org/10.1007/978-3-031-45523-0_12

12.1 Introduction

Biopharmaceuticals are complex therapeutic molecules originating from biological processes and sources including bacteria, fungi, algae, cyanobacteria, and plants. They have revolutionized the field of medicines and therapeutics by providing alternative treatments for cancer, diabetes, autoimmune disorders, cardiovascular diseases, and neurodegenerative diseases. Biopharmaceuticals are well accepted due to their high specificity and sensitivity. They are also appreciated for their potential to rarely cause side effects and their potential to treat patients who are unresponsive to traditional synthetic medicines (Kesik-Brodacka 2018). Since their emergence in 1980, a total of 443 biopharmaceuticals are commercialized solely in European Union and the United States. Sustained growth in the biopharma market has been observed since 2015, with 177 products being approved during 2015–2018 and 197 products during 2018–2022 (Walsh and Walsh 2022). The biopharmaceutical market observed spurred growth as compared to traditional drugs, and it showed tremendous potential to accelerate with an expected market share of >$566 billion by 2032 with a compound annual growth rate (CAGR) of 8.2% (2023).

The biopharmaceutical industry is undergoing sweeping shifts in response to population growth, climate crisis, resource scarcity, financial constraints, and regulatory challenges forcing companies to search for new therapeutic modalities with a focus on process-to-product innovation (Avagyan 2010). Notably, the pharmaceutical industry generates 25–100 kg of waste per 1 kg of product due to process inefficiency that leads to resource scarcity. Indicatively, the emergence of antimicrobial resistance and the shortage of newer antibiotic compounds resulted in a shortage of >400 drugs in the United States (Shaban et al. 2018; Sharma 2015). To mitigate the challenging aspects, the exploration of renewable feedstock such as cyanobacteria for affordable production of biosimilars and next-generation biopharmaceuticals is the trending research area in this domain. This trend is being promoted by the efficacy of cyanobacteria as pharmaceutical agents (Tsolakis et al. 2023).

Cyanobacteria are envisioned as crucial raw materials for third-generation biorefineries with the potential to be implemented in biopharmaceutical industries. They are known for their rich metabolite composition encompassing pigments (phycobilins and carotenoids), lipids, peptides, and secondary metabolites. Easy scale-up, higher growth rate, and environmental sustainability in terms of CO_2 sequestration, adaptation to fluctuating environmental conditions, and high resource recycling efficiency are the added benefits (Shahid et al. 2020, 2023). Currently, cyanobacteria exist in medical value chains as a superfood and nutritional supplements which are either product of whole biomass or their extracted metabolites such as carotenoids, astaxanthin, phycobilins, omega-3 fatty acids, vitamins, etc. (Jha et al. 2017). Although the industrial application of cyanobacterial pharmaceuticals/therapeutics is limited, they are intensively researched from a biorefinery perspective. To leverage the pharmaceutical potential of cyanobacteria, the focus has been on environment-driven metabolite production and development of sustainable and economical processing routes. Being in its infancy, utilization of cyanobacterial biorefinery as a commercial therapeutic is hindered by the higher processing cost and difficult scale-up. The development of potential pharmaceutical value chains in the biorefinery paradigm is imperative from the commercial perspective. This book chapter provides the considerations related to the implementation of cyanobacterial biorefinery for the biopharmaceutical industry. An overview of innovative technologies and their integration with proposed biorefinery is the highlight of the chapter which will provide the basis for real-world implementation of such systems at an industrial scale.

12.2 Considerations for Industrial Implementation of Cyanobacterial Biorefinery

The process of the biopharma industry starts from candidate selection followed by their optimization, process development, and maintenance of commercial requirements. Strain development, upstream-to-midstream processing, and downstream extraction/purification processing are the key step in the cyanobacterial biorefinery (Fig. 12.1). Each step has its own set of positives and negatives which must be dealt with accordingly to ensure continuous supply. For the industrial deployment of cyanobacterial biorefinery, a well-designed and well-managed sustainable supply chain is the key consideration. Indeed, it involves hierarchical steps and is complex to optimize, but integration of multidisciplinary innovation with the cyanopharm

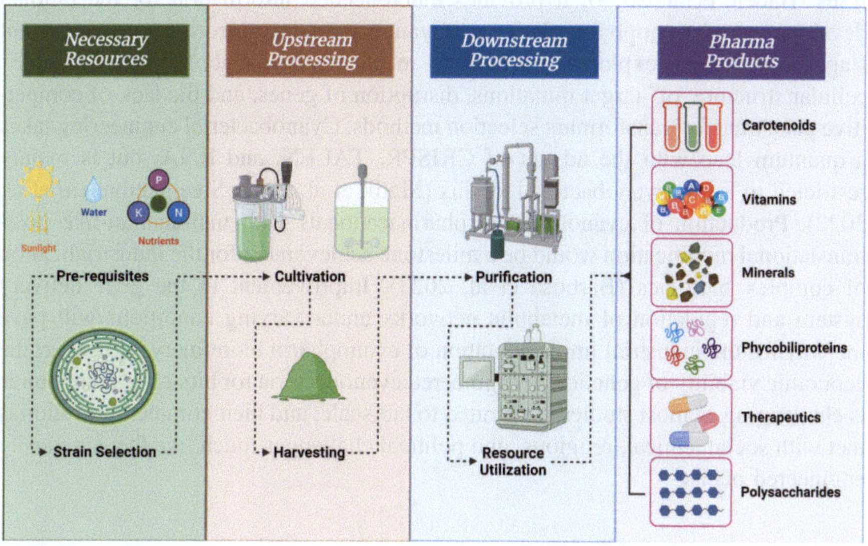

Fig. 12.1 Pictorial representation of the cyanobacterial processing route for the biopharmaceutical industry

(cyanobacterial biorefinery for pharmaceuticals) supply chain will help in process optimization and technology maturation (Espinoza Pérez et al. 2017; Szkodny and Lee 2022).

12.2.1 Strain Development

Commercially, cyanobacterial species, namely, *Oscillatoria*, *Anabaena*, *Spirulina*, *Lyngbya*, *Microcystic,* etc. have been explored as therapeutic tools in the pharmaceutical industry. Efforts must be exerted to amplify the quantities of secondary metabolites. Exposure of strain to varying quantities of macro- and micronutrients, light intensities, temperature ranges, and pH fluctuations has been explored as a viable option to express the desired characteristics during the cultivation stage (Shahid et al. 2019). For this purpose, supply chain designers have to consider the trade-offs such as fluctuating environmental conditions (variation between seasons and weather changes) and facility locations as cyanobacteria, being environment-driven organisms, respond critically to these changes that may restrict their industrial deployment (Tsolakis et al. 2023).

Alternatively, rigorous efforts have been put to identify the discrete bioactive compounds through omics and their manipulation through genetic engineering to achieve the required titers of secondary metabolites (Jeong et al. 2020; Yang et al. 2023). Metabolic flux analysis is the latest intervention in the field of system metabolic engineering that determines the carbon flow in complex biochemical regulatory networks to elucidate the control of metabolic pathways by omics analyses (Babele et al. 2023). It provides much-needed information for the rational development of cyanopharm platforms. Cyanobacterial engineering is at its nascent stages due to underexplored biochemical mechanisms, metabolic drifts, complex cellular structure, off-target mutations, disruption of genes, and the lack of competitive gene transfer/transformant selection methods. Cyanobacterial engineering takes a quantum leap with the advent of CRISPR, TALEN, and RNAi but is mainly restricted to a few cyanobacterial strains (Malik et al. 2022; Sreenikethanam et al. 2022). Production of cyanobacterial pharmaceuticals with mammalian-like post-translational modification would be a milestone achievement for the industrialization of complex biologics (Barbosa et al. 2023). Improvement in the gene delivery system and regulation of metabolic networks under varying conditions will pave the way for the industrial implementation of cyanopharm biorefinery. However, the economic viability of genetically engineered cyanobacteria for bioactive compounds is challenging as most studies are limited to lab-scale, and their commercialization is met with social, ethical, religious, and political challenges much like the genetically engineered plants.

12.2.2 Cultivation Modes for the Cyanopharm Industry

The choice of cultivation mode (batch, fed-batch, or perfusion) is the major consideration of upstream processing that is concomitant with product titer, contamination risks, biotic/abiotic exposure, and safety and stability of the product. Commonly, photoautotrophic batch cultivation is preferred for the industrial production of cyanobacteria as they require minimum input. Indoor/outdoor open shallow raceway ponds are generally used for the mass production of cyanobacteria, yet their effectiveness is compromised due to high incidents of contamination. Alternatively, closed photobioreactors are acclaimed for industrial production due to high biomass density, elevated control of cultivation parameters, and minimal contamination risks, but high operational and maintenance costs are the hindering factors (Hejna et al. 2022; Saha and Murray 2018; Shahid et al. 2019). Specially designed cyanobacterial photobioreactors for the industrial production of pharmaceutics would be beneficial value-addition.

Perfusion cultivation is now primarily being explored by the pharmaceutical industry for denaturation-prone compounds that cannot withstand longer exposure stress. The industrial trend indicated a shift towards perfusion cultivation due to its high versatility, low volumetric requirements, and consistent product quality. Studies indicated the high production cost of the perfusion mode that results in non-viable product economics (Lalor et al. 2019). Implementation of this mode for cyanobacterial pharmaceuticals is underexplored and could be the focus of future research trends. Irrespective of the cultivation mode, the focus should be on optimizing the conditions that promote the co-product of bioactive compounds to enhance the economic viability of biorefinery.

12.2.3 Sustainable Processing for the Cyanopharm Biorefinery

Complex manufacturing of biopharmaceuticals significantly contributes to the process cost as bioprocessing efficiency affects the yield, safety, and ultimately sales/purchase of the product (Rader and Langer 2014). Biomass processing in the biorefinery paradigm is the ultimate solution, but it should be optimized in a way to harness the maximum biomass potential. Co-extraction or sequential extraction of metabolites (usually from high-value to low-value) is the preferred schemes to obtain multiple products from single biomass with minimum-to-zero waste generation. Co-extraction of phycobilins and proteins from *Spirulina* was performed followed by extraction of bioactive compounds such as lipids/fatty acids or polysaccharides (Thevarajah et al. 2022). Co-product processing of antioxidant and anticancerous pigments including phycocyanin, allophycocyanin, and phycoerythrin yielded 93–402 kg pigment/ton-biomass and 25–112 kg-biodiesel/ton-biomass from spent biomass of *Leptolyngbya* sp. (Pekkoh et al. 2023). Similarly, phytochemicals with anticancerous and antioxidant abilities were successfully extracted from the polysaccharide-free *Nostoc* biomass (Pekkoh et al. 2022).

It is imperative to note that the chosen route effectively processes cyanobacteria to its maximum potential, and it must not deteriorate product quality or produce toxic by-products which is the basic necessity for the pharmaceutical processing chain. For this reason, the choice of processing chemicals (Table 12.1), sequential order of extracted metabolites, and processing scheme are of utmost importance. Industrially, chemical-based extraction is common and well-adapted technology, but the milking of cyanobacterial metabolites through green chemicals or chemical-free extraction is gaining dominance to overcome the problems associated with traditional chemical-based extraction. However, before their industrial deployment, optimization of green solvents in terms of cost and environmental impact has to be performed.

In green methods, ionic liquids and supercritical liquids are of utmost importance. Supercritical CO_2 (having both liquid and gaseous properties) has been explored to extract antimicrobial and antioxidant compounds from *S. platensis* (Mendiola et al. 2007) and to extract polyunsaturated fatty acids (Yang et al. 2019). Supercritical CO_2 alone and in combination with ethanol was evaluated to extract carotenoids and fucoxanthin from *Isochrysis galbana*, followed by chemical and water-based extraction of bioactive-rich lipids and proteins (Gilbert-López et al. 2015). Supercritical fluid extraction based on supercritical CO_2 and ethanol extraction was optimized to obtain neuroprotective phytosterol-rich extract from *Phormidium autumnale* (Fagundes et al. 2021). However, complex cell wall structure greatly limits the penetration ability of supercritical fluids. Integration of enzyme-assisted extraction with supercritical fluids helps in obtaining oxidation-sensitive pure extracts with preserved biological properties and improved extraction efficiency (Patil et al. 2021).

Another arising option is the chemical-free extraction of valuable metabolites. Ultrasound-assisted extraction and microwave-assisted extraction are generally employed approaches for this purpose. Microwave-assisted extraction has been optimized for the sequential extraction of 34% fucoxanthin, 23% eicosapentaenoic acid (EPA), and 43.5% of chrysolaminarin from *Phaeodactylum tricornutum* (Zhang et al. 2018). However, heat and pressure produced by these techniques limit their applicability for the extraction of oxidation-sensitive and thermal-sensitive metabolites (Patil et al. 2021).

12.3 Biopharma Projects-Thinking for Future

The core of cyanobacterial biorefinery is sustainability which could only be achieved by mindful processing. Keeping in view SDG 12 of the United Nations "Sensible production and consumption", the biopharma project has to shift towards resource efficiency and cleaner production. Some latest interventions and innovations are being rolled out to aid in the process (Fig. 12.2).

Table 12.1 Comparison of extraction processes in terms of their extraction, efficiency, cost, and environmental impact

Extraction mode	Extracted compounds	Extraction efficiency	Cost analysis	Environmental impact	Recommendations	References
Mechanical disruption	Proteins, pigments	Moderate	Low	Low	Suitable for large-scale operations. Minimal environmental impact	Eppink et al. (2019)
Sonication	Proteins, lipids	High	Moderate	Moderate	Effective for small- to medium-scale operations. Requires careful waste management	Chia et al. (2019)
Enzyme-assisted extraction	Proteins, enzymes	High	High	Low	Efficient but expensive. Use for high-value products. Utilize enzymes from renewable sources	Sinzinger et al. (2022)
Supercritical fluid extraction	Lipids, carotenoids	High	High	Low	Efficient but costly. Utilize green solvents. Ensure proper waste disposal	Pagels et al. (2021)
Solvent extraction (e.g., ethanol, methanol)	Pigments, lipids	Variable	Low to high	High	Cost-effective but environmentally hazardous. Use with caution and consider recycling/recovery methods	Vignesh et al. (2023)
Pressurized liquid extraction	Proteins, carotenoids	High	Moderate	Low	Suitable for medium-scale operations. Choose eco-friendly solvents	Pagels et al. (2021)
Subcritical water extraction	Proteins, polysaccharides	Moderate	Low to moderate	Low	Environmentally friendly. Optimize temperature and pressure conditions	Thiruvenkadam et al. (2015)
Microwave-assisted extraction	Proteins, pigments	High	Moderate	Moderate	Efficient for lab-scale to medium-scale operations. Properly manage energy consumption	Sarma et al. (2022)
Ultrasonic-assisted extraction	Proteins, lipids	High	Low to moderate	Low	Effective for small- to medium-scale operations. Minimize energy use	Pagels et al. (2021)
Aqueous two-phase extraction	Proteins, enzymes	High	High	Low	Efficient for specific applications. Use biocompatible and recyclable phase-forming agents	Porav et al. (2020)

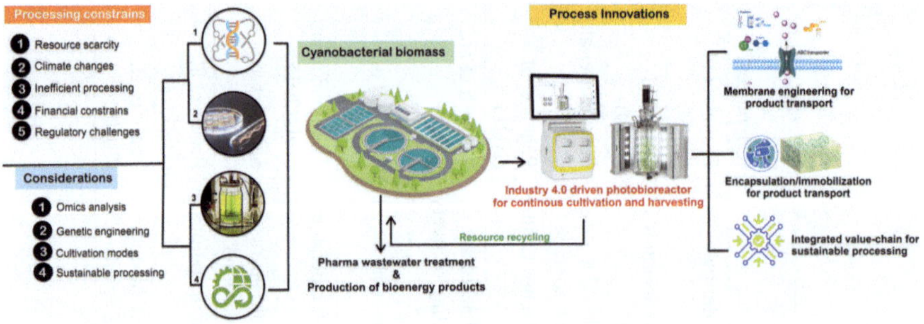

Fig. 12.2 Considerations and innovative processing layout to address constraints in the implementation of cyanobacterial biorefinery for the biopharmaceutical industry

12.3.1 Immobilization and Encapsulation

Recent years have witnessed progress in the development of hybrid molecules, particularly based on microbial living cells. Immobilization and encapsulation are product protection techniques where the living cells are either entrapped in the matrix or are covered in the polymeric sheath to provide an additional layer of protection. These techniques are ideal for product stability, controlled release of drugs, in vivo and in vitro bioavailability, intensifying processes, protection from immune response, and containment to avoid the release of cyanobacteria in the environment (Dawiec-Liśniewska et al. 2022; Li et al. 2022).

Silica-based sol-gel are ideal drug delivery system for the controlled release of drugs as it prevents premature denaturation of drugs (Kamanina et al. 2022). Immobilization of cyanobacteria between modified sol-gel silica layers allows the continuous production of astaxanthin from *Haematococcus pluvialis* (Fiedler et al. 2007). Hydrogel based on phycocyanin and carrageenan cross-linkage exhibited superior hemostatic capabilities that promote rapid wound healing (Dev et al. 2020). In another study, hydrogel compartmentalization was explored for the co-cultivation of *Synechococcus elongatus* and *Yarrowia lipolytica*/*Pseudomonas putida*, resulting in 15- to 22-fold higher indigoidine and β-carotene production with the final titer of 7.5 and 1.3 gL^{-1} hydrogel, respectively (Zhao et al. 2022).

Encapsulation of phycocyanin (pharmaceutical bioactive compound) with 3% sodium alginate and 2.5% calcium chloride improves its stability, especially in a pH environment (Pan-Utai and Iamtham 2019). Chitosan-encapsulated protein hydrolysate of *Spirulina platensis* resulted in the controlled release of the bioactive compound under simulated gastrointestinal conditions with excellent antioxidant properties (Forutan et al. 2022). A similar case was observed with liposome-chitosan-encapsulated *Spirulina* extract, which resulted in elevated antimicrobial properties and preserved antioxidant capacity (Mohammadi et al. 2023). 3D-printing of cyanobacterial hydrogels is the latest and most promising to design biobased devices, drug delivery systems, and biomaterials for surgical sutures and healing dressings (Dawiec-Liśniewska et al. 2022). One possibility is to use the cyanobacterial lipids and polysaccharides as drug-stabilizing agents, or they could be exploited for the controlled release of drugs. Cyanobacterial lipid nanomaterials could encapsulate the drugs and antioxidants to provide structural stability as observed in the case of vitamin B encapsulation with extracellular polymeric substances (EPS) of cyanobacterium *Cyanothece* sp. (Estevinho et al. 2019). Clinical success and industrial production of cyanobacterial therapeutics will help in the emergence of a sustainable biorefinery-based pharmaceutical industry (De Jesus Oliveira Santos et al. 2023). Most of these studies are in preliminary stages and require intensive research and knowledge for the optimization of various immobilization and encapsulation compounds for different bioactive compounds. It is also believed to aid in the product recovery process and needs to explore at length.

12.3.2 Exporter Engineering for Product Recovery

Product export through membrane transporter engineering is an interesting but least explored strategy for metabolite excretion from cyanobacteria without extensive processing. Concurrently it promotes the production of that particular bioactive compound by regulating the metabolic flux. Membrane transporters especially various classes of ABC exporters can secrete pigments, alkanes, polysaccharides, terpenes, organic acids, etc. But they have mainly been explored for the excretion of fatty acids and hydrocarbons (Gonçalves et al. 2021; Lin et al. 2020). Information regarding the pigment exporters and their mechanism remains elusive.

Identification of promiscuous and specific exporters requires a comprehensive understanding of exporter molecules and their heterologous screening for particular metabolites (Malik et al. 2021). Four ABC transporters (Cz08g16130, Cz05g17060, Cz04g21110, and Cz09g27180) were found to form a complex in *Chromochloris zofingiensis* to export astaxanthin from chloroplast (Roth et al. 2017). ABC transporters Eco-MsbA and St-MsbA have been reported to export canthaxanthin and zeaxanthin from *E. coli* (Doshi et al. 2013), while members of the PDR exporter family were able to excrete β-carotene from yeast cells (Chen et al. 2022). The heterologous expression of these exporters has yet to be tested in cyanobacteria to export pharmaceutically important bioactive compounds. However, successful implementation of these efflux pumps in cyanobacteria would open new ways for cyanopharm commercialization by enhancing the yield of biomass and titer of pharmaceutically viable pigments.

12.3.3 Industry 4.0 Technologies

The evolution of modern-day technologies and increased understanding of their integration with industrial processes improves the relationship between processing conditions and product quality which helps in revolutionizing the research scenarios

(Szkodny and Lee 2022). Circular economy 4.0 is the most recent trend that prompts the digital transformation of the supply chain through the interconnection of sensing and computation. It leverages artificial intelligence (AI), big data analysis, and machine learning technologies to automate the cultivation to processing pipeline for continuous monitoring to ensure operational efficiency from R&D to the post-commercialization stage. Although in infancy, digital technologies are now being applied to regulate the drug development process. Implementation of such systems by the biopharmaceutical industry could save time, resources, and expenses as observed by the use of industry 4.0 photobioreactor and electric transducer which streamline the upstream and downstream processing (Kim et al. 2022; Tsolakis et al. 2023).

Parameters obtained from the multi-objective machine learning model were able to enhance the biomass production of *Nostoc* sp. by 90% and phycobilin yield by 61.8% (Saini et al. 2021). Machine learning algorithms are being utilized to control the moisture content of biomass during the vacuum drying process (Pilario et al. 2022), to regulate the flocculant concentration for required floc size (Lopez-Exposito et al. 2019), and to optimize the metabolite extraction parameters (Muhammad et al. 2022; Srivastava et al. 2018). In a real-world case study to extract the anticancerous glycoprotein CSV-u from *Chlorella vulgaris*, a comparison between traditional and industry 4.0-driven biorefinery schemes was done. In the traditional scenario, a centralized facility was considered where biomass was cultivated in the large-scale open pond, and then harvested/dried biomass is transported to the API production facility for the extraction and packaging of CSV-u. Lipid and protein were co-extracted from spent biomass and formulated into aquaculture feed additive, while the remaining biomass was distributed to biofertilizer manufacturers. In the industry 4.0 biorefinery scenario, a distributed system in modular "micro-factories" was designed where industry 4.0-driven photobioreactors were employed for the continuous extraction and production facility followed by utilization of spent biomass according to the traditional approach. The results indicated that the second scenario gains an edge over the traditional route due to low-cost processing, energy saving, and quality improvement. However, significant initial investment and trade-offs related to manufacturing economics are the areas of consideration and research by biopharmaceutical companies (Tsolakis et al. 2023). The major challenge associated with the implementation of Industry 4.0 is data unavailability as the accuracy of implemented models enhances according to the training data sets. Companies will have to go through a certain period of trial and error through collaborative research within open innovation to optimize the process. It is expected that co-patenting will accelerate the digitalization of value-chain in the biopharmaceutical industry.

12.3.4 Integrated Models and Sustainable Value Chains

The main hindrance in the industrial implementation of the cyanobacterial biorefinery is its cost, so it is important to make the process not only economically viable but also sustainable and resource-efficient. A well-defined biorefinery scheme integrates sustainable technologies at both upstream and downstream levels (as discussed in the previous sections) for maximum biomass utilization. High water footprint and inefficient resource utilization during cyanobacterial processing necessitate the implementation of a cyclic bioeconomy for the industrialization of cyanopharm. The use of wastewater is a traditional approach to reducing water footprint due to freshwater scarcity, but it could not be applied for the biopharmaceutical industry due to the need of maintaining a sterile environment. Alternatively, media recycling could be a promising option to reduce water footprint as it could reduce water requirement by 84% and nutrient requirement by 55% (Shahid et al. 2019; Yang et al. 2011) with the additional benefit of cost reduction and positive ecological impact. However, the impact of recycled media on the biomass growth and concentration/nature of produced bioactive compounds is highly species-dependent and is affected by the presence of extracellular metabolites, cell debris, and contamination sources. Therefore, it is imperative to optimize the media replenishment conditions and how many times it could be recycled, particularly at the industrial scale. Life cycle assessment analysis and environmental impact analysis could provide valuable insight into the real-world implementation of such a scheme.

Another possible approach is the two-stage cultivation, where during the first stage finest chemicals will be used under a sterile environment to obtain high-quality cyanobacteria that do not contain any harmful compounds and then subject to co-extraction or sequential extraction of bioactive compounds/secondary metabolites. In the second stage, spent biomass (obtained after the extraction of pigments and bioactive compounds) is applied for the treatment of pharmaceutical industry wastewater, and the produced biomass is then processed for non-food products like polymers and bioenergy compounds (De Jesus Oliveira Santos et al. 2023).

12.4 Conclusion and Prospects

The United Nations' manifesto of "responsible processing and consumption (SDG 12)" propels the transition of the biopharmaceutical industry towards sustainable and circular processing routes. Cyanobacteria have been explored as potential pharmaceutical agents due to their ability to produce an arsenal of metabolites and bioactive compounds. Capital investment, maintenance of product profile, and effective recovery profile are the main challenges in the industrial implementation of cyanopharm biorefinery. Genetic engineering of perspective strain(s), modification of cultivation parameters, and co-extraction or sequential extraction of bioactive compounds (high-to-low value) are potential options to widen the scope of cyanobiorefinery. Encapsulation, immobilization, and exporter engineering promote product recovery in an eco-friendly and energy-efficient manner. It is being observed that the implementation of industry 4.0 technologies and integrated biorefinery models are effective for the development of industrially viable cyanobacterial value chains.

References

Avagyan AB (2010) New design of biopharmaceuticals through the use of microalgae addressed to global geopolitical and economic changes. Are you ready for new development in biopharma? Pharmacol Pharm 1:33–38

Babele PK, Srivastava A, Young JD (2023) Metabolic flux phenotyping of secondary metabolism in cyanobacteria. Trends Microbiol

Barbosa MJ, Janssen M, Südfeld C et al (2023) Hypes, hopes, and the way forward for microalgal biotechnology. Trends Biotechnol 41:452–471

Chen M, Li M, Ye L et al (2022) Construction of canthaxanthin-producing yeast by combining spatiotemporal regulation and pleiotropic drug resistance engineering. ACS Synth Biol 11:325–333

Chia SR, Chew KW, Show PL et al (2019) *Spirulina platensis* based biorefinery for the production of value-added products for food and pharmaceutical applications. Bioresour Technol 289:121727

Dawiec-Liśniewska A, Podstawczyk D, Bastrzyk A et al (2022) New trends in biotechnological applications of photosynthetic microorganisms. Biotechnol Adv 59:107988

De Jesus Oliveira Santos M, Oliveira De Souza C, Marcelino HR (2023) Blue technology for a sustainable pharmaceutical industry: microalgae for bioremediation and pharmaceutical production. Algal Res 69:102931

Dev A, Mohanbhai SJ, Kushwaha AC et al (2020) κ-carrageenan-C-phycocyanin based smart injectable hydrogels for accelerated wound recovery and real-time monitoring. Acta Biomater 109:121–131

Doshi R, Nguyen T, Chang G (2013) Transporter-mediated biofuel secretion. Proc Natl Acad Sci 110:7642–7647

Eppink MHM, Olivieri G, Reith H et al (2019) From current algae products to future biorefinery practices: a review. In: Wagemann K, Tippkötter N (eds) Biorefineries. Springer, Cham

Espinoza Pérez AT, Camargo M, Narváez Rincón PC et al (2017) Key challenges and requirements for sustainable and industrialized biorefinery supply chain design and management: a bibliographic analysis. Renew Sust Energ Rev 69:350–359

Estevinho BN, Mota R, Leite JP et al (2019) Application of a cyanobacterial extracellular polymeric substance in the microencapsulation of vitamin B12. Powder Technol 343:644–651

Fagundes MB, Alvarez-Rivera G, Mendiola JA et al (2021) Phytosterol-rich compressed fluids extracts from *Phormidium autumnale* cyanobacteria with neuroprotective potential. Algal Res 55:102264

Fiedler D, Hager U, Franke H et al (2007) Algae biocers: astaxanthin formation in sol–gel immobilised living microalgae. J Mater Chem 17:261–266

Forutan M, Hasani M, Hasani S et al (2022) Liposome system for encapsulation of *Spirulina platensis* protein hydrolysates: controlled-release in simulated gastrointestinal conditions, structural and functional properties. Materials (Basel) 15

Gilbert-López B, Mendiola JA, Fontecha J et al (2015) Downstream processing of *Isochrysis galbana*: a step towards microalgal biorefinery. Green Chem 17:4599–4609

Gonçalves CF, Lima S, Oliveira P (2021) Product export in cyanobacteria. Cyanobacteria Biotechnol

Hejna M, Kapuścińska D, Aksmann A (2022) Pharmaceuticals in the aquatic environment: a review on eco-toxicology and the remediation potential of algae. Int J Environ Res Public Health 19

Jeong Y, Cho S-H, Lee H et al (2020) Current status and future strategies to increase secondary metabolite production from cyanobacteria. Microorganisms 8

Jha D, Jain V, Sharma B et al (2017) Microalgae-based pharmaceuticals and nutraceuticals: an emerging field with immense market potential. ChemBioEng Rev 4:257–272

Kamanina OA, Saverina EA, Rybochkin PV et al (2022) Preparation of hybrid sol-gel materials based on living cells of microorganisms and their application in nanotechnology. Nanomater 12

Kesik-Brodacka M (2018) Progress in biopharmaceutical development. Biotechnol Appl Biochem 65:306–322

Kim Y, Atukeren E, Lee Y (2022) A new digital value chain model with PLC in biopharmaceutical industry: the implication for open innovation. J Open Innov Technol Market Complex 8:63

Lalor F, Fitzpatrick J, Sage C et al (2019) Sustainability in the biopharmaceutical industry: seeking a holistic perspective. Biotechnol Adv 37:698–707

Li Y, Li X, Liang Z-P et al (2022) Progress of microencapsulated phycocyanin in food and pharma industries: a review. Molecules 27

Lin P-C, Zhang F, Pakrasi HB (2020) Enhanced production of sucrose in the fast-growing cyanobacterium *Synechococcus elongatus* UTEX 2973. Sci Rep 10:390

Lopez-Exposito P, Negro C, Blanco A (2019) Direct estimation of microalgal flocs fractal dimension through laser reflectance and machine learning. Algal Res 37:240–247

Malik S, Shahid A, Liu C-G et al (2021) Developing fourth-generation biofuels secreting microbial cell factories for enhanced productivity and efficient product recovery; a review. Fuel 298: 120858

Malik S, Shahid A, Haider MN et al (2022) Prospects of multiproduct algal biorefineries involving cascading processing of the biomass employing a zero-waste approach. Curr Poll Rep 8:147–158

Market (2023). Biopharmaceutical market research report 2023. In: MARKET.US (ed)

Mendiola JA, Jaime L, Santoyo S et al (2007) Screening of functional compounds in supercritical fluid extracts from Spirulina platensis. Food Chem 102:1357–1367

Mohammadi M, Hamishehkar H, Mcclements DJ et al (2023) Encapsulation of *spirulina* protein hydrolysates in liposomes: impact on antioxidant activity and gastrointestinal behavior. Food Chem 400:133973

Muhammad G, Potchamyou Ngatcha AD, Lv Y et al (2022) Enhanced biodiesel production from wet microalgae biomass optimized via response surface methodology and artificial neural network. Renew Energ 184:753–764

Pagels F, Pereira RN, Vicente AA et al (2021) Extraction of pigments from microalgae and cyanobacteria-a review on current methodologies. Appl Sci 11

Pan-Utai W, Iamtham S (2019) Physical extraction and extrusion entrapment of C-phycocyanin from *Arthrospira platensis*. J King Saud Univ Sci 31:1535–1542

Patil PD, Patil SP, Kelkar RK et al (2021) Enzyme-assisted supercritical fluid extraction: an integral approach to extract bioactive compounds. Trends Food Sci Technol 116:357–369

Pekkoh J, Lomakool S, Chankham J et al (2022) Maximizing biomass productivity of cyanobacterium *Nostoc* sp. through high-throughput bioprocess optimization and application in multiproduct biorefinery towards a holistic zero waste. Biomass Convers Bioref

Pekkoh J, Duangjan K, Phinyo K et al (2023) Turning waste CO_2 into value-added biorefinery co-products using cyanobacterium Leptolyngbya sp. KC45 as a highly efficient living photocatalyst. Chem Eng J 460:141765

Pilario KES, Ching PML, Calapatia AMA et al (2022) Predicting drying curves in algal biorefineries using Gaussian process autoregressive models. Digit Chem Eng 4:100036

Porav AS, Bocăneală M, Fălămaş A et al (2020) Sequential aqueous two-phase system for simultaneous purification of cyanobacterial phycobiliproteins. Bioresour Technol 315:123794

Rader RA, Langer ES (2014) Biopharmaceutical manufacturing: historical and future trends in titers, yields, and efficiency in commercial-scale bioprocessing. Bioprocess J 13

Roth MS, Cokus SJ, Gallaher SD et al (2017) Chromosome-level genome assembly and transcriptome of the green alga *Chromochloris zofingiensis* illuminates astaxanthin production. Proc Natl Acad Sci 114:E4296–E4305

Saha SK, Murray P (2018) Exploitation of microalgae species for nutraceutical purposes: cultivation aspects. Fermentation 4

Saini DK, Rai A, Devi A et al (2021) A multi-objective hybrid machine learning approach-based optimization for enhanced biomass and bioactive phycobiliproteins production in *Nostoc* sp. CCC-403. Bioresour Technol 329:124908

Sarma MK, Saha K, Choudhury R et al (2022) Development of a microwave-assisted solvent extraction process for the extraction of high-value carotenoids from *chlorella* biomass. Biofuels Bioprod Biorefin 16:698–709

Shaban H, Maurer C, Willborn R (2018) Impact of drug shortages on patient safety and pharmacy operation costs. Fed Pract 35:24

Shahid A, Malik S, Alam MA et al (2019) The culture technology for freshwater and marine microalgae. In: Alam MA, Wang Z (eds) Microalgae biotechnology for development of biofuel and wastewater treatment. Springer Singapore, Singapore

Shahid A, Malik S, Khan AZ et al (2020) Multiproduct algal biorefineries: challenges and opportunities. In: Verma P (ed) Biorefineries: a step towards renewable and clean energy. Springer Singapore, Singapore

Shahid A, Khan AZ, Jabeen F et al (2023) Cyanobacteria-based biorefineries for a sustainable future of bioindustry. In: Oncel SS (ed) A sustainable green future: perspectives on energy, economy, industry, cities and environment. Springer, Cham

Sharma V (2015) Applicability of green chemistry in pharmaceutical processes. Fine Chem 5:102–104

Sinzinger K, Schieder D, Rühmann B et al (2022) Towards a cyanobacterial biorefinery: carbohydrate fingerprint, biocomposition and enzymatic hydrolysis of *Nostoc* biomass. Algal Res 65: 102744

Sreenikethanam A, Raj S, J RB, et al. (2022) Genetic engineering of microalgae for secondary metabolite production: recent developments, challenges, and future prospects Front Bioeng Biotechnol 10

Srivastava G, Paul AK, Goud VV (2018) Optimization of non-catalytic transesterification of microalgae oil to biodiesel under supercritical methanol condition. Energy Convers Manag 156:269–278

Szkodny AC, Lee KH (2022) Biopharmaceutical manufacturing: historical perspectives and future directions. Annu Rev Chem Biomol Eng 13:141–165

Thevarajah B, Nishshanka GKSH, Premaratne M et al (2022) Large-scale production of *spirulina*-based proteins and c-phycocyanin: a biorefinery approach. Biochem Eng J 185:108541

Thiruvenkadam S, Izhar S, Yoshida H et al (2015) Process application of subcritical water extraction (SWE) for algal bio-products and biofuels production. Appl Energy 154:815–828

Tsolakis N, Goldsmith AT, Aivazidou E et al (2023) Microalgae-based circular supply chain configurations using Industry 4.0 technologies for pharmaceuticals. J Clean Prod 395:136397

Vignesh K, Atchaya R, Pavan Kumar Rao G et al (2023) A cascaded biorefinery for the sustainable valorization of *Arthrospira maxima* biomass: a circular bioeconomy approach. Bioresour Technol Reports 23:101510

Walsh G, Walsh E (2022) Biopharmaceutical benchmarks 2022. Nat Biotechnol 40:1722–1760

Yang J, Xu M, Zhang X et al (2011) Life-cycle analysis on biodiesel production from microalgae: water footprint and nutrients balance. Bioresour Technol 102:159–165

Yang X, Li Y, Li Y et al (2019) Solid matrix-supported supercritical CO_2 enhances extraction of γ-linolenic acid from the cyanobacterium *Arthrospira* (*Spirulina*) *platensis* and bioactivity evaluation of the molecule in zebrafish. Mar Drugs 17

Yang Y, Hassan SHA, Awasthi MK et al (2023) The recent progress on the bioactive compounds from algal biomass for human health applications. Food Biosci 51:102267

Zhang W, Wang F, Gao B et al (2018) An integrated biorefinery process: stepwise extraction of fucoxanthin, eicosapentaenoic acid and chrysolaminarin from the same *Phaeodactylum tricornutum* biomass. Algal Res 32:193–200

Zhao R, Sengupta A, Tan AX et al (2022) Photobiological production of high-value pigments via compartmentalized co-cultures using Ca-alginate hydrogels. Sci Rep 12:22163

Chapter 13
Cyanobacteria Biotechnology: Challenges and Prospects

Aqib Zafar Khan, Xin-Qing Zhao, Feng-Wu Bai, Hafiz Hassan Mustafa, and Chen-Guang Liu

Abstract The interest in exploiting renewable carbon sources for the sustainable manufacturing of chemicals has increased due to global warming and climatic instability. Cyanobacteria have tremendous potential for using light and CO_2 as the only sources of energy and carbon, respectively, making them excellent cellular factories for manufacturing valuable compounds through carbon-negative processes. Although cyanobacteria biotechnology offers immense potential, however, there are still several challenges in applying cyanobacterial technology starting from lab to industrial scales. Biotechnological innovations are required to address a number challenges including limited biomass productivity, poor tolerance to various abiotic and biotic factors during outdoor cultivation, and difficult harvesting. These issues could be addressed by employing various technologies including synthetic biology, host engineering, and metabolic engineering. Converting cyanobacterial biomass into bioproducts including biohydrogen, bioethanol, biogas, biomethane, and biopolymers has been explored in this chapter. Various methods have also been explored to enhance the ability of the strains to produce a particular amount of biomass or bioproducts through genetic engineering. This chapter also discusses several viewpoints regarding the potential commercialization of cyanobacterial technologies in the future.

Keywords Cyanobacteria · Synthetic biology · Metabolic engineering · Biopolymers · Biohydrogen

A. Z. Khan · X.-Q. Zhao · F.-W. Bai · C.-G. Liu (✉)
State Key Laboratory of Microbial Metabolism, Joint International Research Laboratory of Metabolic & Developmental Sciences, School of Life Sciences and Biotechnology, Shanghai Jiao Tong University, Shanghai, China
e-mail: cg.liu@sjtu.edu.cn

H. H. Mustafa
Bioenergy Research Centre, Department of Bioinformatics and Biotechnology, Government College University Faisalabad, Faisalabad, Pakistan

M. A. Mehmood et al. (eds.), *Pharmaceutical and Nutraceutical Potential of Cyanobacteria*, https://doi.org/10.1007/978-3-031-45523-0_13

13.1 Introduction

Using cyanobacterial biomass is crucial for advancing sustainable development worldwide since it presents prospects for the growth of numerous sectors. Although they may also grow heterotrophically on organic substrates, cyanobacteria can nearly always be found in aquatic bodies that contain inorganic nutrients (such as carbon, nitrogen, phosphorus, and other oligo-elements) and light (for completing oxygenic photosynthesis) (Shahid et al. 2020a, b, 2021a; Khan et al. 2019). Biomass may be transformed into energy products through various approaches including liquefaction, pyrolysis, transesterification, fermentation, and anaerobic digestion (Abomohra et al. 2016; Muhammad et al. 2021; Oliveira et al. 2021). Besides, both microalgae and cyanobacteria are a well-known source of long-chain polyunsaturated fatty acids with antibacterial, anti-cancer, and antioxidant properties which make them suitable candidates for the food and pharmaceutical sectors (Lauritano et al. 2016). Due to their faster growth rate, higher stress tolerance to abiotic factors, acceptability of wastewater as cultivation media, and simplicity of genetic modification, cyanobacteria are favored over terrestrial plants as prospective cell factories. Cyanobacteria have been successfully bred to create a range of high-value products including phycobiliprotein, carotenoids, lipids, carbohydrates, and proteins (Khan et al. 2019).

Synthetic biology has made noteworthy strides, with the development of tools for genome editing and gene silencing like CRISPR, TALEN, and Zinc finger approaches, neutral sites for chromosomal integration, modular gene assembly kits, and several biological components with well-established functions like promoters, terminators, ribosomal binding sites, operon, and gene optimization (Khan et al. 2018, 2019; Shahid et al. 2020c). Parallel to this, computational scientists have been concentrating on cyanobacterial systems biology, which involves genome-scale metabolic models, fluxomic optimization, and other omics approaches to improve the growth and metabolite productivity of model cyanobacteria *Synechocystis* sp. and *Synechococcus* sp.

Contrary to their heterotrophic equivalents, namely, *Escherichia coli* and yeast, cyanobacteria remain relatively unexplored as cell factories despite the abovementioned research developments. For cyanobacteria-based bioprocesses to be used commercially, biomass productivity must be significantly increased. Host engineering and pathway engineering are two main classes into which the toil in this field may be divided. Host engineering refers to creating an organism that serves as a host and has desired traits; this is often accomplished by introducing numerous genetic alterations related to abiotic stresses. On the other hand, pathway engineering or metabolic engineering refers to the genetic alterations to target a specific value-added chemical/metabolite and may be directed by predictions made using metabolic models. While cyanobacteria's genome-scale metabolic models have made great strides, only a small number of their predicted products have undergone experimental testing.

Although cyanobacteria are one of the most sustainable approaches for biotechnological application, many challenges were faced during their cultivation, harvesting, genetic modification, and downstream processing stages, which have been discussed in this chapter.

13.2 Challenges Associated with Cyanobacteria-Based Bioproducts

It has been shown that the manufacturing of products from cyanobacteria grown in wastewater is promising, but issues must be resolved before scaling up. The first is the need to keep cyanobacteria-dominated cultures alive for an extended time. The production of multiproduct such as biofuels, like biodiesel, bioethanol, biogas, and biohydrogen, is thought to be possible using cyanobacteria as a renewable resource (Khan et al. 2022). Both microalgae and cyanobacteria contribute to the production of energy products (Table 13.1), but a few hurdles need to be addressed in terms of biomass productivity (Khan et al. 2022; Malik et al. 2022; Shahid et al. 2021c).

Table 13.1 Valorization of algae/cyanobacteria biomass into bioenergy (Kant Bhatia et al. 2022)

Biofuels	Strains	Fermenting strains	Processes involved	Yields
Biohydrogen	*Chaetomorpha antennina*	Methanogenic bacteria	The addition of surfactant improves the solubilization of organic matter during microwave treatment	74.5 mL/g COD
	U. reticulata	Methanogens	Extraction of volatile fatty acids was almost double than MP only and thus better for bio-H_2 production	63 mL/g COD
	A. platensis and *L. digitata*	Anaerobic sludge	The optimum C/N ratio was 26.2 for optimum H_2 production	85 mL/g VS
Biomethane	*C. protothecoides*	Anaerobic fermentative bacteria	Almost complete utilization of nutrients into multiple products including diesel, methane, and fertilizer	196 $\pm$ 4 Nm^3 CH_4/t VS
	N. gaditana	Anaerobic inoculums	Deoiling affects the structure integrity without cell lysis and improves hydrolysis	360 mL CH_4/g VS
Biogas	*C. protothecoides*	Anaerobic fermentative bacteria	Almost complete utilization of nutrients and conversion into multiple products including diesel, methane, and fertilizer	1.7 $\pm$ 0.1 NL/days

13.2.1 Biopolymers from Cyanobacterial Biomass

The growth of cyanobacteria requires thorough maintenance of the cultivation media that must be monitored regularly against contaminating microbes. Maintaining the dominance of cyanobacteria in mixed cultures and avoiding contamination depend on the regulation of nutrient contents and (molar) ratios as well as ideal culture operating conditions. Keeping the N:P ratios (>32:1) balanced for nutrients in cultivation media is crucial for better growth and yield. Because of their strong affinity for P and ability to store it as polyphosphate by luxury absorption, cyanobacteria may have a greater chance of dominating the green algal species *Chlorella* sp. and *Stigeoclonium* sp. It seems that the use of waste streams with stable properties (like diluted digestate) or low nutrient sources (like secondary effluents) might be the most suitable for cyanobacterial selection in terms of influent sources. Compared to other waste streams with more variable nutrient contents, the nutrient loading from these waste streams can be more readily managed (e.g., municipal wastewater). Operating circumstances are another thing to take into account; research indicates that batch sequencing operations are best suited for choosing cyanobacteria that are simple to settle. However, owing to the emergence of additional easily settleable bacteria such as *Stigeoclonium* sp. or *Pediastrum* sp., sufficient SRT and nutritional loads should be carefully checked (Arias et al. 2020).

Production of carbohydrates and PHB from mixed wastewater-borne cyanobacteria present another solution. Although organic carbon has been employed for PHB synthesis in pure cultures, in mixed cultures this substrate would reduce cyanobacterial dominance and promote heterotrophic bacterial activity (Arias et al. 2020). According to some research findings, mixed cyanobacterial cultures may rival pure cultures. Biopolymers have been seen in mixed cultures that are dominated by different cyanobacterial species. The high production costs that are needed to maintain the sterile conditions for pure cultures, which limits large-scale manufacturing, may be avoided by using mixed cultures. In general, mixed cyanobacteria-bacteria cultures developed in wastewater treatment systems provide a substitute for the manufacturing of polymers and improve the environmental and financial advantages of bioplastics and biofuels. There are just a few controlled experiments that have been done at the laboratory and pilot-scale employing cyanobacteria grown in wastewater to produce PHAs (bioplastics) and carbohydrates (biofuel substrate). Scaling up the technology and carefully examining how outdoor circumstances (e.g., direct sunlight and temperature) affect the cultivation process requires further research and deeper knowledge.

13.2.2 Cyanobacterial Biomass to Biodiesel

Large raceway pond system is required for cyanobacterial development, which results in a significant amount of cyanobacterial biomass for biodiesel generation, which is challenging. Additionally, lipid extraction is also tedious and raises cost of production. Only a few cyanobacteria are abundant in fatty acids or lipids, which will have an impact on output volume and raise the cost of biodiesel production (Khan et al. 2018). Lipids must be converted to biodiesel through transesterification. To examine and generate high-quality biodiesel, it is also required to measure its energy value, including its octane, cetane, calorific value, boiling temperature, freezing point, and carbon emission value (Khan et al. 2022; Malik et al. 2022; Shahid et al. 2021b, c; Haider et al. 2023; Zafar Khan et al. 2022). Current vehicle engines need to be replaced with ones that can run on biodiesel. Vehicle engine replacement is a protracted procedure (Mata et al. 2010).

13.2.3 Cyanobacterial Biomass to Biohydrogen

Biohydrogen production is still a long way from being realized, even though the mechanics behind the initial steps of its synthesis by phototrophic microbial cells are fully understood. The sensitivity of oxygenated enzymes and electron competition across the metabolic pathways that utilized a low number of electrons are major barriers to sustainable hydrogen generation. As a result, hydrogen generation from cyanobacteria has only been observed so far in laboratory settings. To achieve industrial-scale production, it is necessary to increase the rate and duration of biohydrogen generation. Dark fermentation requires more exertion, produces less hydrogen, and requires the removal of CO_2 throughout the process. The cost of producing hydrogen is additional. Utilizing sunlight as an energy source, photo-fermentation is the fermentative conversion of organic substrates into hydrogen and carbon dioxide. $2H_2O$, light, and CH_3COOH yield $4H_2 + 2CO_2$.

Starvation of macronutrients such as nitrogen, sulfur, and phosphorus, pH, temperature, light intensity, carbon, and nitrogen sources play an important role in biohydrogen production Only a minor portion of the process involves releasing H_2 gas at atmospheric pressure. The gas has to be dried out and compressed to a few thousand PSI to obtain a reasonable energy density. Gas compression uses a lot of energy, the majority of which cannot be recovered. Therefore, although H_2 can be created, it does not always indicate the procedure is worthwhile (Saifuddin and Priatharsini 2016). Other difficulties with bio-hydrogen generation include unstable hydrogen synthesis, which may be related to the metabolic change in organisms that produce hydrogen. The best methods for making hydrogen generation affordable and sustainable include genetic modification, metabolic engineering of microalgae, and optimization of critical experimental parameters (Kosourov et al. 2018).

13.2.4 Biomethane from Cyanobacterial Biomass

Due to the absence of an established recovery system of biomass, energy, and other products, the processing of cyanobacterial biomass after cultivation is tedious and energy intensive which makes it costly for replacing fossil fuels. In comparison to theoretical values, microalgae emit less methane in actuality. Experimental findings are now restricted to volatile solids with a CH_4 content of 0.05–0.31 L/g (Sialve et al. 2012). Such a great degree of variety may be attributed to two key factors: (i) the compositional complexity of macromolecules and (ii) the properties of each microalgae species' cell wall. The methane potential of various organic compounds in cyanobacterial cells is what accounts for the differences in anaerobic biodegradability caused by macromolecular composition. To calculate the theoretical methane production, the composition of organic matter may be stoichiometrically transformed into methane. The largest potential methane output is thus found in lipids (1.014 L/g VS), followed by proteins (0.851 L/g VS) and carbohydrates (0.415 L/g VS) (Sialve et al. 2009). Indeed, it has been effectively shown to boost methane output by causing a specific macromolecule accumulation in microalgae cells. Since carbohydrates accumulate as non-structural storage compounds (like starch) rather than structural carbohydrate compounds (like cellulose) in the cell wall, carbohydrate enrichment is a strategy that holds promise for enhancing anaerobic digestion performance. Although carbohydrates have a lower methane potential than lipids and proteins, they may be more accessible to anaerobic bacteria than glycoproteins and lipids that are a component of microalgae cell walls when they are present as non-structural storage components. Since the whole body of knowledge on the approach is centered on the generation of biodiesel, experimental findings on the rise in microalgal methane output following lipid formation are currently lacking. Ammonia toxicity is a significant concern when dealing with cyanobacterial anaerobic digestion. Cyanobacteria cells contain ample protein (around 50–70%); therefore hydrolysis might produce a lot of ammonia, which could be harmful to methanogens.

13.2.5 Cyanobacterial Biomass to Biochemicals

In order to create fuels and chemicals with added value, cyanobacteria have been developed as "green" chassis for many years to produce free fatty acids (FFA), polyhydroxybutyrate (PHB), farnesene, limonene, sucrose, and ethylene (Table 13.2). However, the poor yield of bioproducts makes employing cyanobacteria in this manner difficult despite several successful proof-of-concept experiments. In particular, poor carbon fixation efficiency and insufficient cyanobacterial tolerance to numerous products, including fuels and chemicals, make it difficult to scale up biochemical processes employing cyanobacterial chassis (Kanno et al. 2017; Cui et al. 2021). Numerous initiatives have been attempted to

Table 13.2 The production of different chemicals in commonly used cyanobacterial chassis (Li et al. 2023)

Biochemicals produced	Cyanobacterial species used	Production rate (mgL^{-1})
Free fatty acids	*Synechocystis* 6803	197
	Synechococcus elongatus 7942	640
	Synechococcus 7002	885
	Synechococcus 11901	1540
Limonene	*Synechococcus* 7002	4
	S. elongatus 2973	16.4
	S. elongatus 2973	50.0
Succinic acid	*S. elongatus* 7942	430
	Synechocystis 6803	1800
	Synechocystis 6803	4200
	S. elongatus 11801	930
Sucrose	*S. elongatus* 7942	1250
	Synechocystis 6803	1800
	S. elongatus 2973	8700
	S. elongatus 2973	8000
Glucosylglycerol	*Synechococcus* 7002	13.9
	S. elongatus 2973	80000
Lysine	*S. elongatus* 2973	556
	Synechococcus 7002	455
Polyhydroxybutyrate	*S. elongatus* 7942	80 mgg^{-1}
	S. elongatus 2973	420
Ethylene	*S. elongatus* 11801	338 $\mu moleg^{-1}$
Glutarate	*S. elongatus* 2973	67.5
Cadaverine	*S. elongatus* 2973	55.3

address the challenges posed by cyanobacteria's poor productivity. The growth of sturdy cyanobacterial chassis is the first sign of inadequate production. The fastest-growing cyanobacterial species, *Synechococcus elongatus* 2973, was found and described by Wendt and colleagues as a promising contender for large-scale applications (Wendt et al. 2022) whose replication time is 1.5 h. The newly identified *Synechococcus* sp. PCC 11901 is a robust cyanobacterial strain with high biomass production at high light intensities with a quick doubling time of roughly 2 hours, according to Wlodarczyk and colleagues (Włodarczyk et al. 2020). In addition to the finding of naturally resilient chasses, we can fairly expect that adaptive evolutionary mechanisms might be used to domesticate super cyanobacterial chassis with high photosynthetic efficiency. These chasses can withstand the high-light intensity and salinity. The future photosynthetic efficiencies of cyanobacteria may be greatly enhanced with the development of these practical methods, speeding up commercialization.

Light penetration, a key constraint for the bioproduction of cyanobacteria, is the second factor contributing to poor productivity. By creating semicontinuous algal growth that is informed by machine learning and supported by synthetic technology, Long and colleagues were able to overcome some of this difficulty (Long et al. 2022). This clever marketing tactic draws attention to the trade-offs between economic viability and high harvest costs.

Increasing photosynthesis is the third way to increase production. According to Li and colleagues, plants often exhaust the supply of C5 intermediates that glucose-6-phosphate (G6P) shuns in photosynthesis dark reactions, which compromises the productivity of next photosynthesis light reactions (Li et al. 2019). The presence of a G6P shunt to enhance C5 pools, as established by Shinde and colleagues, permits more dark reaction intermediates in cyanobacteria, resulting in a relatively faster and more productive photosynthesis (Shinde et al. 2019). They showed that fitness and total photosynthetic production depend on a seamless metabolic change from darkness to light. We may confidently expect that large-scale bioproduction applications by cyanobacteria will be possible in the future given the most recent advancements in carbon-negative synthetic biology in cyanobacteria.

13.3 Metabolic Engineering of the Cyanobacteria for Bioproducts

Excellent examples of altered hosts that are often utilized for metabolic engineering are *Saccharomyces cerevisiae*, *Bacillus subtilis*, and *E.* coli. None of the engineered cyanobacterial hosts are commercially available. Thus, wild-type organisms have been employed in pathway engineering investigations of cyanobacteria (Fig. 13.1); the most commonly used strains are *Synechococcus elongatus* PCC 7942, *Synechocystis* sp. PCC 6803, and *Synechococcus* sp. PCC 7002. Additionally, for outdoor cultivation setup, cyanobacteria need to be resistant to intense light, high temperatures, and other abiotic stimuli. Searching for potential strains which could be used as excellent hosts through bioprospecting is one strategy. The cyanobacterial strains *S. elongatus* UTEX 2973 (henceforth *S. elongatus* 2973) and *S. elongatus* PCC 11801 (henceforth *S. elongatus* 11801), which are being thoroughly characterized and have shown promising potential as hosts for several products (Khan et al. 2019), are fast-growing and stress tolerant. A different strategy would be to give slow-growing strains a fast-growth trait; this would work particularly well with the frequently used model organisms *Synechocystis* sp. 6803 and *S. elongatus* 7942. Growth rates and light tolerance increased when single nucleotide polymorphisms (SNPs) from the fast-growing phenotype of *S. elongatus* UTEX 2973 were incorporated into the model cyanobacterium *S.* elongatus 7942 (Zuñiga et al. 2020). *S. elongatus* 11801 and 11802, two closely related fast-growing strains, both have these SNPs. Some nations prohibit the study of genetically engineered microalgae to preserve the natural wild species, which produce less lipid (Shahid et al. 2020c).

Although the outcomes seem to be context-dependent, another strategy has been focused on the discovery and regulation of rate-limiting Calvin-Bassham-Benson

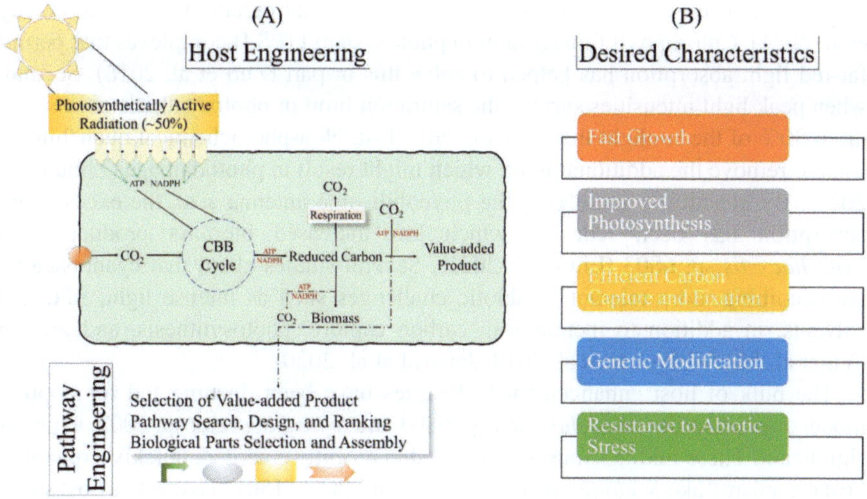

Fig. 13.1 Host and pathway engineering strategies for cyanobacteria. (**a**) Development of a cyanobacterial cell factory can be broadly divided into two parts: (on the left) host engineering or improvement of the cellular properties such as the rates of uptake of CO_2 and conversion to the reduced carbon substrates and (on the right) pathway engineering strategies to maximize the rate, yield, and titer of the product of interest. While host engineering strategies are largely product-independent, pathway engineering involves a search of pathways in databases, ranking available alternatives, selecting enzymes and genetic parts, and being primarily product-specific. Metabolic models can guide pathway engineering by making useful predictions of gene targets. (**b**) (above) A more elaborate scheme of host engineering strategies listing the desired characteristics of an ideal host and current efforts to achieve them via genetic engineering and (below) genome editing and other synthetic biology tools that are an integral part of host engineering

(CBB) cycle enzymes. The bifunctional sedoheptulose-1,7-bisphosphatase/fructose-1,6-bisphosphatase (BiBPase) was ubiquitously expressed in *Synechococcus* sp. 7002 and, interestingly, led to enhanced growth and photosynthetic oxygen evolution, a marker of photosynthetic activity. While *Synechococcus* 7002 did not grow better with the overexpression of ribulose-1,5-bisphosphate carboxylase/oxygenase (RuBisCO) (Weiss et al. 2017), *Synechocystis* sp. 6803 grew better with the overexpression of both BiBPase and RuBisCO (Zhao et al. 2022). The use of a similar approach in combination with route engineering has also been tried. In comparison to the overexpression of a single enzyme, the overexpression of two specific CBB cycle enzymes significantly increased the ethanol titer in *Synechocystis* sp. 6803 (Shinde et al. 2022). The overexpression of carbon transporters, which in turn improved carbon fixing, is a popular method for increasing carbon capture and growth rates. For instance, the marine cyanobacterium *Synechococcus* sp. 7002 had better growth, intracellular glycogen content, and external secretion carbohydrates after overexpressing *SbtA* and *BicA*, the native Na^+-dependent carbon transporters (Wu et al. 2022).

Cyanobacteria have limited photosynthetic efficiency, much like other photosynthesis-dependent species. First off, they can only absorb energy that is photosynthetically active or about 50% of the total solar radiation (Figure a) (Wang et al. 2021). Chlorophyll f integration in photosystem I (PS I) complexes that permit far-red light absorption has helped to solve this in part (Zuo et al. 2018). Second, when peak light intensities surpass the saturation limit of photosynthetic organisms, up to 80% of the sunlight that is absorbed is lost. Non-photochemical quenching is used to remove the additional light, which might result in photodamage (Zhou et al. 2021). By genetically modifying the phycobilisome antenna size, the excess light absorption has been reduced, which has increased biomass production in *Synechocystis* sp. 6803 (Liu et al. 2021). Several studies claim that cyanobacteria are becoming more tolerant to abiotic challenges such as intense light, salt, and solvents, in addition to increases in carbon capture, photosynthesis, and growth (Yu et al. 2015; Ungerer et al. 2018; Jaiswal et al. 2020).

The bulk of host enhancement techniques have been documented for popular model organisms like *Synechocystis* sp. 6803 and *Synechococcus* sp. 7002, which is significant. These findings must be validated in recently emerged, quickly expanding host organisms like *S. elongatus* UTEX 2973 and PCC 11801. To create a host strain for metabolic engineering, various alterations must also be examined to see how they affect one another. Given that cyanobacteria are polyploid, the genetic alterations will most likely take the form of chromosomal integration and necessitate chromosomal segregation. It will be necessary to modify the creature in stages utilizing cutting-edge, marker-free genome editing tools like CRISPR-Cas9. It will also be necessary to thoroughly examine the recombinant host strains' long-term stability. The characterization of metabolic fluxes, the metabolome, transcriptome, and proteome will be necessary to get a thorough knowledge of the host, which is desired. To that goal, various model strains of cyanobacteria have been developed that have genome-scale metabolic models and ^{13}C isotopic tracer-aided flow analysis. Additionally, diurnal alterations in the metabolome have been examined directly (Jaiswal et al. 2020) or through a transcriptome-guided metabolic model (De Porcellinis et al. 2018), which provide more details on the pathways that are active at various times of the light-dark cycle.

13.4 Synthetic Biology Approaches to Develop Cyanobacteria-Based Microbial Platforms

Key synthetic biology elements known as promoters have been investigated for several cyanobacterial species. According to their functional distinctions, the usually defined promoters, such as inducible and constitutive promoters. For convenience of reference, Table 13.3 provides a list of recently described promoters.

Another possible method for making high-value goods directly from CO_2 is via synthetic biology and metabolic engineering of photosynthetic organisms. However, low-flux carbon metabolic pathways like the shikimate pathway, which are found in cyanobacteria, restrict the growth of microbial cell factories for chemicals (Liang

Table 13.3 Impact of promoter engineering on chemical production in cyanobacterial strains [Adapted from (Wang et al. 2020)]

Promoter	Strains	Description
Inducible		
PL03	*Synechocystis* 6803, *Anabaena* 7120	Anhydrotetracycline-inducible promoter. The induction range was 200-fold in *Synechocystis* 6803. When tested in *Anabaena* 7120, the PL03-driving expression improved by 7% of the total protein
Psca6-2	*Synechocystis* 6803	A variant based on *E. coli* Ptac with approximately tenfold induction ratios
PO2	*Synechocystis* 6803	Promoter activated under dark or anaerobic conditions
PEZtet	*Synechococcus* 7002	Hybrid of PcpcB and two tet operators with a 32-fold dynamic range
PcptOO-cLac143	*Synechococcus* 7002	IPTG-inducible promoter with a 48-fold dynamic range
PompC	*Synechocystis* 6803	Promoter activated under dark conditions
Pvan	*S. elongatus* 7942	Vanillate-inducible promoter
PBAD	*Synechocystis* 6803, *S. elongatus* 7942	Promoter involved in activation of L-arabinose
Constitutive		
PpsbA	*Synechocystis* 6803, *S. elongatus* 7942	The activity of limonene synthase was enhanced 100-fold under PpsbA than that under Ptrc in *S. elongatus* 7942
Pcpc560	*Synechocystis* 6803	Pcpc560-driving expression of proteins results in about 15% of the total soluble proteins
PR-PS	*S. elongatus* 7942	Proteins generated by PR-PS could account for about 12% of the total extracted proteins
Ptrc	*Synechocystis* 6803, *S. elongatus* 2973	An *E. coli*-derived promoter was used to drive the expression of yfp
Psca3-2	*Synechocystis* 6803	A Ptac-variant promoter with moderate activity
Plac	*S. elongatus* 2973	An *E. coli*-derived promoter was used to drive the expression of *cscB* for sucrose production
PpsbA2S	*Synechocystis* 6803	A derivative of PpsbA2 promoter with shorter sequence. It shows fourfold higher strength when compared to its original version
PA2520	*Synechococcus* 7002	PA2520 showed about eightfold higher strength than Prbc of *Synechococcus* 7002

and Lindblad 2016). Redirecting carbon flow may improve the synthesis of chemicals and biofuels by metabolic network reprogramming (Roussou et al. 2021). Many research efforts have been put into altering carbon flow. In *S. elongatus*, for instance, Ni and colleagues created a potent metabolic sink called a synthetic feedback-inhibition-resistant cassette, which sped up electron transport and decreased energy waste in the shikimate pathway (Ni et al. 2018).

Cyanobacterial consortium engineering may potentially enable high productivity and titers by removing several of the metabolic barriers that are unique to cyanobacteria. For instance, Zuiga and associates (Gupta et al. 2020) presented a solid modeling framework for the deliberate design of synthetic heterotrophic and phototrophic communities with optimum growth sustainability. By using cyanobacteria, Kruyer and colleagues were able to transform CO_2 into sugars that were then improved into 2,3-BDO by modified *E. coli* (Kruyer et al. 2021). Synthetic light-driven cyanobacteria and heterotrophic bacteria were reported by Weiss and colleagues to produce stable PHB (Weiss et al. 2017). By putting this focus on it, cyanobacteria can export photosynthetically fixed carbon in the form of sucrose, which heterotrophic bacteria use to produce large amounts of compounds. Therefore, it is possible to guide carbon flow into the intended bioproduct pathways, such as riboswitch-mediated biosensors, by using fine-tuned synthetic biology toolkits and supporting technologies. This would increase the economic viability of cyanobacteria for large-scale bioproduction.

Because the energy and material metabolisms of cyanobacteria are two largely distinct systems, conducting metabolic engineering and control is difficult. A novel concept of metabolic network reprogramming has been put forth in light of the development of synthetic biology. It starts with the node components of carbon metabolism and utilizes synthetic biology regulator components, such as RBS and promoter cassettes, to optimize metabolic flux in synthetic pathways. For better metabolic flux management, we can also combine two or three advanced metabolic reprogramming design toolkits. For instance, Ni and colleagues redirected $>30\%$ of the carbon to the shikimate pathway for the synthesis of 2-phenyl ethanol in *S. elongatus* by combining "metabolic sink" biotechnology with an artificial feedback-inhibition-resistant cassette (Ni et al. 2018).

In addition to flux redirection, metabolic network modeling-based methodologies are also important for enhancing the economics to obtain high bioproduction by cyanobacteria. The most recent advances in modeling have been focused on applying mathematical modeling to pinpoint metabolic bottlenecks to enhance bioproduction. In order to pinpoint characteristics linked to high photosynthetic efficiency in NADP-malic enzyme-type C4 photosynthesis, Zhao and colleagues created a dynamic system model of C4 photosynthesis based on maize (Zhao et al. 2022). These results provide potential improvements to C4 photosynthesis. After defining the key sigma factor in *S. elongatus* PCC 7942 using computational modeling and wet lab analyses, Shinde and colleagues boosted the limonene titer to 19 mgL^{-1} in 7 days (Kirst et al. 2014), offering a fresh approach for upcoming terpenoid engineering in phototrophs. Frameworks built on machine learning have opened up new possibilities for modeling-based work. Wu and associates created a machine-learning-based system for ^{13}C-fluxomics, emphasizing improvements in the synthetic biology cycle of design-build-test-learn (Cui et al. 2020). Through comparative proteomics and experimental validation, Wang and colleagues increased our understanding of the metabolic processes involved in natural guanidine breakdown and showed how biological applications might promote the bioproduction of ethylene (Wang et al. 2021). These structures provide cyanobacteria the ability to alter carbon flow and improve the carbon metabolic pathway. By doing this, operating duties will be greatly reduced, and cyanobacteria will be more effective at producing high-value chemicals.

13.5 Conclusion and Prospects

Biodiesel, bioethanol, and valuable chemicals are the principal applications of cyanobacterial biotechnology. They are being used to produce PUFAs, pigments, carbohydrate-derived compounds, and more. However, cyanobacterial biorefineries have untapped prospects and obstacles. Wastewater nutrient removal and recycling using cyanobacterial-based technology seem sustainable. Cyanobacteria may absorb nutrients from water naturally, and also cultivation has removed nutrients in lab trials. Species selection, culture system design, wastewater source assessment, and pilot and large-scale system performance have all been studied extensively. The algae-based wastewater treatment (WWT) method still faces various obstacles.

Due to technological and economic limitations, cyanobacterial-based WWT is not widely used. The technical problem is maintaining viable cyanobacterial culture for long-term performance. Cyanobacterial culture techniques for large-scale wastewater treatment are required. To complement nutrient removal, the best method is to co-locate cyanobacterial cultivation with a wastewater treatment facility. This method would minimize process disruption. Using digester biomass CO_2 to boost cyanobacterial growth is another advantage of co-location. Life cycle analysis, carbon balancing, and economic analysis have been disregarded in the literature, especially in pilot and large-scale studies. These assessments should be done concurrently or early in the projects. Species selection would improve economics, for instance, fast-growing, high-photosynthetic, environmental-tolerant species with easy downstream processing (e.g., harvesting), and high-value biomass should be selected. A large biobank is available to choose cyanobacterial strains, a varied collection of microorganisms (0.2–1 million species) which might be improved by genetically altering organisms and/or developing consortiums. Utilizing native algae should be prioritized. Wild cyanobacteria have a lower organic percentage but greater ash content than modified cyanobacteria. Natural cyanobacterial production and composition can alter daily and seasonally. However, wild cyanobacteria do not need a complicated cultivation system and expensive nutrients. Water blooms of natural cyanobacteria (algae) in lakes and coastal regions endanger human health and the coastal ecology owing to their toxicity. However, these algal blooms may create essential lipids, carbohydrates, and proteins (Zuo et al. 2018). Thus, using such cyanobacteria would benefit both the industry and the environment. Natural cyanobacteria feedstock is expensive to harvest. This requires energy- and cost-efficient algae harvesting technology. Magnetic nanoparticles (Fe) and flocculation using chemicals, microbiological agents, or electric fields are efficient and cost-effective algae harvesting methods (Malik et al. 2022; Shahid et al. 2020b). Therefore, using natural algae with these collecting techniques may be cost-effective.

Cyanobacterial fractionation, or selective conversion of one ingredient (lipids, carbs, or proteins), seems promising due to cyanobacterial biomass' complexity. Therefore, building a "lipid-, carbohydrate-, or protein-first biorefinery" is crucial since cyanobacterial culture is usually optimized for one component. For instance, lipid-rich algae may be harvested and converted into biofuel or its derivatives.

Carbohydrate-rich algae may create HMF and levulinic acid. High-protein cyanobacteria may be used as food or transformed into amino acids or other nitrogenous substances. Insoluble dark-brown humin is produced during algae carbohydrate conversion. Condensed soluble molecules like HMF create insoluble humin that is seldom converted or used (Zhou et al. 2021). Therefore, humin, which lowers product carbon balance, should be avoided. Catalysts, reaction solvents, and circumstances have been tried to reduce humin production. However, further research is needed to understand humin generation and construct reaction systems for carbohydrate-based value chemical synthesis.

The metabolic pathway for the biosynthesis of protein needs to be improved for human or animal nutrition. Valorizing cyanobacterial proteins with high-value products is promising. Proteins are a prospective feedstock for N-containing chemical or polymer synthesis as one of the primary biologically fixed nitrogen sources. Fatty amides, fatty nitriles, or N-heterocycles are discovered in liquid cyanobacterial products after thermochemical conversion (Liu et al. 2021). These N-containing compounds, with excellent biodegradability and low toxicity, are valuable in surfactant and lubricant production. Therefore, techniques to increase their yields or effectively separate them must be devised.

Based on the foregoing viewpoints, future research might concentrate on using natural cyanobacteria with effective collecting techniques (magnetic nanoparticles, flocculation), improving the carbon balance during conversion, and scaling up the process for industrialization. Algae fractionation-based products include (1) algal lipids that can be directly recovered or converted to oleochemicals; (2) algal carbohydrates which can be converted into value-added building block chemicals (HMF, succinic acid, lactic acid, polyols); and (3) algal proteins that can be thermochemically converted into fatty amides, fatty nitriles, or N-heterocycles.

References

Abomohra AE-F, Jin W, Tu R et al (2016) Microalgal biomass production as a sustainable feedstock for biodiesel: current status and perspectives. Renew Sust Energ Rev 64:596–606

Arias DM, García J, Uggetti E (2020) Production of polymers by cyanobacteria grown in wastewater: current status, challenges and future perspectives. New Biotechnol 55:46–57

Cui J, Sun T, Li S et al (2020) Improved salt tolerance and metabolomics analysis of *Synechococcus elongatus* UTEX 2973 by overexpressing Mrp antiporters. Front Bioeng Biotechnol 8

Cui J, Sun T, Chen L et al (2021) Salt-tolerant *Synechococcus elongatus* UTEX 2973 obtained via engineering of heterologous synthesis of compatible solute glucosylglycerol. Front Microbiol 12

De Porcellinis AJ, Nørgaard H, Brey LMF et al (2018) Overexpression of bifunctional fructose-1,6-bisphosphatase/sedoheptulose-1,7-bisphosphatase leads to enhanced photosynthesis and global reprogramming of carbon metabolism in *Synechococcus* sp. PCC 7002. Metab Eng 47:170–183

Gupta JK, Rai P, Jain KK et al (2020) Overexpression of bicarbonate transporters in the marine cyanobacterium *Synechococcus* sp. PCC 7002 increases growth rate and glycogen accumulation. Biotechnol Biofuels 13:17

Haider MN, Khan AZ, Usman M et al (2023) Impact of seasons and wastewater cultivation on the biomass and biodiesel production by the Plectonema terebrans BERC10 as a candidate for a multiproduct algal biorefinery. Fuel 332:125987

Jaiswal D, Sengupta A, Sengupta S et al (2020) A novel *cyanobacterium Synechococcus* elongatus PCC 11802 has distinct genomic and Metabolomic characteristics compared to its neighbor PCC 11801. Sci Rep 10:191

Kanno M, Carroll AL, Atsumi S (2017) Global metabolic rewiring for improved CO_2 fixation and chemical production in cyanobacteria. Nat Commun 8:14724

Kant Bhatia S, Ahuja V, Chandel N et al (2022) Advances in algal biomass pretreatment and its valorisation into biochemical and bioenergy by the microbial processes. Bioresour Technol 358: 127437

Khan AZ, Bilal M, Rasheed T et al (2018) Advancements in biocatalysis: from computational to metabolic engineering. Chin J Catal 39:1861–1868

Khan AZ, Bilal M, Mehmood S et al (2019) State-of-the-art genetic modalities to engineer cyanobacteria for sustainable biosynthesis of biofuel and fine-chemicals to meet bio–economy challenges. Life (Basel) 9:54

Khan AZ, Malik S, Mehmood MA et al (2022) Two-stage algal cultivation for the biotransformation of urban wastewater's pollutants into multiple bioproducts in a circular bioeconomy paradigm. Energy Convers Manag 273:116400

Kirst H, Formighieri C, Melis A (2014) Maximizing photosynthetic efficiency and culture productivity in cyanobacteria upon minimizing the phycobilisome light-harvesting antenna size. Biochim Biophys Acta 1837:1653–1664

Kosourov S, Jokel M, Aro E-M et al (2018) A new approach for sustained and efficient H2 photoproduction by *Chlamydomonas reinhardtii*. Energy Environ Sci 11:1431–1436

Kruyer NS, Realff MJ, Sun W et al (2021) Designing the bioproduction of Martian rocket propellant via a biotechnology-enabled in situ resource utilization strategy. Nat Commun 12:6166

Lauritano C, Andersen JH, Hansen E et al (2016) Bioactivity screening of microalgae for antioxidant, anti-inflammatory, anticancer, anti-diabetes, and antibacterial activities. Front Mar Sci 3

Li J, Weraduwage SM, Preiser AL et al (2019) A cytosolic bypass and G6P shunt in plants lacking peroxisomal hydroxypyruvate reductase. Plant Physiol 180:783–792

Li Z, Li S, Chen L et al (2023) Fast-growing cyanobacterial chassis for synthetic biology application. Crit Rev Biotechnol 2023:1–15

Liang F, Lindblad P (2016) Effects of overexpressing photosynthetic carbon flux control enzymes in the cyanobacterium *Synechocystis* PCC 6803. Metab Eng 38:56–64

Liu L, Zhou Y, Guo L et al (2021) Production of nitrogen-containing compounds via the conversion of natural microalgae from water blooms catalyzed by ZrO_2. ChemSusChem 14:3935–3944

Long B, Fischer B, Zeng Y et al (2022) Machine learning-informed and synthetic biology-enabled semi-continuous algal cultivation to unleash renewable fuel productivity. Nat Commun 13:541

Malik S, Shahid A, Betenbaugh MJ et al (2022) A novel wastewater-derived cascading algal biorefinery route for complete valorization of the biomass to biodiesel and value-added bioproducts. Energy Convers Manag 256:115360

Mata TM, Martins AA, Caetano NS (2010) Microalgae for biodiesel production and other applications: a review. Renew Sust Energ Rev 14:217–232

Muhammad G, Alam MA, Mofijur M et al (2021) Modern developmental aspects in the field of economical harvesting and biodiesel production from microalgae biomass. Renew Sust Energ Rev 135:110209

Ni J, Liu H-Y, Tao F et al (2018) Remodeling of the photosynthetic chain promotes direct CO_2 conversion into valuable aromatic compounds. Angew Chem Int Ed 57:15990–15994

Oliveira CYB, D'alessandro EB, Antoniosi Filho NR et al (2021) Synergistic effect of growth conditions and organic carbon sources for improving biomass production and biodiesel quality by the microalga *Choricystis minor* var. minor. Sci Total Environ 759:143476

Roussou S, Albergati A, Liang F et al (2021) Engineered cyanobacteria with additional overexpression of selected Calvin-Benson-Bassham enzymes show further increased ethanol production. Metab Eng Commun 12:e00161

Saifuddin N, Priatharsini P (2016) Developments in bio-hydrogen production from algae: a review. Res J Appl Sci Eng Technol 12:968–982

Shahid A, Malik S, Khan AZ et al (2020a) Multiproduct algal biorefineries: challenges and opportunities. In: Verma P (ed) Biorefineries: a step towards renewable and clean energy. Springer Singapore, Singapore

Shahid A, Malik S, Zhu H et al (2020b) Cultivating microalgae in wastewater for biomass production, pollutant removal, and atmospheric carbon mitigation: a review. Sci Total Environ 704:135303

Shahid A, Rehman AU, Usman M et al (2020c) Engineering the metabolic pathways of lipid biosynthesis to develop robust microalgal strains for biodiesel production. Biotechnol Appl Biochem 67:41–51

Shahid A, Khan AZ, Malik S et al (2021a) Advances in green Technologies for the Removal of effluent organic matter from the urban wastewater. Curr Pollut Rep 7:463–475

Shahid A, Malik S, Liu C-G et al (2021b) Characterization of a newly isolated cyanobacterium *Plectonema terebrans* for biotransformation of the wastewater-derived nutrients to biofuel and high-value bioproducts. J Water Process Eng 39:101702

Shahid A, Usman M, Atta Z et al (2021c) Impact of wastewater cultivation on pollutant removal, biomass production, metabolite biosynthesis, and carbon dioxide fixation of newly isolated cyanobacteria in a multiproduct biorefinery paradigm. Bioresour Technol 333:125194

Shinde S, Zhang X, Singapuri SP et al (2019) Glycogen metabolism supports photosynthesis start through the oxidative pentose phosphate pathway in cyanobacteria. Plant Physiol 182:507–517

Shinde S, Singapuri S, Jiang Z et al (2022) Thermodynamics contributes to high limonene productivity in cyanobacteria. Metab Eng Commun 14:e00193

Sialve B, Bernet N, Bernard O (2009) Anaerobic digestion of microalgae as a necessary step to make microalgal biodiesel sustainable. Biotechnol Adv 27:409–416

Sialve B, Bernet N, Steyer J (2012) Impact of microalgae characteristics on their conversion to biofuel, Part II: Focus on biomethane production. Biofuels Bioproducts Biorefining 6(2): 205–218

Ungerer J, Wendt KE, Hendry JI et al (2018) Comparative genomics reveals the molecular determinants of rapid growth of the cyanobacterium *Synechococcus elongatus* UTEX 2973. Proc Natl Acad Sci 115:E11761–E11770

Wang F, Gao Y, Yang G (2020) Recent advances in synthetic biology of cyanobacteria for improved chemicals production. Bioengineered 11:1208–1220

Wang B, Xu Y, Wang X et al (2021) A guanidine-degrading enzyme controls genomic stability of ethylene-producing cyanobacteria. Nat Commun 12:5150

Weiss TL, Young EJ, Ducat DC (2017) A synthetic, light-driven consortium of cyanobacteria and heterotrophic bacteria enables stable polyhydroxybutyrate production. Metab Eng 44:236–245

Wendt KE, Walker P, Sengupta A et al (2022) Engineering natural competence into the fast-growing cyanobacterium *Synechococcus elongatus* strain UTEX 2973. Appl Environ Microbiol 88

Włodarczyk A, Selão TT, Norling B et al (2020) Newly discovered *Synechococcus* sp. PCC 11901 is a robust cyanobacterial strain for high biomass production. Commun Biol 3:215

Wu C, Yu J, Guarnieri M et al (2022) Computational framework for machine-learning-enabled 13C Fluxomics. ACS Synth Biol 11:103–115

Yu J, Liberton M, Cliften PF et al (2015) *Synechococcus elongatus* UTEX 2973, a fast growing cyanobacterial chassis for biosynthesis using light and CO_2. Sci Rep 5:8132

Zafar Khan A, Haider MN, Zhao X-Q et al (2022) Evaluation of resource recovery potential of the *Pseudoscillatoria coralii* BERC01 under variable compositions of wastewater to produce biomass for cyanobacterium biorefinery. Sustain Energy Technol Assess 54:102804

Zhao H, Wang Y, Lyu M-JA et al (2022) Two major metabolic factors for an efficient NADP-malic enzyme type C4 photosynthesis. Plant Physiol 189:84–98

Zhou Y, Li M, Chen Y et al (2021) Conversion of polysaccharides in *Ulva prolifera* to valuable chemicals in the presence of formic acid. J Appl Phycol 33:101–110

Zuñiga C, Li T, Guarnieri MT et al (2020) Synthetic microbial communities of heterotrophs and phototrophs facilitate sustainable growth. Nat Commun 11:3803

Zuo Z, Ni B, Yang L (2018) Production of primary metabolites in *Microcystis aeruginosa* in regulation of nitrogen limitation. Bioresour Technol 270:588–595

Chapter 14
Global Research Trends in Cyanobacteria: Bioproducts and Culture Collection

Mahwish Amin, Aqsa Mushtaq, Hira Ashfaq, Hesham R. El-Seedi, and Ning Wang

Abstract Cyanobacteria, a diverse group of photosynthetic organisms, have garnered significant attention in recent years due to their promising biotechnological potential and their pivotal role in global ecosystems. Cyanobacteria are proficient in converting solar energy into biomass, making them highly attractive for biofuel production. Several studies have demonstrated the viability of cyanobacteria as feedstock for bioethanol, biodiesel, and biohydrogen, paving the way for a more sustainable energy future. The rise of cyanobacterial research can be attributed to the increasing demands for sustainable and eco-friendly alternatives to traditional resources. In addition to biofuels, cyanobacteria have emerged as promising producers of various high-value bioproducts. These include pigments such as phycocyanin and carotenoids, which find applications in the food, cosmetic, and pharmaceutical industries. Moreover, cyanobacteria have shown potential as sources of bioactive compounds, such as antimicrobial peptides and antioxidants, further enhancing their value in modern biotechnology. The global trend towards utilizing cyanobacteria as bioproduct factories is accompanied by an increased focus on establishing and expanding cyanobacteria culture collection banks. These banks serve as repositories of diverse cyanobacterial strains, ensuring their preservation, accessibility, and distribution to researchers worldwide. They play a crucial role in supporting fundamental research, bioprospecting endeavors, and biotechnological

M. Amin · A. Mushtaq · H. Ashfaq
Bioenergy Research Center, Department of Bioinformatics and Biotechnology, Government College University Faisalabad, Faisalabad, Pakistan

H. R. El-Seedi
International Research Center for Food Nutrition and Safety, Jiangsu University, Zhenjiang, China

Jiangsu Education Department, International Joint Research Laboratory of Intelligent Agriculture and Agri-Products Processing, Jiangsu University, Nanjing, China

Pharmacognosy Group, Department of Pharmaceutical Biosciences, Biomedical Centre, Uppsala University, Uppsala, Sweden

N. Wang (✉)
School of Bioengineering, Sichuan University of Science and Engineering, Zigong, China

M. A. Mehmood et al. (eds.), *Pharmaceutical and Nutraceutical Potential of Cyanobacteria*, https://doi.org/10.1007/978-3-031-45523-0_14

applications. This chapter provides a comprehensive overview of the current global trends concerning cyanobacteria, with a particular focus on their bioproducts and the importance of cyanobacteria culture collection banks.

Keywords Cyanobacteria · Culture Collection Bank (CCB) · Bioproducts · Publication · Bibliography

14.1 Introduction

Cyanobacteria, also known as blue-green algae, are a diverse group of photosynthetic microorganisms that play a crucial role in aquatic ecosystems and the Earth's oxygen production. They are among the oldest organisms on the planet, dating back 3500 million years, and have contributed significantly to shaping the Earth's atmosphere and ecosystem (Mehdizadeh Allaf and Peerhossaini 2022). These remarkable microorganisms are capable of harnessing sunlight to carry out photosynthesis, converting CO_2 into oxygen and organic matter (Wang et al. 2020). Cyanobacteria thrive in a wide range of environments, from freshwater lakes and oceans to extreme habitats like hot springs and deserts (Pawlowski and Bergman 2007). Some cyanobacterial species can form distinctive colonies and mats, imparting vibrant color to water bodies (Paerl et al. 2000).

Spirulina is a rich source of proteins (55-65%) and vitamins (particularly Vitamin B12 and provitamin β-carotene) (Seghiri et al. 2019). Phycobilins especially phycocyanin are the major health-promoting protein-based pigments produced by the *Spirulina* which are known for their health-promoting and disease-protection properties (Arslan et al. 2021; Stanic-Vucinic et al. 2018). Additionally, *Spirulina* is composed of 20% carbohydrates and 5% essential fatty acids-rich lipids. The antioxidant, anti-inflammatory, antiviral, antimicrobial, and immune-protectant activities of these metabolites render the use of *Spirulina* in nutraceuticals and pharmaceuticals (Han et al. 2021; Guldas et al. 2020).

Moreover, cyanobacteria also offer a wealth of bioproducts with diverse applications across various industries including biofuels, biofertilizer, antibiotic, pharmaceutical, and bioplastic industries (Agarwal et al. 2022). Biofuels such as biodiesel and bioethanol hold the potential to lessen our dependence on fossil fuels and reduce greenhouse gas emissions (Khanna et al. 2011). Additionally, high-value compounds of cyanobacteria including pigments, antioxidants, and vitamins have significant applications in food, cosmetic, and pharmaceutical industries (Rashi et al. 2023). Furthermore, these versatile microorganisms also play a crucial role in bioremediation by absorbing excess nutrients and pollutants, aiding in the restoration of polluted ecosystems (Abatenh et al. 2017).

Globally, culture collection banks have been developed to preserve and manage these versatile photosynthetic microorganisms (Yuorieva et al. 2023). These collection banks play a vital role in the field of cyanobacterial research, environmental studies, and biotechnological applications as they maintain a diverse array of cyanobacterial strains isolated from different habitats (Yuorieva et al. 2023). By safeguarding and disseminating cyanobacterial resources, these culture banks contribute significantly to advancing our understanding of cyanobacteria and promoting sustainable innovations for the benefit of society and the environment (Pathak et al. 2018).

The objective of this chapter was to comprehensively analyze the worldwide cyanobacterial research trends in the last two decades. This bibliometric analysis provided insight into research direction in different fields. It also provided an insight into global culture collection banks and cyanobacterial bioproducts which are a wealth of opportunities for industries, research, and innovation to address pressing global challenges.

14.2 Methodology

The complete information about global trends in cyanobacterial research was extracted from the Elsevier Scopus database (https://www.scopus.com) by using the [TITLE-ABS-KEY: Cyanobacteria, AND Nutraceuticals, AND Pharmaceuticals, AND Bioproducts] as a search query. The search resulted in the top 10 documents, and the research period from 2003 to 2023 was selected for analysis. It was observed that the results may vary based on search query and parameters as new data is being added to the database daily. The information related to cyanobacteria-based bioproducts, and culture collection banks were obtained by the freely available information on the Internet mainly Google and Google Scholar.

14.3 Results and Discussion

14.3.1 Global Publications by Year

A global publication by year refers to analyzing and categorization of worldwide dissemination of information research or content, with the focus on different years to observe trends and developments in global publishing activities over time, and Fig. 14.1 shows the evolution in global scientific publications from 2003 to 2023. It was observed that the total number of publications during the period of 2003–2023 was 610, and there was only one paper published in 2003, which reached 192 publications in the year 2022. The number of publications in the year 2023 is unavailable. It was observed that cyanobacteria, nutraceutical, pharmaceutical, and cyanobacterial bioproducts-based publications rose from 2012 to 2022 which indicate higher research activity in the last decade.

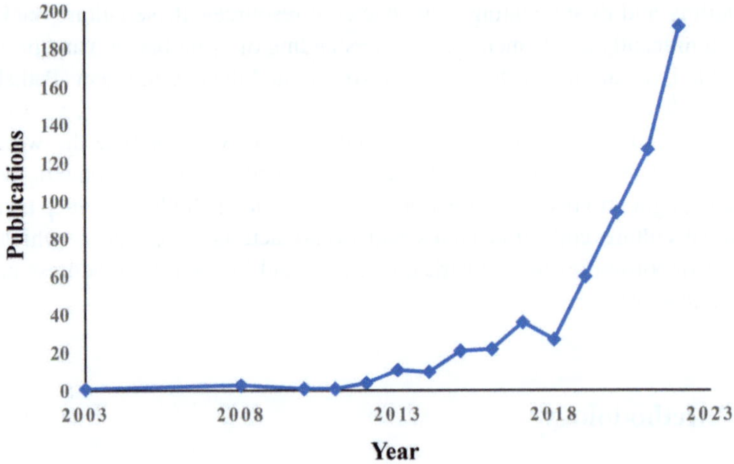

Fig. 14.1 Research publication trend in cyanobacteria for nutraceutical and pharmaceutical and cyanobacterial bioproducts, based on research from 2003 to 2023 (Source: SCOPUS)

14.3.2 Global Production by Country/Territory

Global production by country refers to the total publications produced by a specific country. The top 28 countries/territories, indicating the major contributors (807 publications) in the field of cyanobacteria for nutraceutical and pharmaceutical bioproducts, were listed (Fig. 14.2). India, China, Brazil, Portugal, and Spain are top 5 publishing countries with 141, 70, 66, 50, and 46 publications, respectively. Among these top 28 countries, 10 are Asian including India, China, Malaysia, South Korea, Taiwan, Thailand, Japan, Pakistan, Iran, and Saudi Arabia, 2 are American including the United States and Brazil, 6 are European mainly Portugal, Spain, France, Italy, United Kingdom, and Germany, and 3 are transcontinental countries including Turkey, Egypt, and Spain. The results showed that Asian countries lead the cyanobacterial publications which indicate higher cyanobacteria-based research activity in these countries.

14.3.3 Global Publications by Source per Year

The global publications focused on cyanobacterial bioproducts for nutraceutical and pharmaceutical bioproducts, based on research published in top 7 journals from 2003 to 2023, are represented in Fig. 14.3. The results showed that *Marine Drugs* is the fastest-growing journal leading with 32 publications; however, its trend started to decline after 2020 and become stable in 2022. Whereas *Bioresource Technology* and *Science of the Total Environment* are the second leading journals focusing on cyanobacterial research where 17 and 9 articles were published, respectively.

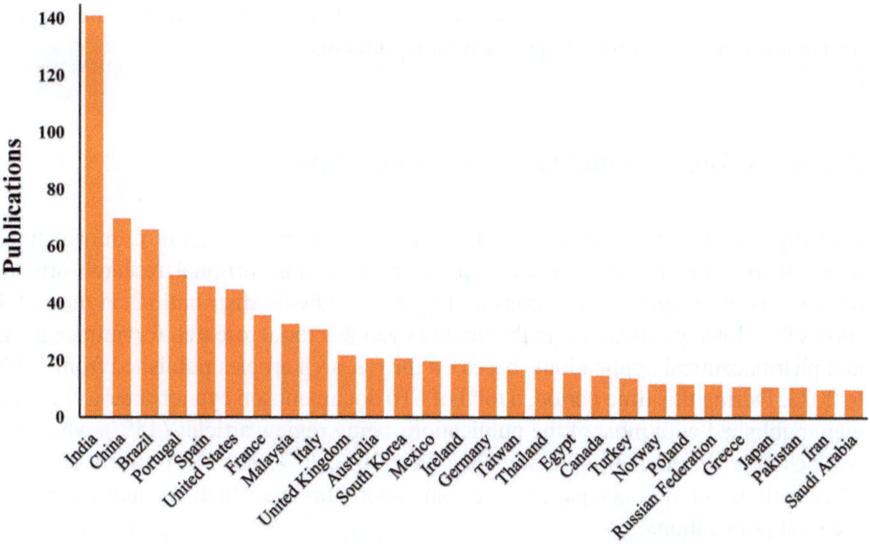

Fig. 14.2 Country-wise publications focused on cyanobacterial bioproducts for nutraceutical and pharmaceutical bioproducts, from 2003 to 2023 by countries/territories (Source: SCOPUS)

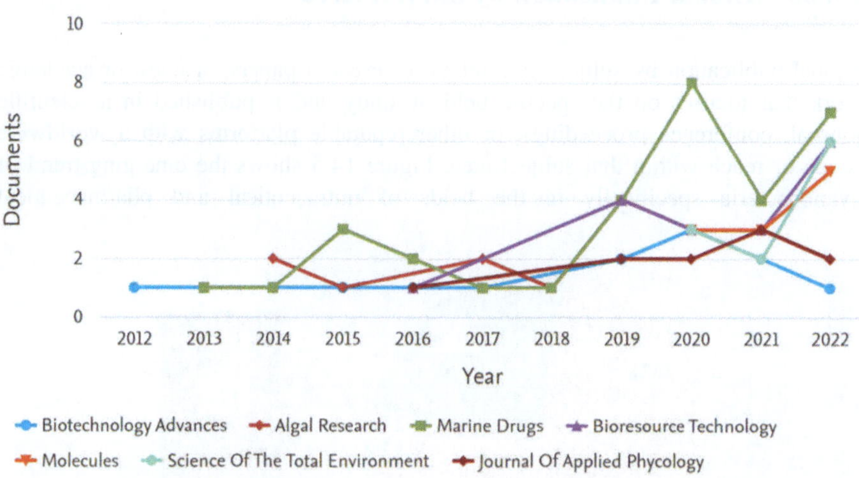

Fig. 14.3 Leading journals publishing research focused on cyanobacterial bioproducts for nutraceutical and pharmaceutical applications, based on the data collected from SCOPUS published during the years 2003 to 2023

MDPI's journal *Molecules* started gaining popularity during the last two years where 11 articles were published. Moreover, the *Journal of Applied Phycology* has been another prominent journal for cyanobacterial publication with ten documents, and it tends to decline after 2021. In the case of Biotechnology Advances, 14 publications were observed, and it tends to decline after 2020. The *Algal Research* is the highest unstable journal with only six publication documents.

14.3.4 Global Publications by Article Type

Global publication by type refers to the analysis and categorization of information, research, or content based on the type of publications (original research article, reviews, book chapters, etc.), considering its worldwide distribution. Figure 14.4 shows the global publications in the field of cyanobacterial research for nutraceutical and pharmaceutical applications, based on the research articles published from 2003 to 2023. During this time period, a total of 610 documents on cyanobacteria research were published with most of the publications being review articles (45%), while the book chapter, research articles, and book were 26%, 26%, and 1%, respectively. The other includes conference paper, notes, and short surveys which account for 1% of the total publications.

14.3.5 Global Publication by Subject Area

Global publication by subject area refers to research papers, articles, or academic work that focuses on the specific field of study and is published in a scientific journal, conference proceedings, or other reputable platforms with a worldwide scope or reach within that subject area. Figure 14.5 shows the emerging trend of cyanobacteria specifically in the fields of nutraceutical and pharmaceutical

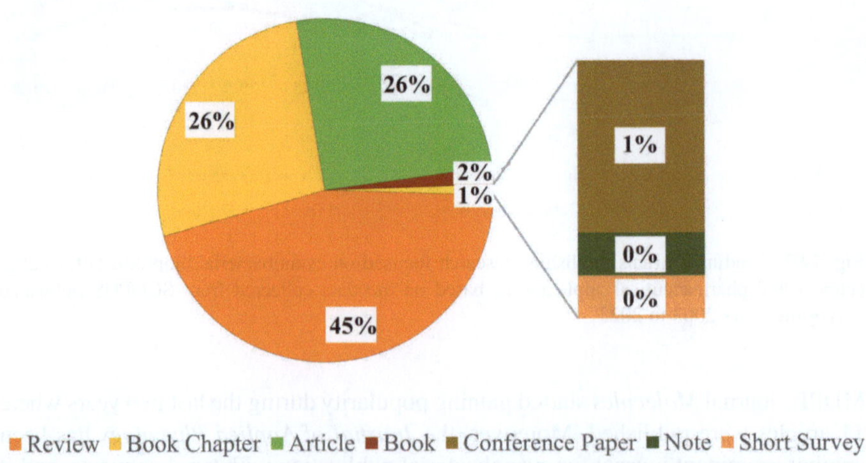

Fig. 14.4 Global research focusing cyanobacteria for pharmaceutical and nutraceutical applications, based on the article types published during 2003–2023 and indexed in SCOPUS

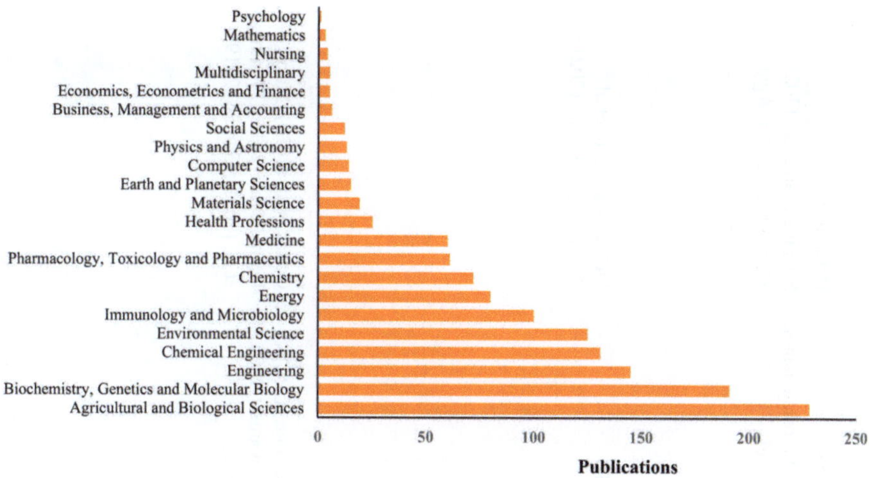

Fig. 14.5 Subject-wise research papers focused on cyanobacterial research for pharmaceutical and nutraceutical applications published from 2003 to 2023 and indexed in SCOPUS

bioproducts-based research in top 22 research areas. It depicted that the highest number of publications were in the field of Agricultural and Biological Sciences (228), Biochemistry, Genetics and Molecular Biology (191), Engineering (145), Chemical Engineering (131), and Environmental Science (125). Other includes Immunology and Microbiology (100), Energy (80), Chemistry (72), Pharmacology, Toxicology and Pharmaceutics (61), Medicine (60), Health Professions (25), Materials Science (19), Earth and Planetary Sciences (15), Computer Science (14), Physics and Astronomy (13), Social Sciences (12), Business Management and Accounting (6), Economics, Econometrics and Finance (5), Multidisciplinary (%), Nursing (4), Mathematics (3), and Psychology (1). These numbers clearly indicated that cyanobacterial research has purely become an interdisciplinary research discipline for a myriad of applications.

14.4 Biotechnological Potential of Cyanobacteria for Diverse Bioproducts

Cyanobacteria are a fascinating group of photosynthetic microorganisms with immense biotechnological potential. These ancient organisms have evolved over billions of years, developing diverse metabolic pathways that enable them to produce a wide array of valuable bioproducts (Mehdizadeh Allaf and Peerhossaini 2022). Table 14.1 summarizes the bioproducts of cyanobacteria. From biofuels and bioplastics to nutraceuticals and pharmaceutical compounds, cyanobacteria offer a sustainable and eco-friendly approach to meet the growing demands of the bioeconomy (Rashi et al. 2023). Their ability to fix atmospheric nitrogen, capture

Table 14.1 Cyanobacteria-based bioproducts and their mechanism of production

Bioproducts	Cyanobacterial species	Mechanism of production	Bioproducts	References
Biofuels	*Synechococcus* sp., *Synechocystis* sp., *Nostoc* sp., *Anabaena* sp.	Overproduction and transesterification of fatty acids to generate fatty acid methyl esters, fatty alcohols, or fatty alkanes are the two processes involved in the production of fatty acid-based biofuels	Biogas, cellulosic ethanol, and biodiesel	Silambarasan et al. (2021), Mund et al. (2022), Agarwal et al. (2022)
Bioplastics	*Synechocystis, Spirulina, Anabaena,* and *Nostoc muscorum*	Cyanobacteria convert CO_2 to organic carbon compounds during the photosynthesis process, and β-ketothiolase effectuates the formation of two acetyl-CoA molecules. The acetyl-CoA molecule sustains a NADPH-dependent reaction that catalyzes the addition of hydrogen to impart hydroxybutyryl-CoA molecules, and PHB synthase facilitates the chain formation	PHA, PHB	Mohanasundaram et al. (2023), Samrot et al. (2021)
Pigments	*Arthrospira platensis, Anabaena variabilis*	Heme group is the precursor molecule for phycobilins which is converted to biliverdin, and phycobilin lyases convert biliverdin to phycobilins. The precursor molecules for carotenoid biosynthesis are isopentenyl pyrophosphate (IPP) and dimethylallyl pyrophosphate (DMAPP), and these are enzymatically converted to carotenoids	C-phycocyanin (CPC), allophycocyanin (APC), carotenoids	Cottas et al. (2020), Mary Leema et al. (2010), Montoya et al. (2021), Gandhi (2019)

Antibiotics	*Oscillatoria angustissima, Calothrix parietina, Nostoc* sp., *Anabaena* sp.	Cyanobacteria possess gene cluster for biosynthesis of antibiotic, and these cluster express under stress conditions. The primary metabolic pathways are the precursor such as glycolysis, pentose phosphate and citric acid cycle. Under stress conditions, gene cluster regulate the biosynthesis of antibiotics	Nostopeptolide, anabaenopeptins	Issa (1999), Erlich et al. (2020)
Nutritional supplements	*Spirulina platensis* and *Chlorella vulgaris*	Cultivation of cyanobacteria under ideal conditions and harvesting of desired products after desired period	Antioxidants (phycobilins, chlorophyll, carotenoids), vit-B12, vit-C, vit-A	Gurney and Spendiff (2022)
Biofertilizers	Diazotrophs, *Nostoc linckia, Anabaena variabilis, Aulosira fertilissima, Calothrix* sp., *Tolypothrix* sp., *Scytonema* sp.	Wastewater-grown nitrogen-fixing cyanobacterial strain's biomass is a biofertilizer	Whole wastewater-grown cyanobacterial biomass is a biofertilizer	Chittora et al. (2020)
Pharmaceuticals	*Dunaliella salina, Chlorella vulgaris, Arthrospira platensis*	The secondary metabolites which are produced during oxidative stress are mainly the pharmaceutical products	Antioxidants (phycobilins, carotenoid, chlorophyll), vit-B12, vit-C, vit-A	Jha et al. (2017)
Bioremediation agents	*Anabaena flos-aquae* UTCC64, *Phormidium autumnale* UTEX1580, and *Synechococcus* sp. PCC7942	Efficient absorbance of heavy metals, organic nutrient removal from water bodies, industrial dye degradation, atmospheric carbon sequestration	Exopolysaccharides (EPS)	Dellamatrice et al. (2017)

CO_2 (Wang et al. 2020), and thrive in various environmental conditions further enhances their appeal as promising candidates for green biotechnology and bioremediation (Abatenh et al. 2017). As research and innovation continue to advance, cyanobacteria hold the promise of contributing significantly to the development of novel bioproducts that address global challenges while fostering a more sustainable future.

14.5 Global Culture Collection Banks for Cyanobacteria Preservation and Sharing

Global cyanobacterial culture collection banks (Table 14.2) play a vital role in preserving and sharing the diverse genetic resources of cyanobacteria from around the world (Yuorieva et al. 2023). These collections serve as repositories for living cyanobacterial strains, encompassing various species and genotypes with unique traits and capabilities. Researchers, scientists, and biotechnologists rely on these culture banks to access authenticated and well-characterized cyanobacterial strains for diverse applications such as biotechnology, bioremediation, and pharmaceutical research (Yuorieva et al. 2023).

Moreover, these culture collections facilitate taxonomic studies, biodiversity assessment, and the exchange of valuable scientific information among different institutions and countries. By safeguarding the genetic diversity of cyanobacteria, these culture banks contribute to sustainable innovation and the advancement of knowledge for the benefit of both scientific research and the broader bioeconomy (Pathak et al. 2018).

14.6 Conclusion and Prospects

A great evolution of cyanobacteria for nutraceutical, pharmaceutical, and other bioproducts has been observed, based on the research focused on these topics during previous two decades (2003–2023). It was shown that a total of 610 publications were published in different research journals with an interdisciplinary way during 2003 to 2023. In addition to the trend regarding the number of publications in the world, other aspects under study are the number of publications by country/territory, source, type, subject area, different bioproducts, and worldwide Culture Collection Banks (CCB) of cyanobacteria. The countries publishing most on cyanobacteria-based bioproducts for nutraceutical and pharmaceutical applications included India (141 publications), China (70 publications), and Brazil (66 publications). Most of the publications are from two well-reputed journals, namely, Marine Drugs (32 publications) and Bioresource Technology (17 publications). Among these, most of the publications are review articles (45%) and to a lesser extent are research articles

Table 14.2 Worldwide Culture Collection Banks (CCB) of cyanobacteria and microalgae

Culture Collection Bank (CCB)	Country	Cyanobacterial genus available	Number of strains available	URL
Culture Collection of Algae and Protozoa (CCAP)	UK	*Anabaena, Anacystis, Aphanizomenon, Aphanothece, Calothrix,* Trichormus, *Tychonema, Limnospira, Nostoc*	154	www.ccap.ac.uk
Pasteur Culture Collection (PCC)	France	*Synechococcus, Nostoc, Calothrix, Oscillatoria, Tolypothrix, Cyanobium, Spirulina, Kamptonema, Leptolyngbya,* etc.	473	catalogue-crbip.pasteur.
University of Texas at Austin Culture Collection of Algae (UTEX)	USA	*Chlorella, Spirulina, Scytonema, Haematococcus, Anabaena, Nannochloropsis, Plectonema, etc.*	NNA	utex.org
Belgian Coordinated Collections of Microorganisms (BCCM)	Belgium	*Nostoc, Microcystis, Heteroleibleinia, Calothrix, Chroococcus, Anabaena, Arthrospira, Geitlerinema, Oscillatoria,* etc.	246	bccm.belspo.be/catalogues
German Collection of Microorganisms and Cell Cultures (DSMZ)	Germany	*Phormidium, Synechocystis, Chroococcus, Spirulina, Nodularia, Calothrix, Pseudanabaena, Synechococcus, Nostoc*	117	www.dsmz.de
American Type Culture Collection (ATCC)	USA	*Anabaena, Nostoc, Synechocystis, Arthrospira, Calothrix, Trichormus, Synechococcus*	25	www.atcc.org
Culture Collection of Algae at Gottingen University (SAG)	Germany	*Spirulina, Chlorella, Nostoc, Synechococcus, Pseudanabaena, Synechococcus,* etc.	NNA	sagdb.uni-goettingen.de
North East Pacific Culture Collection (NEPCC)	Canada	*Tetraselmis, Dermocarpa, Synechococcus, Oscillatoria, Anabaena, Tolypothrix, Fischerella, Spirulina*	19	db.botany.ubc.ca
Freshwater Algal Culture Collection (FWAC)	Canada	*Scytonema, Synechococcus, Oscillatoria, Anabaena, Tolypothrix, Fischerella, Spirulina*	835	algae.ihb.ac.cn
Roscoff Culture Collection	France	*Oscillatoria, Cyanophyceae XX*	28	roscoff-culture-collection.org
	USA	*Spirulina, Nostoc, Oscillatoria, Scytonema, Tolypothrix,*	NNA	gcm.wdcm.org

(continued)

Table 14.2 (continued)

Culture Collection Bank (CCB)	Country	Cyanobacterial genus available	Number of strains available	URL
National Center for Marine Algae and Microbiota (NCMA)		*Scytonema, Haematococcus,* etc.		
NBRC Microalgal Collection	Japan	*Acaryochloris, Dermocarpa, Phormidium, Pseudanabaena, Synechococcus, Thermosynechococcus,* etc.	53	www.nite.go.jp
PhycolBank	Pakistan	*Spirulina* sp., *Oscillatoria* sp., *Phormidium* sp., *Tribonema* sp., *Plectonema* sp., *Chlorella* sp., *Tetraselmis* sp., *Scenedesmus* sp., *Bracteacoccus* sp.	27	www.gcuf.edu.pk

NNA, number not available

(26%), book chapters (26%), and conference papers (1%). This study also provides basic information about the research direction of cyanobacteria. It is observed that most of the publications are achieved in the field of Agricultural and Biological Sciences (228). This study also shows that cyanobacteria have a promising future as a number of bioproducts are continuously produced from it. Moreover, cyanobacteria culture collection banks are valuable resources for studying biodiversity, taxonomy, and ecological aspects of cyanobacteria. By preserving strains from different geographic locations and environments, they facilitate the investigation of cyanobacterial adaptability to changing environmental conditions and their role in maintaining ecological balance. The increased understanding of cyanobacterial physiology and the implementation of sustainable production practices will drive the transformation of cyanobacteria into essential components of a green and bio-based economy.

References

Abatenh E, Gizaw B, Tsegaye Z et al (2017) The role of microorganisms in bioremediation-A review. Open J Environ Biol 2:038–046

Agarwal P, Soni R, Kaur P et al (2022) Cyanobacteria as a promising alternative for sustainable environment: synthesis of biofuel and biodegradable plastics. Front Microbiol 13:939347

Arslan R, Eroglu EC, Aksay S et al (2021) Determination of bioactive properties of protein and pigments obtained from Spirulina platensis. J Food Process 45:e15150

Chittora D, Meena M, Barupal T et al (2020) Cyanobacteria as a source of biofertilizers for sustainable agriculture. Biochem Biophys Rep 22:100737

Cottas AG, Cunha WR, Ribeiro EJ et al (2020) Influence of medium composition on the production of Phycocyanin from Anabaena variabilis. Ind Biotechnol 16:45–49

Dellamatrice PM, Silva-Stenico ME, de Moraes LAB et al (2017) Degradation of textile dyes by cyanobacteria. Braz J Microbiol 48:25–31
Erlich JR, To EE, Liong S et al (2020) Targeting evolutionary conserved oxidative stress and immunometabolic pathways for the treatment of respiratory infectious diseases. Antioxid Redox Signal 32:993–1013
Gandhi SG (2019) Chapter 8 - Synthetic biology for production of commercially important natural product small molecules. In: Singh SP, Pandey A, Du G, Kumar S (eds) Current developments in biotechnology and bioengineering. Elsevier
Guldas M, Ziyanok-Demirtas S, Sahan Y et al (2020) Antioxidant and anti-diabetic properties of Spirulina platensis produced in Turkey. Food Sci Technol 41(3)
Gurney T, Spendiff O (2022) Algae supplementation for exercise performance: current perspectives and future directions for spirulina and chlorella. Front Nutr 9
Han P, Li J, Zhong H et al (2021) Anti-oxidation properties and therapeutic potentials of spirulina. Algal Res. 55:102240
Issa AA (1999) Antibiotic production by the cyanobacteria Oscillatoria angustissima and Calothrix parietina. Environ Toxicol Pharmacol 8:33–37
Jha D, Jain V, Sharma B et al (2017) Microalgae-based pharmaceuticals and nutraceuticals: an emerging field with immense market potential. ChemBioEng Rev 4:257–272
Khanna M, Crago CL, Black M (2011) Can biofuels be a solution to climate change? The implications of land use change-related emissions for policy. Interface Focus 1:233–247
Mary Leema JT, Kirubagaran R, Vinithkumar NV et al (2010) High value pigment production from Arthrospira (spirulina) platensis cultured in seawater. Bioresour Technol 101:9221–9227
Mehdizadeh Allaf M, Peerhossaini H (2022) Cyanobacteria: model microorganisms and beyond. Microorganisms 10:696
Mohanasundaram Y, Natarajan V, Arumugam N (2023) Cyanobacteria as a source of biodegradable plastics. In: Archana T (ed) Cyanobacteria. IntechOpen, Rijeka
Montoya EJO, Dorion S, Atehortua-Garcés L et al (2021) Phycobilin heterologous production from the Rhodophyta Porphyridium cruentum. J Biotechnol 341:30–42
Mund NK, Liu Y, Chen S (2022) Advances in metabolic engineering of cyanobacteria for production of biofuels. Fuel 322:124117
Paerl H, Pinckney J, Steppe T (2000) Cyanobacterial-bacterial mat consortia: examining the functional unit of microbial survival and growth in extreme environments. Environ Microbiol 2:11–26
Pathak J, Rajneesh MPK et al (2018) Cyanobacterial farming for environment friendly sustainable agriculture practices: innovations and perspectives. Front Environ Sci 6
Pawlowski K, Bergman B (2007) Chapter 11 Plant symbioses with Frankia and cyanobacteria. In: Bothe H, Ferguson SJ, Newton WE (eds) Biology of the nitrogen cycle. Elsevier, Amsterdam
Rashi T, Pankaj Kumar S, Archana T (2023) Cyanobacteria as the source of antioxidants. In: Archana T (ed) Cyanobacteria. IntechOpen, Rijeka
Samrot AV, Samanvitha SK, Shobana N et al (2021) The synthesis, characterization and applications of polyhydroxyalkanoates (PHAs) and PHA-based nanoparticles. Polymers (Basel) 13(19):3302
Seghiri R, Kharbach M, Essamri AJ (2019) Functional composition, nutritional properties, and biological activities of Moroccan Spirulina microalga. J Food Qual. 2019:3707219
Silambarasan S, Logeswari P, Sivaramakrishnan R et al (2021) Cultivation of Nostoc sp. LS04 in municipal wastewater for biodiesel production and their deoiled biomass cellular extracts as biostimulants for Lactuca sativa growth improvement. Chemosphere 280:130644
Stanic-Vucinic D, Minic S, Nikolic MR et al (2018) Spirulina phycobiliproteins as food components and complements. In: Microalgal biotechnology. IntechOpen, London, pp 129–149

Wang L, Chen L, Yang S et al (2020) Photosynthetic conversion of carbon dioxide to oleochemicals by cyanobacteria: recent advances and future perspectives. Front Microbiol 11
Yuorieva N, Sinetova M, Messineva E et al (2023) Plants, cells, algae, and cyanobacteria in vitro and Cryobank collections at the Institute of Plant Physiology, Russian Academy of Sciences: A platform for research and production center. Biology 12:838

MIX
Papier aus verantwortungsvollen Quellen
Paper from responsible sources
FSC® C105338

If you have any concerns about our products,
you can contact us on
ProductSafety@springernature.com

In case Publisher is established outside the EU,
the EU authorized representative is:
Springer Nature Customer Service Center GmbH
Europaplatz 3, 69115 Heidelberg, Germany

Printed by Libri Plureos GmbH
in Hamburg, Germany